CATZ-JPTC

(18)

Water resources and reservoir engineering

Water resources and reservoir engineering

Proceedings of the seventh conference of the British Dam Society held at the University of Stirling, 24-27 June 1992

Edited by Noel M. Parr, J. Andrew Charles and Susan Walker

Thomas Telford, London

Conference organized by the British Dam Society, the British Hydrological Society, the Scottish Hydrological Group and the Institution of Water and Environmental Management

Organizing Committee
- N. M. Parr, *Chairman*
- Dr J. A. Charles
- R. M. Jarvis
- A. I. B. Moffat
- A. C. Robertshaw
- Dr Susan Walker
- N. Tyler

First published 1992

A CIP catalogue record for this book is available from the British Library

ISBN 0 7277 1692 1

© The British Dam Society, 1992, unless otherwise stated

All rights, including translation, reserved. Except for fair copying, no part of this publication may be reproduced, stored in a retrieval system or transmitted in any form or by any means electronic, mechanical, photocopying, recording or otherwise, without the prior written permission of the Publications Manager, Publications Division, Thomas Telford Services Ltd, Thomas Telford House, 1 Heron Quay, London E14 9XF.

Papers or other contributions and the statements made or the opinions expressed therein are published on the understanding that the author of the contribution is solely responsible for the opinions expressed in it and that its publication does not necessarily imply that such statements and or opinions are or reflect the views or opinions of the organizers or publishers.

Published on behalf of the organizers by Thomas Telford Services Ltd, Thomas Telford House, 1 Heron Quay, London E14 9XF.

Printed in England by Redwood Press Ltd, Melksham, Wilts.

Contents

RESERVOIR PLANNING, OPERATION AND ENVIRONMENTAL ASPECTS

The reliability of single, historic estimates of reservoir capacity. A. J. ADELOYE	1
Optimal operation of local reservoir sources. A. H. BUNCH	11
Control rules for Tayside's linked system of reservoirs supplying two demand zones. J. A. COLE and A. L. GORDON	19
The South East Wales Conjunctive Resource Scheme. S. A. GALLIMORE and J. C. MOSEDALE	33
Comparison of the yields of variants of the Broad Oak Water scheme. J. K. HALL and D. E. MACDONALD	41
An investigation into the maintenance of high summer water levels at recreationally important reservoirs using a simulation model of the Tees reservoir system. A. I. HILL	53
Operational yields. W. L. JACK and A. O. LAMBERT	65
The Balquhidder research catchments: development of the results for application to water resources. R. C. JOHNSON	73
The water resources of Madras. M. C. D. LA TOUCHE and P. SIVAPRAKASAM	81
Estimation of river flow requirements to meet electricity demands in the Cameroon. P. E. ROBINSON	93
Evaluation impacts on environment of Narmada Sagar project, India. H. SAHU	101
Reassessment of reservoir yields in North-West England in the light of possible climate change. H. SMITHERS	107
Computer software for optimizing the releases from multiple reservoirs operated for flow regulation. T. WYATT, E. V. HINDLEY and T. C. MUIR	117

CONTENTS

RESERVOIR DESIGN AND CONSTRUCTION

Modelling the uncertainty of sediment deposition upstream of flood control dams. G. W. ANNANDALE — 125

The construction of a cut-off in a volcanic residual soil using jet grouting. L. J. S. ATTEWILL, J. D. GOSDEN, D. A. BRUGGEMANN and G. C. EUINTON — 131

Two embankment dams on alluvial foundation: Durlassboden and Eberlaste. H. CZERNY — 141

Reservoir competency study made in respect of limestone formations found upstream of Kodasalli Dam, India. H. V. ESWARAIAH, V. S. UPADHYAYA and C. R. RAMESH — 147

The instrumentation, monitoring and performance of Roadford Dam during construction and first filling. J. D. EVANS and A. C. WILSON — 157

Recent examples of reinforced grass spillways on embankment dams based on CIRIA report 116. R. FREER — 167

Reservoir construction development for irrigation in the United Kingdom 1960-92. S. M. HAWES — 175

The use of stepped blocks for dam spillways. H. W. M. HEWLETT and R. BAKER — 183

A review of spillway flood design standards in European countries, including freeboard margins and prior reservoir level. F. M. LAW — 191

The flood control works for the Cardiff Bay Barrage. P. J. MASON, S. A. BURGESS and T. N. BURT — 203

Numerical modelling of reservoir sedimentation. I. C. MEADOWCROFT, R. BETTESS and C. E. REEVE — 211

Triggers to severe floods: extreme rainfall and antecedent wetness. D. W. REED — 219

A statistical perspective on reservoir flood standards. D. W. REED and C. W. ANDERSON — 229

Trapping efficiency of reservoirs. C. E. REEVE — 241

Small embankment-type reservoirs for water supply and amenity use. B. H. ROFE, C. G. HOSKINS and M. F. FLETCHER — 253

Rock for dam face protection and the CIRIA/CUR manual on rock in coastal engineering. J. E. SMITH and J. D. SIMM — 259

Design, construction and performance of Mengkuang Dam. E. H. TAYLOR, C. M. WAGNER and J. H. MELDRUM — 271

RESERVOIR MONITORING AND MAINTENANCE

Implementation of the Reservoirs Act 1975 and monitoring of dams. D. C. BEAK	287
Response of a clay embankment to rapid drawdown. C. J. A. BINNIE, D. J. SWEENEY and M. W. REED	301
The role of instrumentation and monitoring in safety procedures for embankment dams. J. A. CHARLES, P. TEDD and K. S. WATTS	311
Investigation, monitoring and remedial works at Tiga Dam, Nigeria. H. S. EADIE, D. J. COATS and N. LEYLAND	321
Vegetation and embankment dams. C. G. HOSKINS and P. R. RICE	329
Resume of maintenance contracts on hydroelectric reservoirs. C. K. JOHNSTON and N. M. SANDILANDS	339
Performance of blockwork and slabbing slope protection subject to wave action. H. T. LOVENBURY and R. A. READER	353
Surveillance and monitoring methods for Italian dams. D. MORRIS	361
Site investigation of existing dams. J. M. REID	373
Dam ageing. G. P. SIMS	383
Development of a three-dimensional computer system for dam surveillance data management. D. M. STIRLING and G. L. BENWELL	395
The BRE dams database. P. TEDD, I. R. HOLTON and J. A. CHARLES	403
The reservoir safety research programme of the Department of the Environment. C. E. WRIGHT, D. J. COATS and J. A. CHARLES	411

The reliability of single, historic estimates of reservoir capacity

A. J. ADELOYE, PhD, MIWEM, Imperial College of Science, Technology and Medicine

SYNOPSIS. Direct-supply reservoirs are the oldest form of artificial, man-made surface water impoundments. Hydrological analysis to determine the capacity-yield relationships of such reservoirs is often carried out empirically, using a record of observed streamflow data at the site with one of several techniques which come under the general rubric of "critical period methods". The major problem with this approach is that it does not provide, explicitly, an estimate of the reliability of the reservoir system. Information on the reliability is extremely important for a system whose failure to meet targets is often accompanied by huge economic losses. In this paper an attempt is made, using Monte Carlo simulation techniques, to quantify the reliability of single historic estimates of reservoir capacity. The results show that the reliability is not unique but varies depending on the length of data record, the statistical characteristics of the inflow series and the yield level.

INTRODUCTION

1. The analysis of direct-supply reservoirs using one of the many critical period techniques (ref. 1) with a historic data record at the project site gives a single estimate of reservoir capacity. However, by their nature, traditional critical period techniques, e.g. the mass curve and its automated version the sequent peak algorithm (SPA) (ref. 2), when used in this way do not give a measure of reliability for the capacity estimate; they merely design for the worst drought on record.

2. The fact that no explicit estimate of the reliability is produced by these techniques does not mean the capacity estimate has no reliability. In fact, analysts have made different assumptions regarding the inherent reliability of this single estimate. For example it is sometimes assumed that, since the critical period respresents a period of extreme low flows in the record, the probability of the single capacity estimate will be that of the lowest, i.e. rank 1, event (ref. 1). Another common assumption is that the single estimate is the median of the probability distribution of reservoir capacity, i.e. the capacity with a probability of 50% (refs 3-4).

3. There are other assumptions, and this divergence of opinion has resulted from the fact that no systematic assessment of the inherent reliability of single estimates of reservoir capacity has been carried out. Yet the practice (of using single historic records) remains popular with hydrology practitioners, and often is the only one that can be used in certain situations, particularly in developing countries where paucity of

data is commonplace.

4. In this work, the results of an operational hydrology method for assessing the reliability of capacity estimates obtained from single historic records are presented. For ease of tractability, only systems dominated by over-year storage requirements have been considered. For such systems, annual streamflow sequences can be analysed to derive the required information (ref. 5).

METHOD OF RESERVOIR STORAGE-YIELD ANALYSIS

5. The technique of reservoir storage-yield analysis adopted is the double cycling sequent peak algorithm, SPA (ref. 2). This technique has been chosen for two main reasons: first it is the automated version of the traditional mass curve analysis which, because it is sequential, takes into account all the pertinent streamflow statistics characterising the data record; and second, the marginal distribution of capacity estimates from the technique has been found to be the Extreme Value Type 1 (EVI) (ref. 6).

6. The SPA can be formulated thus (ref. 2)
Let K_t = storage capacity required at the beginning of period t,
Y_t = required release during the period,
Q_t = corresponding inflow during period t.
If we define

$$K_t = \begin{cases} Y_t - Q_t + K_{t-1} & ; \text{ if positive} \\ 0 & ; \text{ otherwise} \end{cases} \quad \text{for } t \subseteq 2T \quad (1)$$

then the reservoir capacity C required is given by

$$C = \max (K_t); \quad t = 1, 2, ..T, T+1, ..2T \quad (2)$$

7. To implement eq. (1), the reservoir is assumed to be initially full, i.e. $K_0 = 0.0$. Also, eq. (1) is applied over 2T, where T is the length of data record, to take care of a situation in which the critical period is close to the end of record T. Hence eq. (1) is termed the double cycling SPA and the capacity C thus obtained represents the minimum storage required over the T-year period to supply the desired yield Y without shortages (ref. 7). In our simulations, T was also assumed to be equal to the useful life of the reservoir.

8. If eq. (1) is applied to several T-year records, then a distribution can be fitted to the reservoir capacities; this will ultimately enable the determination of reservoir capacity with a desired reliability (i.e. probability of non-exceedance) p over the T-year period. In this work, this distribution was assumed, following Burges and Linsley (ref. 6), to be EVI. The density function and the cdf of the EVI are respectively (ref. 8)

$$f_X(x) = \alpha \exp(-\alpha(x-u)) \exp(-\exp(-\alpha(x-u))) \quad (3a)$$

$$F_X(x) = \exp(-\exp(-\alpha(x-u))) \quad (3b)$$

where α is the scale parameter and u is the location parameter. Moment estimates of the parameters can be obtained from

$$\mu = u + \gamma/\alpha \qquad (4a)$$

$$\sigma^2 = \pi^2/6\alpha^2 \qquad (4b)$$

where γ = Eulers constant = 0.577216,
 μ = sample estimate of the mean of the distribution of capacity, and
 σ = sample estimate of the standard deviation of the distribution of capacity.

MONTE CARLO SIMULATION

9. Synthetic annual streamflow data records of various lengths were generated assuming alternately the normal, 2-p log-normal and the gamma density functions. These flows were assumed to be generated by a Markov (AR-1) persistence process thus

$$Q_{t+1} = \mu_q + r(Q_t - \mu_q) + z\sigma_q \sqrt{(1-r^2)} \qquad (5)$$

where Q_{t+1}, Q_t are the flows in periods t+1 and t respectively, μ_q is the mean flow (MAF), σ_q is the standard deviation of flow, r is the lag-1 serial correlation coefficient and z is the random standard normal variate. This simple model is adequate for our purpose since, according to previous studies, the results of the analysis are not expected to be very sensitive to moderate departures from the true model.

10. In the analysis, annual streamflow statistics used in eq. (5) were pre-specified and were assumed to be known with certainty; this eliminated the problem of parameter estimation variability. The parameters μ_q, σ_q and r specified are directly applicable to the normal model. For the log-normal model, the specified parameters were transformed using moment transformation equations, while for the gamma model, the normal variates, z, were transformed using the Wilson-Hilferty transformation (ref. 1).

11. The mean annual flow μ_q was taken throughout to be unity; this was to simplify the analysis. r was varied between 0.0-0.4 and CV values were varied between 0.2-0.6. These ranges cover typical values often encountered in most European streams. For the gamma model, estimates of the skewness coefficient are also required; these were obtained as (ref. 2)

$$S_k = 2CV \qquad (6)$$

12. The simulation experiment was similar in all cases. Using eq. (5) and the combinations of streamflow statistics, 1000 synthetic traces of annual streamflow data record of length T years were generated. In the study, T values of 5, 10, 20, 50, 80, 90 and 100 years were considered. Each of the 1000 traces was routed through the reservoir using the SPA (eqs 1&2) and the capacity required to satisfy the yield Y without failure over the T-year period was determined. To ensure that over-year storages predominate for the assumed combinations of streamflow statistics, very high yields (ranging from 0.7 to 0.98 of μ_q) were considered. Using all

RESERVOIR PLANNING AND OPERATION

Table 1. Summary of results for normal, log-normal and gamma functions

cv	r	T (Yrs)	Y	μ	σ	Hist	p-Hist	C_{99}	C_{95}	C_{90}
1	2	3	4	5	6	7	8	9	10	11
0.2	0.0	5	0.7	0.019[a] / 0.017[b] / 0.010[c]	0.015 / 0.024 / 0.031	0.054 / 0.020 / 0.028	0.796 / 0.430 / 0.764	0.177 / 0.087 / 0.107	0.113 / 0.053 / 0.068	0.085 / 0.040 / 0.051
			0.8	0.058 / 0.048 / 0.048	0.084 / 0.062 / 0.065	0.154 / 0.100 / 0.128	0.879 / 0.752 / 0.892	0.322 / 0.218 / 0.250	0.215 / 0.147 / 0.168	0.168 / 0.115 / 0.132
			0.9	0.159 / 0.132 / 0.141	0.115 / 0.095 / 0.099	0.254 / 0.183 / 0.228	0.849 / 0.752 / 0.832	0.511 / 0.410 / 0.453	0.364 / 0.309 / 0.327	0.299 / 0.256 / 0.271
		10	0.7	0.042 / 0.017 / 0.025	0.074 / 0.035 / 0.045	0.226 / 0.100 / 0.101	0.977 / 0.348 / 0.938	0.276 / 0.125 / 0.166	0.182 / 0.082 / 0.109	0.141 / 0.062 / 0.084
			0.8	0.119 / 0.081 / 0.094	0.108 / 0.072 / 0.081	0.326 / 0.278 / 0.201	0.953 / 0.553 / 0.901	0.457 / 0.307 / 0.349	0.320 / 0.216 / 0.246	0.259 / 0.175 / 0.200
			0.9	0.261 / 0.251 / 0.224	0.142 / 0.114 / 0.117	0.426 / 0.307 / 0.301	0.881 / 0.515 / 0.765	0.708 / 0.527 / 0.601	0.527 / 0.432 / 0.452	0.447 / 0.389 / 0.386
		50	0.7	0.166 / 0.074 / 0.099	0.102 / 0.054 / 0.066	0.268 / 0.215 / 0.216	0.864 / 0.106 / 0.944	0.482 / 0.244 / 0.306	0.352 / 0.175 / 0.222	0.294 / 0.144 / 0.185
			0.8	0.296 / 0.198 / 0.226	0.127 / 0.076 / 0.091	0.368 / 0.315 / 0.316	0.762 / 0.106 / 0.855	0.694 / 0.435 / 0.511	0.533 / 0.395 / 0.395	0.462 / 0.368 / 0.344
			0.9	0.551 / 0.456 / 0.479	0.195 / 0.170 / 0.170	0.481 / 0.441 / 0.477	0.410 / 0.761 / 0.566	1.161 / 1.048 / 1.013	0.914 / 0.797 / 0.797	0.805 / 0.805 / 0.702
		100	0.7	0.216 / 0.104 / 0.135	0.094 / 0.062 / 0.062	0.288 / 0.168 / 0.149	0.811 / 0.874 / 0.661	0.511 / 0.268 / 0.328	0.391 / 0.301 / 0.249	0.339 / 0.172 / 0.215
			0.8	0.369 / 0.242 / 0.277	0.122 / 0.077 / 0.089	0.388 / 0.262 / 0.249	0.633 / 0.671 / 0.433	0.752 / 0.483 / 0.558	0.597 / 0.385 / 0.442	0.528 / 0.342 / 0.394
			0.9	0.685 / 0.572 / 0.596	0.211 / 0.175 / 0.177	0.546 / 0.591 / 0.515	0.271 / 0.614 / 0.364	1.347 / 1.119 / 1.151	1.079 / 0.898 / 0.926	0.960 / 0.799 / 0.863
0.4	0.0	5	0.7	0.204 / 0.111 / 0.139	0.214 / 0.119 / 0.136	0.129 / 0.028 / 0.212	0.414 / 0.250 / 0.753	0.876 / 0.483 / 0.565	0.604 / 0.332 / 0.393	0.484 / 0.266 / 0.316
			0.8	0.312 / 0.219 / 0.254	0.245 / 0.164 / 0.193	0.329 / 0.227 / 0.312	0.598 / 0.591 / 0.695	1.081 / 0.733 / 0.796	0.769 / 0.525 / 0.576	0.632 / 0.433 / 0.479
			0.9	0.460 / 0.378 / 0.413	0.272 / 0.208 / 0.213	0.529 / 0.422 / 0.412	0.667 / 0.661 / 0.568	1.313 / 1.029 / 1.081	0.967 / 0.766 / 0.811	0.815 / 0.649 / 0.691
		10	0.7	0.358 / 0.209 / 0.249	0.241 / 0.130 / 0.152	0.237 / 0.147 / 0.311	0.343 / 0.716 / 0.716	1.113 / 0.609 / 0.727	0.807 / 0.443 / 0.534	0.672 / 0.370 / 0.448
			0.8	0.514 / 0.368 / 0.416	0.289 / 0.197 / 0.197	0.437 / 0.477 / 0.411	0.449 / 0.632 / 0.560	1.391 / 1.033 / 1.033	1.035 / 0.788 / 0.783	0.878 / 0.673 / 0.673
			0.9	0.742 / 0.628 / 0.674	0.329 / 0.260 / 0.264	0.692 / 0.622 / 0.511	0.506 / 0.598 / 0.289	1.772 / 1.443 / 1.502	1.355 / 1.166 / 1.166	1.179 / 1.018 / 1.018
		50	0.7	0.757 / 0.419 / 0.514	0.271 / 0.172 / 0.172	0.704 / 0.682 / 0.436	0.484 / 0.940 / 0.367	1.607 / 1.054 / 1.054	1.263 / 0.803 / 0.835	1.111 / 0.739 / 0.739
			0.8	1.067 / 0.749 / 0.842	0.374 / 0.265 / 0.278	1.004 / 0.921 / 0.834	0.498 / 0.783 / 0.559	2.239 / 1.580 / 1.713	1.764 / 1.244 / 1.359	1.554 / 1.095 / 1.204
			0.9	1.634 / 1.387 / 1.477	0.547 / 0.462 / 0.481	1.576 / 1.454 / 2.276	0.526 / 0.627 / 0.936	3.349 / 2.836 / 2.985	2.655 / 2.249 / 2.374	2.348 / 1.989 / 2.104
		100	0.7	0.913 / 0.509 / 0.629	0.287 / 0.161 / 0.193	0.738 / 0.801 / 0.908	0.293 / 0.947 / 0.916	1.814 / 1.012 / 1.234	1.449 / 0.808 / 0.989	1.287 / 0.718 / 0.881
			0.8	1.296 / 0.920 / 1.046	0.399 / 0.291 / 0.326	0.940 / 1.101 / 1.108	1.712 / 0.776 / 0.644	2.549 / 1.834 / 2.069	2.042 / 1.464 / 1.655	1.817 / 1.300 / 1.472

cv	r	n	p							
0.2	0.2	5	0.7	0.015 / 0.000 / 0.000	0.042 / 0.022 / 0.028	0.000 / 0.000 / 0.000	0.410 / 0.456 / 0.448	0.147 / 0.075 / 0.095	0.094 / 0.035 / 0.059	0.070 / 0.035 / 0.044
			0.8	0.054 / 0.035 / 0.039	0.078 / 0.056 / 0.061	0.042 / 0.028 / 0.052	0.505 / 0.512 / 0.669	0.299 / 0.210 / 0.232	0.200 / 0.139 / 0.154	0.156 / 0.103 / 0.119
			0.9	0.144 / 0.123 / 0.130	0.117 / 0.098 / 0.102	0.142 / 0.128 / 0.155	0.562 / 0.564 / 0.664	0.512 / 0.424 / 0.451	0.363 / 0.307 / 0.321	0.297 / 0.248 / 0.263
		10	0.7	0.042 / 0.013 / 0.023	0.068 / 0.038 / 0.045	0.000 / 0.000 / 0.097	0.291 / 0.884 / 0.934	0.254 / 0.129 / 0.165	0.168 / 0.084 / 0.108	0.129 / 0.064 / 0.082
			0.8	0.120 / 0.095 / 0.095	0.110 / 0.078 / 0.089	0.062 / 0.195 / 0.197	0.330 / 0.884 / 0.884	0.466 / 0.325 / 0.366	0.326 / 0.225 / 0.256	0.264 / 0.172 / 0.208
			0.9	0.268 / 0.238 / 0.247	0.154 / 0.129 / 0.136	0.245 / 0.187 / 0.297	0.508 / 0.508 / 0.705	0.759 / 0.623 / 0.673	0.555 / 0.457 / 0.500	0.469 / 0.424 / 0.424
		50	0.7	0.164 / 0.126 / 0.105	0.107 / 0.062 / 0.075	0.117 / 0.069 / 0.373	0.372 / 0.516 / 0.994	0.501 / 0.389 / 0.342	0.364 / 0.191 / 0.246	0.304 / 0.194 / 0.204
			0.8	0.330 / 0.325 / 0.262	0.152 / 0.106 / 0.121	0.217 / 0.210 / 0.573	0.234 / 0.592 / 0.979	0.807 / 0.557 / 0.642	0.614 / 0.423 / 0.488	0.528 / 0.373 / 0.419
			0.9	0.663 / 0.556 / 0.596	0.258 / 0.211 / 0.227	0.522 / 0.530 / 0.786	0.323 / 0.519 / 0.826	1.474 / 1.218 / 1.307	1.145 / 1.040 / 1.019	1.003 / 0.881 / 0.892
		100	0.7	0.230 / 0.110 / 0.149	0.113 / 0.093 / 0.077	0.186 / 0.202 / 0.515	0.397 / 0.915 / 0.999	0.585 / 0.308 / 0.389	0.441 / 0.221 / 0.292	0.378 / 0.193 / 0.249
			0.8	0.426 / 0.383 / 0.335	0.168 / 0.111 / 0.134	0.286 / 0.382 / 0.715	0.195 / 0.834 / 0.985	0.954 / 0.672 / 0.755	0.740 / 0.477 / 0.585	0.646 / 0.433 / 0.509
			0.9	0.855 / 0.713 / 0.765	0.299 / 0.251 / 0.261	0.524 / 0.880 / 0.960	0.098 / 0.874 / 0.812	1.793 / 1.483 / 1.584	1.412 / 1.232 / 1.232	1.245 / 1.073 / 1.105
0.4	0.2	5	0.7	0.200 / 0.173 / 0.126	0.224 / 0.137 / 0.363	0.369 / 0.363 / 0.363	0.807 / 0.929 / 0.941	0.904 / 0.458 / 0.554	0.619 / 0.323 / 0.381	0.493 / 0.241 / 0.304
			0.8	0.306 / 0.234 / 0.234	0.262 / 0.181 / 0.181	0.469 / 0.463 / 0.463	0.777 / 0.895 / 0.895	1.127 / 0.802 / 0.802	0.798 / 0.572 / 0.572	0.647 / 0.497 / 0.470
			0.9	0.453 / 0.394 / 0.394	0.291 / 0.206 / 0.223	0.569 / 0.606 / 0.563	0.714 / 0.808 / 0.808	1.364 / 1.094 / 1.094	0.995 / 0.816 / 0.816	0.832 / 0.685 / 0.685
		10	0.7	0.373 / 0.203 / 0.264	0.278 / 0.159 / 0.191	0.559 / 0.429 / 0.409	0.789 / 0.810 / 0.810	1.244 / 0.701 / 0.862	0.891 / 0.498 / 0.619	0.735 / 0.410 / 0.513
			0.8	0.594 / 0.390 / 0.452	0.332 / 0.235 / 0.260	0.659 / 0.685 / 0.509	0.692 / 0.899 / 0.656	1.592 / 1.128 / 1.268	1.169 / 0.829 / 0.937	0.983 / 0.687 / 0.791
			0.9	0.797 / 0.674 / 0.725	0.397 / 0.325 / 0.343	0.759 / 0.995 / 0.609	0.531 / 0.898 / 0.422	2.042 / 1.805 / 1.803	1.537 / 1.282 / 1.366	1.315 / 1.099 / 1.173
		50	0.7	0.902 / 0.507 / 0.660	0.384 / 0.214 / 0.282	0.783 / 1.183 / 1.074	0.434 / 0.990 / 0.918	2.197 / 1.177 / 1.544	1.619 / 0.905 / 1.186	1.403 / 0.785 / 1.028
			0.8	1.302 / 0.998 / 1.095	0.524 / 0.302 / 0.435	0.951 / 1.479 / 1.474	0.265 / 0.852 / 0.832	2.944 / 2.083 / 2.459	2.279 / 1.618 / 1.818	1.985 / 1.413 / 1.663
			0.9	2.005 / 1.442 / 1.859	0.725 / 0.680 / 0.680	1.444 / 2.230 / 1.988	0.220 / 0.819 / 0.644	4.281 / 3.749 / 3.994	3.359 / 2.934 / 3.129	2.952 / 2.579 / 2.747
		100	0.7	1.146 / 0.850 / 0.861	0.419 / 0.239 / 0.311	1.015 / 0.270 / 1.211	0.433 / 0.894 / 0.876	2.459 / 1.401 / 1.836	1.927 / 1.097 / 1.441	1.692 / 0.962 / 1.267
			0.8	1.678 / 1.203 / 1.418	0.578 / 0.422 / 0.472	1.470 / 1.370 / 1.611	0.410 / 0.713 / 0.718	3.492 / 2.528 / 2.898	2.757 / 1.991 / 2.298	2.432 / 1.754 / 2.033
			0.9	2.672 / 2.308 / 2.308	0.875 / 0.783 / 0.825	2.242 / 1.925 / 2.146	0.349 / 0.344 / 0.372	5.417 / 4.780 / 5.081	4.305 / 3.783 / 4.039	3.813 / 3.342 / 3.578

[a] = normal; [b] = log-normal; [c] = gamma; Hist = historic capacity; p-Hist = probability of historic capacity; C_p = capacity with probability p; cv = coefficient of variation of annual flows; r = serial correlation; μ = mean of capacity; σ = standard deviation of capacity.

the 1000 traces gave 1000 such capacity estimates to which was fitted the EVI density function.

13. Next, the 1000 traces were averaged to obtain a pseudo-historic sequence of length T. The averaging preserved the mean of the flows but not the standard deviation as expected. A correction was thus applied to the averaged flows so that the annual CV was preserved using

$$Q_E = (Q - \bar{Q})E_\sigma + \bar{Q}E_\mu \qquad (7)$$

where Q_E is the corrected annual flow, Q is the original annual flow (uncorrected), $\bar{Q}$ is the uncorrected mean flow, E_σ is (100+% change in the standard deviation as a result of averaging)/100, E_μ is (100+% change in the mean flow as a result of averaging)/100.

14. The pseudo-historic data record of length T obtained as above was then routed through the reservoir and its capacity obtained. This is the single estimate of reservoir capacity which will be obtained in a typical analysis, but with no knowledge at all of its reliability over the T-year period. In this work, the reliability of this single historic estimate was assessed by referring to the parent EVI distribution function.

RESULTS AND DISCUSSIONS

15. Given the combinations of streamflow and yield characteristics considered, a large volume of results was generated which cannot all be presented here for lack of space. Therefore only a section of the results are presented.

16. Table 1 is a summary of some of the results. The behaviours of μ, σ, Hist, C_{99}, C_{95}, C_{90} are as expected and are in accordance with previous studies (ref. 6). The one novel information of the Table is in column 8, which contains the estimated probability of the single historic capacity estimate, p-Hist. Contrary to earlier assumptions about the inherent reliability of this single estimate as discussed in sections 2 and 3, the probability values in column 8 of Table 1 are not unique but vary widely depending on the streamflow characteristics, the yield level and the probability distribution of the streamflows.

17. For the normal distribution, there was a general tendency for the reliability, given a particular yield, to decrease as the record length (or planning period) increases; for a given planning period, however, there was a much reduced variation across yield levels. The reduction at the longer planning periods was most pronounced for a yield of 0.9 of the mean flow. The reliability, as remarked earlier, is over the entire planning period and what these results suggest is that over a very short planning period, e.g 5 years, the reservoir capacity obtained with a single data record having a Gaussian density function could have an acceptable reliability. However, for the more practical medium to longer-term planning horizons, the single capacity estimate can be extremely unreliable. Because of the huge capital investments often involved, most reservoir systems are planned for a useful life of over 50 years. For such systems, the reliability for a single capacity estimate can be as low as 10%, yet this is an approach which has been used, and still being used, for the design of large water resources projects.

18. The normal distribution may be difficult to justify as a

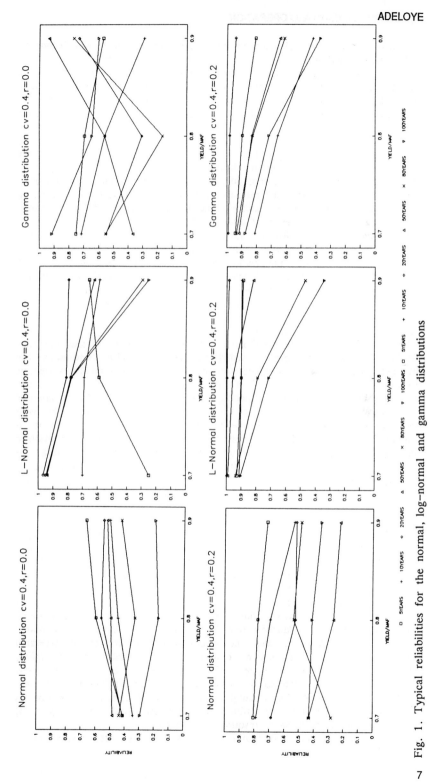

Fig. 1. Typical reliabilities for the normal, log-normal and gamma distributions

distribution hypothesis for streamflow since most hydrological processes have been known to exhibit some skewness. The two skewed distributions considered, i.e the log-normal and the gamma, are perhaps more plausible distribution hypotheses for describing the annual streamflow data. For these two distributions, there is also a tendency for the reliability to decrease as the planning horizon increases, although this trend is less discernible than that observed for the normal distribution. Also here, the reliabilities for the single historic estimate of capacity were generally higher than those obtained for the normal distribution. One possible reason for this is the inherent skewness in the log-normal and gamma distributions, whose effect would be to produce a higher capacity for a given yield; this will in turn translate to a higher reliability. The influence of this factor was most felt for a yield of 0.7 of mean flow where the reliabilities were very high, almost approaching unity, irrespective of the planning horizon. Increasing the sustainable yield from the reservoir caused the reliability to reduce, sometimes by a factor of 2 between yields of 0.8 and 0.9 of the mean annual flow. Fig. 1 is a typical comparison of the behaviour of reliability for all the three distributions considered, which helps to illustrate some of the points made above.

19.   Thus the above results have again demonstrated another limitation of analysing reservoir storage-yield relationships using a single data record, i.e. that the undisclosed reliability of the capacity is not unique but varies widely. The problem is obviously more serious for long planning horizons and for very high yields. For these conditions, the inherent reliability of the single estimate can be so low and unreliable as to make a nonsense of the entire exercise. The use of stochastic hydrology with an appropriate technique for reservoir storage-yield analysis has been advocated for a long time (ref. 9) and represents the best approach by which the analyst can ensure that, over a planning period, a reservoir capacity with the desired reliability is chosen. For streamflows which follow certain distribution functions, the task of accomplishing this has been made easy (refs 7 and 10).

REFERENCES
1. MCMAHON T.A. and MEIN R.G. River and reservoir yield. Water Resources Publications, Littleton, CO, 1986.
2. LOUCKS D.P. et al. Water resource systems planning and analysis. Prentice Hall Inc., Englewood Cliffs, NJ, 1981.
3. VOGEL R.M. and STEDINGER J.R. The value of stochastic streamflow models in overyear reservoir design applications. WRR, 1988, vol. 24(9), 1483-1490.
4. KENDALL D.R. and DRACUP J.A. A comparison of index-sequential and AR(1) generated hydrologic sequences. Journal of Hydrology, 1991, vol. 122, 335-352.
5. HOSHI K. et al. Reservoir design capacities for various seasonal operational hydrologic models. Proc. Jpn. Soc. Civ. Eng., 1978, vol. 273, 22-35.
6. BURGES S.J. and LINSLEY R.K. Some factors influencing reservoir storage. Journal of Hydraulics Div., ASCE, 1971, vol. 97, 977-991.
7. BAYAZIT M. and BULU A. Generalized probability distribution of reservoir capacity. Journal of Hydrology, 1991, vol. 126, 195-205.
8. ANG A.H-S. and TANG W.H. Probability concepts in engineering

planning and design. John Wiley, 1984.
9. MAASS A. et al. Design of water resource systems. Harvard University Press, Cambridge, Mass., 1962.
10. VOGEL R.M. and STEDINGER J.R. Generalized storage-reliability-yield relationships. Journal of Hydrology, 1987, vol. 89, 303-327.

Optimal operation of local reservoir sources

A. H. BUNCH, BSc, DMS, MICE, North West Water Ltd

SYNOPSIS
This paper addresses the way in which simulation programs with graphical output routines have been developed, and are used within North West Water to allow proactive management of reservoir systems. This approach ensures that the risk of failure of water supply may be spread across the region, thus reducing the likelihood of individual source failures. The user friendly presentation and operation of the program allows a rapid analysis and subsequent appraisal of the whole system, allowing a dynamic approach to be taken to reservoir yields as a drought event unfolds.

INTRODUCTION
1. North West Water (NWW) is a Water Services Company, serving the needs of approximately 7 Million people in the North-West region of England. It operates 168 impounding reservoirs which are used conjunctively in certain instances with river abstractions and boreholes to supply water at a rate of approximately 2,400 Mld.

HISTORICAL PERSPECTIVE
2. Within NWW the facility to determine reservoir yields and develop appropriate control curves to operate virtually all of its reservoir sources to a defined reliability for a defined yield has been available for many years (ref 1). The development of a probabilistic approach to reservoir inflows and their subsequent use has meant that individual managers have used control curves, as shown in Fig. 1, which allows them to operate their sources to an ascribed yield at minimum cost. If the reservoir is above the control curve the source can be overdrawn, but when the reservoir volume goes below the line the demand is cut back to its ascribed yield, and water is imported from the regional sources.
3. However with the integrated network of sources and aqueducts that exists within NWW, part of which is shown in Fig. 2, there is the possibility of managing resources in a much more proactive way. Recent experience within the United Kingdom has shown that different regions can experience

RESERVOIR PLANNING AND OPERATION

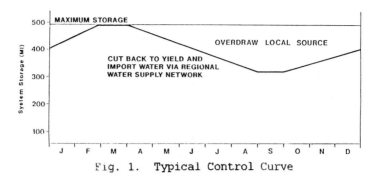

Fig. 1. Typical Control Curve

different levels of drought intensity, and that is also the case within regions. Thus if the 1989 experience is considered, the variation in rainfall was such that while certain areas of North-West England at the end of September had experienced rainfall with a return period of approximately 1 in 50 years for an event of 5 months duration, other areas had return periods of 1 in 5 years. Virtually all sources had been below their control curves for a number of months, all capable of withstanding a return period of 1 in 50 years at the beginning of the drought. Yet at the end of the summer some areas had sources with average contents of 24%, whereas other areas had sources that were 53% full.

4. It would seem unreasonable to apply a hose pipe ban in one location alone just because the criteria of a 1 in 10 years event had been met, particularly where regional supplies are available to supplement the local sources. If another area had experienced a much less severe event, then it may be appropriate to reduce the amount of support it received from regional supplies. The water thus made available could be used to supply areas that were worse off. The difficulty in carrying out the necessary transfer of regional supplies would be in determining what risk should be taken with the area that is giving up part of its support. A solution to this may be to ensure that the operation of all sources are modified to achieve the same level of future risk, notwithstanding the severity of weather already experienced. Again, the techniques to carry out the necessary analysis have been available in NWW for a number of years, and were used extensively in 1984 and 1989.

5. The system is mounted on one of the company's mainframe computers, and provides a numerical output which requires further processing to provide easily understood information for managers to act upon. The whole process could quite easily take three days of lapsed time, from beginning the file input to actually providing a report that could be acted upon. The actual time to carry out the program runs and subsequent manual analysis and information preparation could take up to three man-days, depending upon the number of

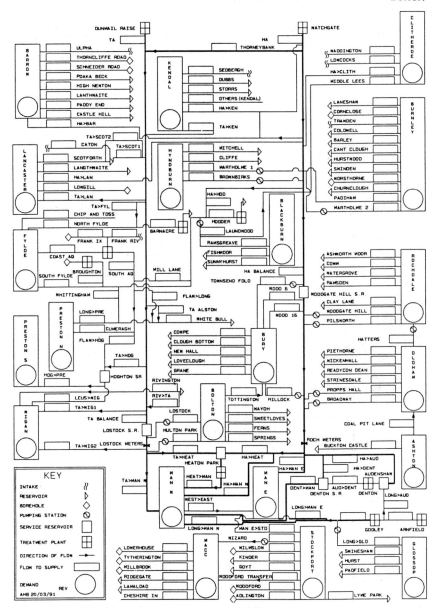

Fig. 2. Schematic of Northern Sources and Aqueducts

systems being analysed (at peak times of analysis there were as many as eighty reservoir systems and sub-systems being investigated in the 1984 drought). Any further investigations, following a "what-if" type of approach would require a further 3 days of lapsed time, meaning that the program could only be used as a guide to the actions that could be taken, rather than as a real management tool.

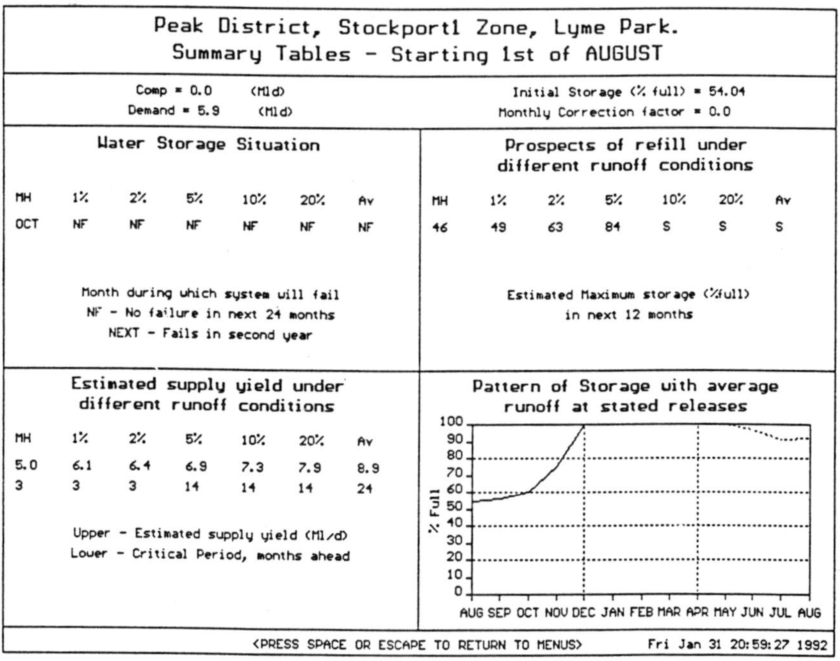

RESERVOIR PLANNING AND OPERATION

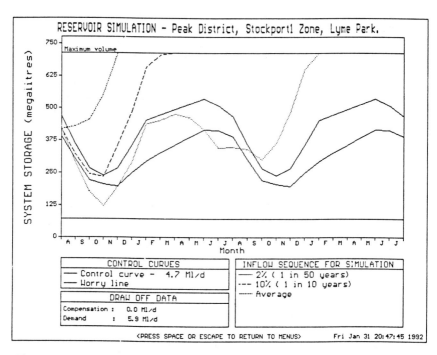

Fig. 3. Risk to supply for varying runoff conditions

Fig. 4. Reservoir Simulation for various inflows

Reservoir	Comp Demand Ml/d	Demand Ml/d	Initial Storage %Full	MH	1%	2%	5%	10%	20%	Av
Peak District, Stockport1 Zone — Estimated supply yield under different inflow conditions — Starting 1st of AUGUST										
Lyme Park	0.0	5.9	54	5.0 / 3	6.1 / 3	6.4 / 3	6.9 / 14	7.3 / 14	7.9 / 14	8.9 / 24
Kinder	1.1	14.2	76	18.0 / 14	19.0 / 14	19.8 / 14	21.2 / 14	22.5 / 14	24.2 / 14	26.9 / 24
Goyt	13.6	34.5	54	28.6 / 14	30.5 / 14	32.3 / 14	35.1 / 14	37.8 / 14	40.4 / 24	45.6 / 24
Lamaload	1.8	5.5	59	4.7 / 13	5.0 / 24	5.2 / 24	5.5 / 24	5.8 / 24	6.2 / 24	6.9 / 24
Langley	2.5	2.0	23	2.3 / 2	2.9 / 3	3.4 / 3	4.0 / 2	4.7 / 2	5.7 / 2	7.7 / 24
Starting 1st of OCTOBER										
Lyme Park	0.0	5.9	21	5.8 / 14	5.8 / 12	6.1 / 12	6.6 / 12	7.0 / 12	7.6 / 12	8.6 / 24
Kinder	1.1	14.2	62	18.8 / 14	19.1 / 12	20.0 / 12	21.4 / 12	22.7 / 12	23.9 / 24	** / **
Goyt	13.6	25.5	22	25.2 / 14	25.7 / 12	27.5 / 12	30.3 / 12	33.0 / 12	36.5 / 12	41.9 / 24
Lamaload	1.8	5.5	38	4.1 / 12	4.5 / 24	4.7 / 24	5.0 / 24	5.3 / 24	5.7 / 24	6.4 / 24
Langley	2.5	2.0	6	1.2 / 1	1.7 / 1	2.1 / 1	3.0 / 1	4.0 / 1	5.7 / 1	7.4 / 24

Fig. 5. Variation in yield during drought

6. With the development of a pc mounted version of the mainframe programs in 1989/90 (ref 2) the means to carry out quick and accurate assessments of probabilistic reservoir yields for a given start date and storage volume was available. Thus it has become possible to consider yields on a dynamic basis during a drought event, giving the ability to quantify risks across all reservoir resources in an easily applicable way, as shown in Fig. 3. The behaviour of the reservoir system can be modelled as shown in Fig. 4.

7. The package allows the reservoir systems that are to be assessed to be modelled in advance, taking into account reservoir size, historical inflows and geographical location. Some pre-processing is carried out on the inflow data so that the actual run time in carrying out a yield analysis for a number of partially drawn-down reservoirs is made as short as practicable. The whole process of carrying out analyses and providing the results in an easily understood form for eighty sources can be completed within a day, both in terms of lapsed time and man-days effort. A single source or small group of sources can be analysed within minutes, including the provision of a paper copy of the results. The other benefit in carrying out the pre-processing is that the input screens are already provided, and only require the user to input the month for which the analysis is being carried out, demand, storage and for certain analysis the choice of the probabilities of inflows to be considered. The program has

been in use on an operational basis during 1990 and 1991, when assessments of reservoir yields were carried out on a fortnightly basis during significantly dry periods. Provided the analysis is begun early enough in the event, and the suggested changes to reservoir outputs can be implemented, then there is the possibility of gradual modifications to outputs being made, so that all of the reservoir systems are supplying water at the same risk, in terms of probabilistic inflows. This can be seen in Fig. 5, which shows the dynamic change in the various probabilistic yields that occur as a drought develops.

8. What is apparent from these particular results is that the ratio of the yields, when compared to the demand, for one of the systems, stays significantly higher than the other ones. This is because of restrictions on the output, such that the recommended abstraction cannot be introduced into the distribution system. However it provides a useful reminder that must always be borne in mind, ie that system constraints must always be taken into account. Provided that restrictions such as these are recognised in advance the overall assessment can take them into account. Fortunately the events in 1991 were not so severe that the first line of demand management, namely publicity to heighten the public's awareness and encourage water saving, were required, even though technically some local areas had met the criteria to justify a further stage of demand management, namely hose-pipe bans.

DISCUSSION

9. The concept of a dynamic yield through a drought situation is not something that is easily visualised, and yet its logic is apparent when the following is considered. If the yield of a reservoir was based on minimum historic inflows, and the reservoir suffered a worse than minimum historic event, then it would fail unless its output was reduced to below its original yield. Therefore on the basis of minimum historic yields, a reservoir yield can be dynamic, particularly at extreme events.

10. The difficulty will be in deciding, once a reservoir has moved below its control line, what probabilistic inflow record should be used to control its output. This is an indeterminate problem, without the availability of highly accurate weather forecasts, extending for months in advance. It is where hydrological judgement must be exercised, taking into account recent weather, current levels of demand, weather forecasts and many more factors. However, the decision support system developed by North West Water will help to determine the risks that are being taken, and should therefore lead to improved management and operation of resources in times of drought.

REFERENCES

1 Pearson D and Walsh PD. The derivation and use of control curves for the regional allocation of water resources. Optimal allocation of water resources, MJ Lowing (ed), IAHS Pub No 135.
2 Walker S, Jowitt P and Bunch AH. Development of a decision support system for drought management within North West Water. IWEM meeting at Dawson House, Warrington on 4/12/91.

Control rules for Tayside's linked system of reservoirs supplying two demand zones

J. A. COLE, WRc plc, and A. L. GORDON, Tayside Regional Council

SYNOPSIS. Backwater and Lintrathen reservoirs are major sources of raw water which together supply Tayside with 102 Ml/d of potable water, following treatment. There are two large supply zones, the City of Dundee and parts of Angus; Dundee is normally supplied from Lintrathen reservoir via the Clatto treatment works and the remainder is normally supplied from Backwater reservoir via the Lintrathen treatment works. Exceptionally the raw water sources can be interchanged, at the cost of some extra pumping and loss of some hydropower generation, or either reservoir can be used on its own to supply both demand zones.

In order to comprehend the action of this complex system, WRc created a computer model and set up a hydrological basis for simulating the system's operation. Sixteen years of monthly inflows, compensation releases, spills and draw-offs were simulated, based on the period 1968-1983. The reliable yield was re-assessed, revealing the nearness of present demands to the system's ultimate capabilities.

Control rules were tried, based on setting a limiting contents level in each reservoir, below which draw-off is curtailed and taken over by the other reservoir. A rationing rule was also postulated, to cut back on draw-off rates in a balanced fashion, when both reservoirs fall below critical levels.

The simulation program has been set up on a PC for day-to-day use by the Tayside Regional Council Water Services Department, who have made progressive improvements to the control rules and have had experience of using them in severe drought situations during the autumns of 1989 and 1990.

To make the system more readily used as an operational tool, the PC package was configured to read dates from and output results to the standard spreadsheet used by the Water Department, for ease of display and assimilation.

Extensive refurbishment of the flow measuring stations was undertaken, along with automation of data gathering, so that frequent checks could be made on the model to allow for its use for predictive analysis of possible operating plans.

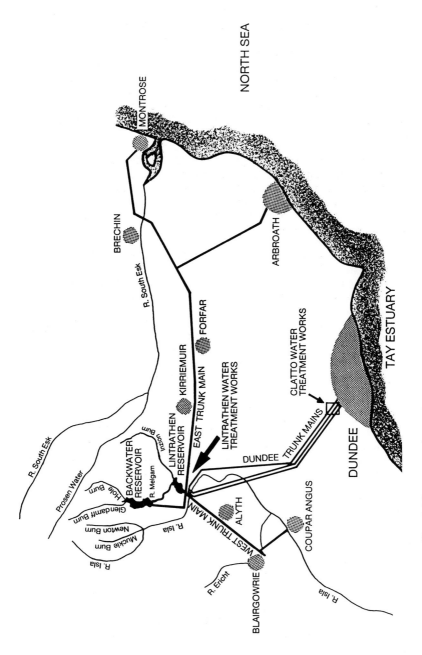

Fig. 1. The reservoirs and trunk mains supplying Tayside

INTRODUCTION
1. Tayside Region obtains most of its raw water supplies from two upland reservoirs, Lintrathen (operating since 1873) and Backwater (operating since 1968). The reservoirs are interconnected and supply raw water via several pipelines to the Clatto treatment works supplying Dundee and adjacent areas; see Figure 1. Also the new Lintrathen water treatment works (WTW) has recently (1990) been commissioned, adjacent to Lintrathen reservoir, to deliver a potable supply to rural areas of Angus and Perthshire previously served by uneconomic local works.
2. In 1988 WRc was approached by Tayside Regional Council Water Services Department (Tayside RCWSD) to collaborate in developing operating policies for their reservoir sources, incorporating the new treatment works at Lintrathen and allowing for anticipated demands in the 1990s. This paper presents some of the fruit of this collaboration, including operational experience with the control rules and their progressive development.
3. The basic tool has been a computer simulation of water storage in and draw-off from the Backwater and Lintrathen system. The prime objective has been that of ensuring a high reliability of water supply. Subsidiary objectives included exploitation of the existing hydropower unit and consideration of how to allocate water on times of shortage.
4. The following tasks were planned:-
   a. Defining the System Configuration
   b. Assessment of Hydrological and Draw-off Data
   c. Creation of Standard Data Sets for Simulation
   d. Forming Control Rules
   e. Flow-charting and Coding for Computer
   f. Performing Simulations
   g. Assessing Yield Reliability vs Control Policies
   h. Assessing Other Variables
5. The investigation tackled all the above tasks, but within the time and cost constraints, tasks c. and h. originally had less than the desired attention. Simulations were confined to use of a rainfall-derived inflow series for the years 1968-1983 only; extension back to 1929 and calibration with good river flow measurements remain as future goals.

SYSTEM CONFIGURATION
The Sources and Demands
6. Figure 2 shows the system of the two reservoirs, Backwater and Lintrathen, their inflows, draw-offs and spillages. A prominent feature of this system is cross-linkage, whereby each demand zone C and R can be supplied by either reservoir. Normally the **rural demand R** will be met from Backwater, via the hydropower turbine T and Lintrathen WTW. Alternatively R can be met by pumping from Lintrathen reservoir, to the WTW directly, not via the turbine. Normally the **city demand C** is met by Lintrathen

RESERVOIR PLANNING AND OPERATION

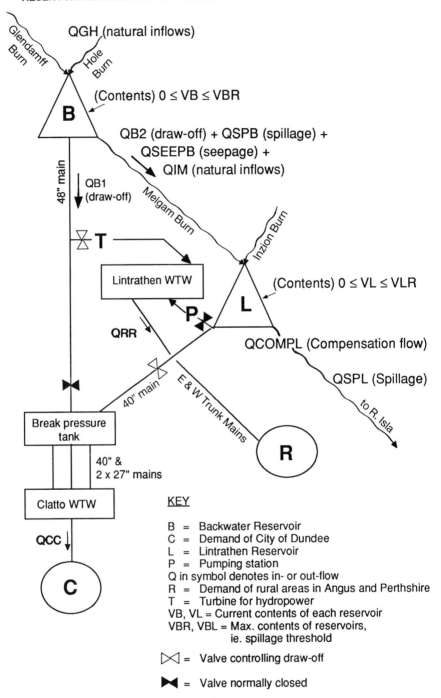

Fig. 2. Schematic diagram of Backwater and Lintrathen reservoirs and their links to demand areas

reservoir via the 40" main, but if that source is cut off for any reason, Backwater draw-off through the 48" main can be increased and deployed to demand C.

7. Spillage from Backwater flows down the Melgam Burn and is thus totally intercepted by Lintrathen reservoir. Seepage through the Backwater dam is similarly intercepted. A constant compensation discharge is legally required from Lintrathen; this and any spillage from the reservoir are lost to the system.

Table 1. Reservoir and Catchment Data

	Backwater catchment	Lintrathen catchment
Reservoir capacity (m^3)	23.15×10^6	9.72×10^6
Catchment area (km^2)	34.2	38.6
Annual rainfall (mm) [1941-1970 Met. Office data]	1029	888
Annual ET (mm)	342	393
River gauging	Glendamff and Hole Burns (since 1990)	Melgam and Inzion Burns (since 1929)
Normal supply (m^3/d)	42000 to zone R	54000 to zone C
Compensation (m^3/d)	Nil	19000
Seepage (m^3/d)	say 1500	say 300

HYDROLOGICAL AND DRAW-OFF DATA
Rainfall and Run-off Records

8. There exists a continuous data set, in a hand-written ledger held by Tayside RCWSD, of rainfalls and river gaugings in the Lintrathen Reservoir catchment for the years 1929 to 1968 inclusive. The data are for calendar months and relate to the period before Backwater Reservoir was commissioned.

9. Some empirical relationships were sought between the monthly rainfalls in the Melgam catchment and monthly (rainfall minus run-off) losses, but in view of the scatter in the points for the drier months (due no doubt to variable soil moisture conditions) no hard and fast relationship was obtained. This attempt was abandoned in favour of a simpler water-balance calculation, as explained in section 3.2 below. A further reason for so doing was the inconsistency apparent in some of the more recent run-off gaugings, when comparing the concurrent data for the Inzion Burn and Melgam Burn. Inspection of the gauging stations revealed the continual problem of siltation behind the thin-plate weirs, which has undoubtedly impaired the run-off record's accuracy over many

years, to an extent which is indeterminate.

10. Rainfall data for the period 1968 to 1983 were available in computer-readable form. The Theissen polygon method was used to obtain areally-weighted rainfalls for the Backwater catchment and for that of the Lintrathen Reservoir (excluding the Backwater catchment).

Run-off Estimated from Rainfall

11. Because of various inconsistencies in the run-off record of the two principal gauged streams, the Inzion Burn and the Melgam Burn, it was decided initially to employ a conceptual model to estimate run-offs for subcatchments directly from rainfall and evaporation data. Subcatchment rainfall was well gauged and Thiessen polygons were used to weight the point measurements. The Met. Office Advisory Service, Edinburgh, supplied MORECS calculations of monthly evapotranspiration (ET) for the 20 years 1956-75, which served to identify a cycle of evaporation, repeated annually. The run-off model was simply:-

Run-off = Rainfall - ET ; **if positive, otherwise;**
Run-off = 0

12. In the event of a negative (Rainfall - ET) value, this deficit is accumulated so as to enhance the next month's evaporation debit, and so on if necessary in succeeding months.

13. The model served as the basis for a 16 year run-off sequence, for the years 1968-83, on which the operating policies for the reservoirs could be given a preliminary test.

14. As more reliable stream gauging data are now becoming available, due to the improved maintenance of the gauging

Table 2. Operational discharges

Source or use of water	Code in Fig. 2	Discharges and demands (Ml/month)
Backwater releases	QB2 to L; QB1 via 48" main	up to 15,000 for scouring purposes
Backwater seepage	QSEEPB	542.0
Lintrathen compensation	QCOMPL	552.0
Untreated supply from Lintrathen to Clatto WTW, Dundee	QCC	1671.0
Treated supply from Lintrathen WTW via E & W trunk mains to Angus and Perthshire	QRR	(Aug 1989) 1256.2 (Feb 1991) 1358.2

weirs, it will be possible to re-assess the model. Then the model should be employed over the whole period of rainfall records, from 1929 to the present day, so giving a considerably more representative basis for simulating the operation of the system.

Draw-off, Seepage and Compensation Discharge

15. Tayside RCWSD provided the figures in Table 2 as the basis for calculating the system operation as it existed in August 1989 and was anticipated for 1991.

16. In course of the simulation runs carried out for this report, a range of QCC and QRR values was explored, extending above 2000 Mlitres per month. Based on a separate statistical study, on the combined outflow data for the five years 1984-88, no significant seasonality of demand was identifiable, so demands QCC and QRR were held constant in all the simulations. It would, however, be easy to incorporate seasonal fluctuations if ever required.

THE 'TAYSIDE' COMPUTER PROGRAM
Simulation of Reservoir Operation

17. Returning to the schematic diagram, Figure 2, this shows labels for most of the quantities entering the calculation. In principle one may use data of any time increment, eg monthly, weekly or daily inflow data, as available or appropriate. To date only monthly data have been employed, which means that spillages are rather crudely estimated.

18. The 'TAYSIDE' program was written in Fortran by B J Cox (WRc) on the basis of a control strategy and flow diagram devised by the first author. Broadly the program takes each reservoir in turn, adds the new inflow and checks on the resulting disposable water. Three cases arise:-
   1. Spillage, even after meeting demand in full
   2. Adequacy, so that demand can be met without going below shortage level, neither is full capacity exceeded
   3. Shortage, deemed to be when disposable water is less than some preset amount

In case 3 the program permits the demand to be transferred to the other reservoir, provided that still has an adequacy of water in store.

Control of Inter-zonal Transfers

19. The parameters G and H define the 'shortage level' in each reservoir, as a fraction of full capacity. A third parameter F defines the split of draw-off from Backwater reservoir. When Backwater reservoir is less than F times its full contents all of its releases go down the 48" pipeline to Lintrathen works, generating useful hydro-power en route. Above the F threshold there is the opportunity to transfer water from Backwater down the Melgam Burn into Lintrathen reservoir, space permitting.

RESERVOIR PLANNING AND OPERATION

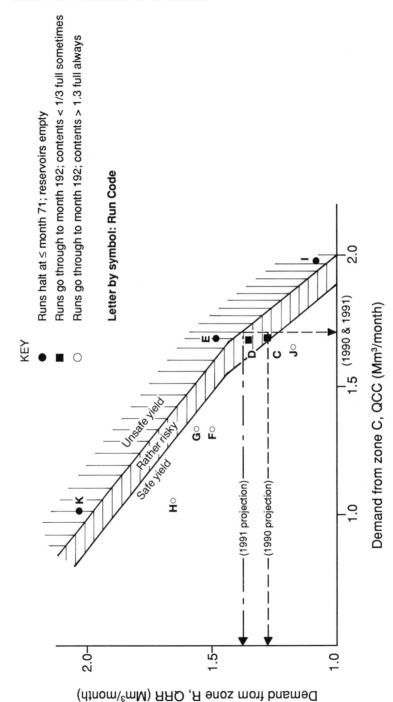

Fig. 3. Yield reliability of Tayside reservoir system assessed on 1968-83 data

Rationing Rule

20. At present the program stops running if both reservoirs reach their shortage levels. This serves as a quick way to weed out extravagant patterns of demand. It is intended as a refinement at a future stage of the study to program in a rationing rule, so that draw-off may continue at a reduced rate, if the contents go below a set threshold.

Program Testing

21. An imaginary series of inflows, draw-offs and compensation discharges was devised, incorporating a drought spell, so that the program logic could be verified. This had the effect of detecting some minor faults, which showed up by simple arithmetic and which might have been undetected later on.

'TAYSIDE' RUNS USING 1968-1983 DATA
Various Combinations of Zonal Demands

22. Numerous combinations of demands, compensation flows and control parameters have been simulated for a 192-month period, using the 1968-1983 rainfall-minus-evaporation calculation described in paragraphs 11 and 12. Most of the runs kept the control parameters F, G and H fixed and ranged over various combinations of demand from zones C and R.

23. Several runs are plotted on Figure 3, showing the region of safe yield, in which demands are met in full throughout the simulation of 192 months, and a rather more risky zone, where yields are met only by invoking transfer of demand on 6% of occasions. Note that the 1989 and projected 1990 demands fall in this 'rather risky' zone. At even higher demands the simulations do not run their full course, despite frequent use of demand transfer opportunities, so this is declared as an 'unsafe' zone.

24. In the format adopted for printing out the test simulations there is an initial list of the preset variables, including control parameters F, G and H and the starting contents. Columns show volumes of water stored at the beginning of each month. Spillages are shown as monthly volumes, when appropriate; these are inevitably crude estimates as no account can be made of within-month behaviour. Transfers of demand occur when the usual draw-off QRR from Reservoir B is temporarily met from Reservoir L, or conversely when Reservoir B takes on QCC in addition to its normal draw-off. Volumes of shortfall and surplus are not calculated every month, only when transfers are called for.

25. In subsequent development of the program for personal computer use, thought was given to the derivation of further statistics from the simulation runs, such as annual sub-totals and run totals of the amounts drawn off, spilt and transferred.

Identification of Control Rule

26. As an exercise in optimisation for the parameters F, G

RESERVOIR PLANNING AND OPERATION

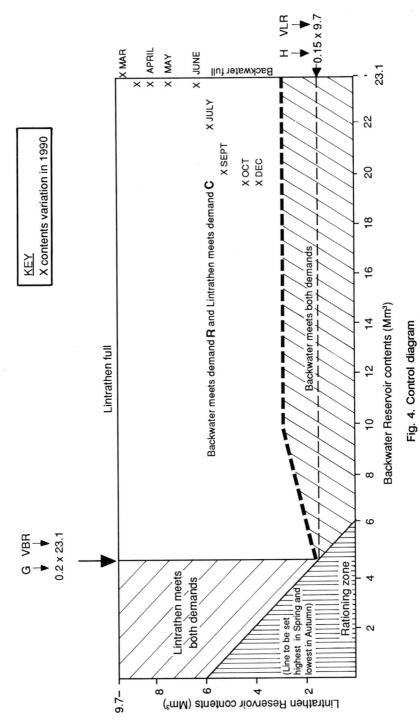

Fig. 4. Control diagram

and H, some variations were made around a case for which QCC = 1.8 Mm³/month, QRR = 1.6 Mm³/month and capacities, starting volumes and compensation flows remain as before.

27. On the criteria of choosing the combination with the least risk, viz least extent of draw-down, the parameters proving to yield the best result, are: F = 0.3 G = 0.2 H = 0.15 The sensitivity to parameter setting is not great though, which is doubtless due to the system running in its normal transfer-free mode for most of the time.

28. Incorporating the above 'best result' and guided also by noting the levels at which demand transfers occurred in other simulations, a control diagram can be drawn up as in Fig 4. This has axes corresponding to current contents of each reservoir. The upper right portion of the diagram is the normal area of working. The upper left and lower right areas are where demand transfers operate, rectifying temporary imbalance of reservoir contents.

29. There will have to be a region of the control diagram, provisionally shown as a triangle at the bottom left of Fig 4, where supplies from both reservoirs are rationed. This feature has not been explored as yet, but undoubtedly a seasonal variation in the boundary of the rationing zone should be allowed for.

OPERATIONAL CONSIDERATIONS

30. As can be seen from the earlier part of the paper there was some doubt about the results from the gauging facilities at Lintrathen and no facilities at all at Backwater. A regular cleaning regime was introduced for the gauging weirs on the Inzion and Melgam Burns near their inflows to Lintrathen Reservoir and new gauging weirs have been constructed at Backwater on its two main inlet streams, the Hole and Glendamff Burns.

31. Telephone loggers were installed at each weir with pressure transducers calibrated to give millimetre accuracy on the level readings. The data is retrieved on a weekly basis, converted to flow using a look up table in the logger software summed to give weekly inflow totals for each catchment, then set up in the required format for the 'SYMRES' package to handle using the 'SYMPHONY' database package which is in use in the Water Department for various purposes.

32. The reservoir simulation was originally set up to carry out the analysis on a monthly basis, partly to reduce the amount of data to be entered manually, but also because some of the rainfall stations are only read on a monthly basis. It was evident that this time interval was too coarse, but with the automation of the data collection, a weekly interval was adopted although any suitable interval can be used provided it is kept constant throughout.

33. Because only the two main streams are measured in each catchment, a comparison is being carried out between the estimated runoff from the catchment using the

RESERVOIR PLANNING AND OPERATION

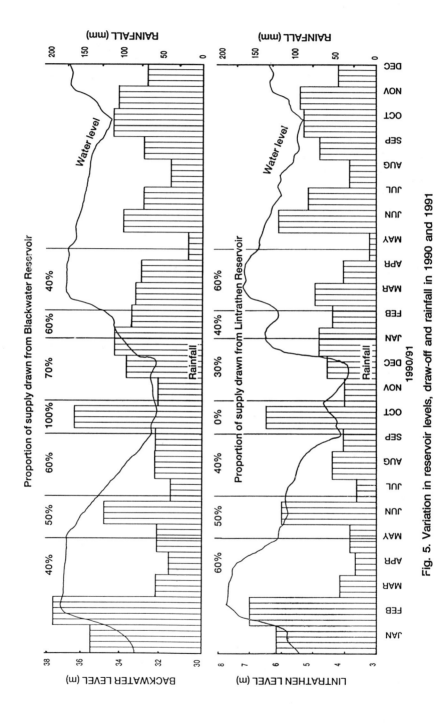

Fig. 5. Variation in reservoir levels, draw-off and rainfall in 1990 and 1991

rainfall-minus-evaporation estimate and the actual flow gaugings to see if any trends emerge which would need to be taken into account. Part of this work has also been automated using tipping bucket rain gauges linked to dataloggers. These were introduced to provide more information for the Meteorological Office and the Tay River Purification Board. They have the added advantage of reducing the overtime worked to maintain the readings.

34. The operational aim has been to equalise the rates of decline at the reservoirs as far as possible; that this is easier said than done is clear from the 1990 data plotted in Figs 4 and 5. In the past the balance of contents has been attempted empirically by drawing 60% of the total demand from Lintrathen and 40% from Backwater in the spring and autumn, 100% from Backwater in the summer and 100% from Lintrathen in the winter. Now, by using the computer simulation, the effect of any changes in operating regime can be rehearsed theoretically ahead of implementation. This is essential, given that the catchment is now operating at its design capacity with the new Lintrathen treatment works on stream.

35. The simulation package is used in three ways:
   1. To compare actual level with predicted levels using

Table 3. Lintrathen catchment rainfall (monthly and annual totals, mm)

Year	Jan	Feb	Mar	Apr	May	Jun	Jul	Aug	Sep	Oct	Nov	Dec	Annual total
1980	98	118	132	10	21	93	71	107	90	74	96	79	989
1981	63	50	112	17	68	40	48	19	190	76	98	73	854
1982	82	89	81	29	41	34	40	121	201	211	124	118	1171
1983	95	38	66	63	120	79	9	31	94	99	28	125	847
1984	157	88	130	12	22	37	44	9	126	105	229	88	1047
1985	93	21	51	72	67	71	102	198	133	39	94	104	1045
1986	101	53	61	54	134	32	50	70	8	49	84	147	843
1987	54	39	102	57	41	95	55	88	100	135	49	70	885
1988	124	65	100	61	48	24	166	156	112	159	74	28	1117
1989	66	92	106	34	25	47	20	119	59	77	39	56	740
1990	127	152	49	30	36	120	26	58	44	140	40	62	874
11-year average (1980-90)	96	74	90	40	57	61	57	89	105	106	87	86	[Annual average] [947]
31-year average (1941-71)	83	60	56	49	71	65	82	91	82	78	83	88	[888]
Change	+13	+14	+34	-9	-14	-4	-25	-2	+23	+28	+4	-2	[+59]

RESERVOIR PLANNING AND OPERATION

measured rainfall and run-off to check the accuracy of the model.
2. To evaluate the effects of any changes in operation.
3. To simulate "what if?" situations when the reservoirs are operating at extremes of supply and demand.

36. The latter can be done using 'Dry Year' or long term average time series and it is possible to allow for increase in evaporation by increasing demand for parts of the year. These adjustments were used when during the summer of 1990 the reservoir showed signs of stress when it was found that an adjustment to the compensation would be sufficient to allow demands to be met. The situation in fact resolved itself while the legal preparations were being made for this.

37. An interesting offshoot from the reservoir studies is that we have noted a change in the annual rainfall distribution over the past few years. More of the rain appears to be falling in the spring and autumn although the same total rainfall is being recorded (see Table 3). It is still too early to say that this change is significant, but it indicates that we must keep the reliable yield under review.

CONCLUSIONS AND RECOMMENDATIONS

38. Based on simulations made to date using the rescaled 1968-1983 rainfall-minus-evaporation data, the Backwater and Lintrathen reservoirs are just able to sustain the forecast 1991 demands, provided the system is operated flexibly, using a control rule of the type offered in Fig 4.

39. The control rule diagram should be refined be development of realistic rationing rules for low-contents states.

40. A more extensive evaluation of the hydrology of the reservoired catchments, coupled with use of statistically generated data sets, will now be required in order to arrive at a definitive yield/reliability relationship; Fig 3. has been a interim guide to that relationship, but it will need validation.

41. Arrangements are already in hand to improve the accuracy of the river gauging in the catchments. Once these are satisfactory a check can be made on the rainfall-evaporation model, which itself could be used as a criterion of selection for past river gauging data, from which some good runs may be salvaged.

42. The maintenance and operation of all reservoir plant and the periodic monitoring of reservoir parameters need to be carried out on an unfailingly regular basis. Quality Assurance for such activities should incorporate a routine reminder system.

43. The computer model has proved a useful tool for reservoir management, reinforcing traditional decision making. It should therefore continue to be developed to ensure that maximum benefit is gained from the Lintrathen-Backwater complex.

The South East Wales Conjunctive Resource Scheme

S. A. GALLIMORE, BSc, Welsh Water, and J. C. MOSEDALE, BSc, MIWEM, National Rivers Authority, Welsh Region

SYNOPSIS. The South East Wales Conjunctive Resources Group, comprising water resources, water treatment and fishery staff from the National Rivers Authority and Welsh Water, was established to optimise the combined use of the water resources serving south east Wales. This paper identifies the current resources and briefly considers the historical operation of the sources and the changes instigated following the droughts of 1976 and 1984. More recent changes, particularly to the regulation of the River Usk, are discussed.

DESCRIPTION OF SOUTH EAST WALES CONJUNCTIVE RESOURCE SYSTEM
1. The South East Wales Conjunctive Resource area is approximately 2400 sq km and affords a supply of 502 Ml/d to a population of 1.25 million people concentrated in the conurbations of Newport and Cardiff and the industrial valleys of Gwent and Glamorgan and the Usk valley. The resources available to meet this demand are shown schematically in Figure 1.

OPERATION BEFORE 1976
2. Prior to 1976, the available water resources comprised:
a) Six upland direct supply reservoirs located in the Heads of the Valleys area (roughly mid-way between Cardiff and Brecon) and constructed between 1884 and 1939. These are, from west to east, the Taf Fawr group (Beacons, Cantref and Llwynon reservoirs), the Taf Fechan group (Upper Neuadd and Pontsticill reservoirs) and Talybont.
b) Usk reservoir, situated in the headwaters of the River Usk near Trecastle used for both direct supply to Swansea valley and regulation of the River Usk.
c) A pumped storage reservoir at Llandegfedd, near Pontypool, relying on abstraction from the River Usk at Rhadyr near to the tidal limit.
d) An abstraction from the River Usk at Llantrisant, also near to the tidal limit.
e) A number of springs and small reservoirs collectively

RESERVOIR PLANNING AND OPERATION

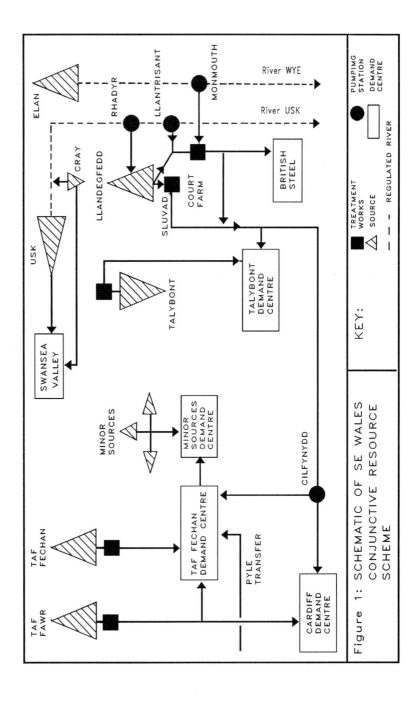

Figure 1: SCHEMATIC OF SE WALES CONJUNCTIVE RESOURCE SCHEME

known as the Minor Sources mostly located in the Heads of the Valleys area near Merthyr Tydfil.

3. The raw water from these sources was supplied to a number of major treatment works. There were two works for the Taf Fawr sources (Cantref and Llwynon), two works for the Taf Fechan sources (Neuadd and Pontsticill), a separate works at Talybont, and two works (Sluvad and Court Farm) which are effectively one unit and were associated with Llandegfedd and the River Usk abstractions at Rhadyr and Llantrisant.

4. There were two other upland sources in the area, but neither were available for water supply purposes in south east Wales:

a) The Elan Valley reservoir complex in mid-Wales provided direct gravity supply to Birmingham and had a minor river regulation role on the River Wye.

b) Cray Reservoir, situated near Usk reservoir, which provided a direct supply to the upper Swansea valley.

5. Each of the above reservoirs was constructed to provide direct supply to a specific demand area. There was minimal provision for the transfer of water between sources and demand areas. The demand areas for each source and the usable storage is detailed below:

Source	Storage (Ml)	Original Demand Centre
Talybont	10,340	Newport
Taf Fawr	7,869	Cardiff
Llandegfedd	23,650	Cardiff
Taf Fechan	15,913	Rhymney and Taff valleys and Llantrisant
Elan	99,106	Birmingham
Usk	11,990	Swansea Valley
Cray	4,204	Swansea Valley

6. The lack of any real flexibility led to major problems in 1976. Some sources were virtually empty whilst others were not. Most of Cardiff and Merthyr Tydfil were subject to 18 hour supply rota cuts and supplies were only maintained by obtaining drought orders to permit abstraction from the River Usk, further depleting the already low flows. However, Newport's supplies were not interrupted.

SYSTEM CHANGES 1976 - 1989

7. Following the drought of 1976, Welsh Water Authority made substantial investment to increase both the resources available to south east Wales and the flexibility of their use. Initially progress was made by developing the embryonic River Wye Regulation Scheme for water supply purposes. This started in 1985 by redeploying the Elan Valley reservoirs' compensation water and licensing an associated abstraction from the River Wye at Monmouth. This abstraction is pumped

to Court Farm in the lower Usk catchment for treatment. In addition, powers were obtained to use Cray reservoir for regulation purposes to support the Rhadyr abstraction to Llandegfedd.

8. Operational yields were derived for each of the reservoirs and the regulation systems using the methods discussed in Reference 1.

9. Extensive analysis of inflow data for each of these sources has shown that the majority are single season critical with the exception of Usk Reservoir. Taf Fechan and Talybont reservoirs are borderline.

10. The Conjunctive Resource area can be subdivided into a number of demand centres which are listed below together with their current demand values:

Cardiff Demand Centre	223 Ml/d
Talybont Demand Centre	62 Ml/d
Taf Fechan Demand Centre	120 Ml/d
Minor Sources Demand Centre	23 Ml/d
British Steel (Llanwern)	74 Ml/d
Total	502 Ml/d

11. The total demand of 502 Ml/d is matched by an operational yield of 510 Ml/d. Overall, there are just sufficient resources to meet the demands under drought conditions.

12. In the Heads of the ValLeys area, the small, elevated Minor Sources have a particular problem - their output has to be reduced early in drought conditions to maintain certain minimum supplies. Operationally, this shortfall can only be met from the Taf Fechan reservoirs which are already unable to satisfy their demand area fully. The solution has come from the additional resource from the regulated River Wye via Court Farm which is now able to supply the Cardiff and British Steel demand centres. This in turn allows resources in the Taf Fawr reservoirs to be redeployed to support the Taf Fechan sources up to a maximum of 42 Ml/d. An extra 10 Ml/d can be supplied into the Taf Fechan demand centre from south west Wales (originating from Llyn Brianne reservoir) via the Pyle Transfer together with a further 20 Ml/d from Sluvad and Court Farm by pumping up the Taff valley at Cilfynydd.

13. This has required construction of a number of interconnections, small pumping stations, and a large increase in treatment capacity at Court Farm. Additional minor interconnections allow Sluvad/Court Farm water to support certain parts of the Talybont supply area.

OPERATIONAL MANAGEMENT DEVELOPMENTS AFTER 1989

14. Since the droughts of 1989 and 1990, the emphasis has

been on refining the operation of the existing resources to maximise their use, rather than on major engineering capital investment. The small surplus of yield over demand requires careful management of the resources to minimise the risk of supply failure and the need to apply for drought orders. This requires particular care early in the year to ensure that the upland, direct supply sources are not overdrawn too soon. Security of supply only exists if there is sufficient water retained in Taf Fawr to support Taf Fechan. This in turn requires Llandegfedd to replace the Taf Fawr water normally provided to the Cardiff demand centre.

15. Operational control rules have been agreed by the Group. The underlying premise of the control rules is that the system is operated to maximise resource and supply security during times of drought or when a resource shortfall threatens rather than to minimise costs.

OPERATIONAL CONTROL RULES

16. As an example of the control rules used, this section examines the operation of the Cray, Usk and Llandegfedd Reservoirs - the River Usk Regulation Scheme.

17. Prior to 1991, the Rhadyr abstraction had two main conditions. Firstly, below a particular river flow, water had to be released from Usk Reservoir to match the abstraction on a 'put and take' basis. Secondly, releases had to be made to the river to leave a minimum river flow of 227 Ml/d below the abstraction point. Under such conditions, resources should have been 'donated' to the river which then could not be abstracted. In practice, this never happened.

18. In practice during the 1976 and 1984 droughts, the regulation water was held in Usk Reservoir until late in the drought. By that time, the river flows had recessed to the point where the maintained flow condition came into force. Since overall resources were depleted, drought orders were obtained to remove the maintained flow condition and the less stringent 'put and take' condition prevailed. Once the regulation storage was exhausted, further drought orders were obtained authorising abstraction from the natural flow.

19. During the 1989 and 1990 droughts, drought orders were avoided by transferring Usk Reservoir storage to Llandegfedd early in the season and therefore before the maintained flow condition became effective.

20. Whilst avoiding drought orders and fully using the Usk Reservoir resource, the early season transfer does have its' drawbacks. Usk Reservoir has poor refill characteristics. Drawing down the reservoir early in the summer may prove unnecessary, and could result in the reservoir not being full at the start of the next summer and compromise its operation in that second year. In addition, the direct supply to Swansea Valley would be restricted to the minimum 5 Ml/d throughout the summer and possibly the following winter. These problems have been overcome by the agreed conrtol

RESERVOIR PLANNING AND OPERATION

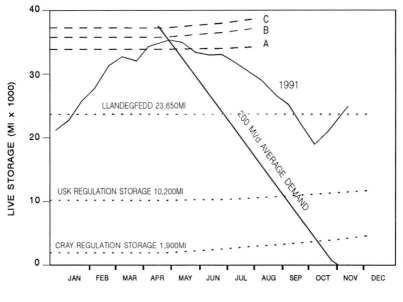

FIGURE 2: USK REGULATION AGGREGATED STORAGE

KEY:
A LLANDEGFEDD + USK
B LLANDEGFEDD + USK + CRAY
C LLANDEGFEDD + USK + CRAY + 1500 Ml FROM DROUGHT ORDER

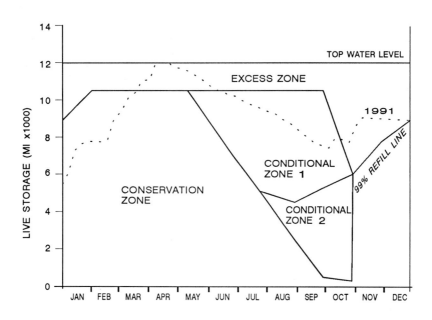

FIGURE 3: USK RESERVOIR CONTROL DIAGRAM

diagrams (Figures 2 and 3) and by varying the Rhadyr abstraction licence to remove the maintained flow condition, initially on a trial basis.

21. Both Cray and Usk reservoirs are still able to provide a direct supply to the Swansea valley. This can be at rates of 43.2 Ml/d and 36 Ml/d respectively when storage allows. However the majority of the storage in these reservoirs is reserved initially for river regulation purposes to support supplies in the Conjunctive Resource area. As the year progresses the total quantities in storage are matched against demands on the Llandegfedd/Court Farm system to establish how much of the Cray and Usk storage needs to be retained for regulation purposes.

22. This is illustrated in Figure 2. This shows the available storage of Cray, Usk and Llandegfedd singly and aggregated on the vertical axis and the average demand line of 200 Ml/d plotted against the calendar on the horizontal axis. It should be noted that the demand curve terminates on the 1st November, the agreed drought end date. After that date, flows in the River Usk are assumed to have recovered sufficiently to allow unsupported direct abstraction.

23. Figure 2 demonstrates that the combined usable storage of the three reservoirs cannot guarantee to meet the average demand (assuming the onset of drought conditions) unless storage is at a maximum on the 5th May.

24. As the summer progresses, the storage required to meet the south east Wales demand reduces. Any surplus storage can be directed to the Swansea valley as direct supply. Thus if the total storage in Usk and Llandegfedd is in excess of 34,000 Ml on the 12th May it is possible to reallocate the remaining Cray storage for direct supply. If the Llandegfedd storage is in excess of 23,500 Ml after the 5th July the Usk Reservoir storage can be reallocated for direct supply.

25. Study of the River Usk drought flow records shows that the earliest date on which the 'put and take' condition could ever come into operation is the 10th May (1976). Using the flow recession characteristics of the River Usk, it is possible to forecast the minimum quantity that can be abstracted from the natural river flow up to 20 days in advance. Hence, with suitable flow conditions, the dates on which Cray and Usk can be reallocated to direct supply may be earlier than indicated in paragraph 25.

26. Figure 3 illustrates the control diagram for Usk Reservoir, which is used in conjunction with the curve shown in Figure 2 and has been agreed by the Consultative Group.

27. This control diagram was constructed using the '10-Component Method' (Reference 2). Detailed explanation of the construction of the control rules is not given here. However the control rule diagram clearly defines how the reservoir should be operated. The EXCESS ZONE defines the storage position where refill is guaranteed and maximum direct supply can be taken. The CONSERVATION ZONE defines

where the priority must be to conserve storage for regulation against the onset of a drought and allows only the minimum direct supply abstraction. The CONDITIONAL ZONE defines the area where storage can be allocated between regulation and/or direct supply (Zone 1) or regulation with minimum direct supply only (Zone 2) as the summer progresses and the overall resource position unfolds.

FUTURE DEVELOPMENTS

28. An additional feature of Figure 3 which is currently being examined is the ability to make specific releases to the river purely for environmental and fishery considerations. If in the Autumn the storage is above the 99% refill curve and the full direct supply entitlement is being achieved, releases can be made for fishery purposes. This may benefit the full length of the river in Autumn and the operators can be secure in the knowledge that refill will not be compromised.

29. Other issues are the conditions for the River Usk abstractions, the relation of the River Usk regulation to control diagrams for the Taf Fawr and Taf Fechan sources, and the potential to vary current statutory compensation discharges.

CONCLUSION

30. The Consultative Group has brought together staff from both the NRA and Welsh Water to optimise existing resources and resolve matters of potential conflict. Progress to date is evidence of the positive results that can be achieved by close co-operation between the two organisations. These include improved supply security, operational flexibility and benefits to the river flows for migratory fish and the river in general

31. There remain many opportunities to operate this complex Conjunctive Resource Scheme in different ways to improve resource availability, improve river flows or improve resource security. The organisational framework is in place to enable these developments to be progressed.

ACKNOWLEDGEMENTS

The authors acknowledge the efforts of their colleagues on the Consultative Group. The views expressed here are the authors, and not those of their respective organisations.

REFERENCES

1. JACK W.L and LAMBERT A.O. 'Operational Yield'. Water Resources and Reservoir Engineering Conference, 1992.
2. LAMBERT A.O. 'An Introduction to Operational Control Rules Using the 10-Component Method.' BHS Occasional Paper No.1, 1989.

Comparison of the yields of variants of the Broad Oak Water scheme

J. K. HALL, BSc, MIWEM, and D. E. MACDONALD, BSc, MSc, MIWEM, Binnie & Partners

---

SYNOPSIS. A reservoir simulation model has been constructed on a 386 PC computer to help plan and design a pump storage reservoir near Canterbury in East Kent. The model has been used to assess the yield and environmental impact of alternative scheme configurations. The structure of the model, the derivation of input sequences of daily flow and the results of the simulations are described.

INTRODUCTION

1. Broad Oak Water is one of a number of options being considered to meet the demand for increased supplies of water to Kent. The proposed site is close to the village of Broad Oak in the Sarre Penn valley, north of Canterbury (see Fig. 1). It is the first reservoir scheme to be planned in Great Britain for more than 10 years. It therefore provides one of the few recent examples of the application of up-to-date hydrological and computer simulation techniques in the field of reservoir planning. The scheme has also had to take account of the latest environmental legislation and guidelines.

2. The construction of a reservoir in this locality has been under consideration since the site was first identified by Lapworth in 1946 (ref. 1). Detailed plans and outline designs were drawn up by Binnie & Partners in the late 1970s but the proposed scheme was rejected at Public Enquiry on the grounds that the need for a new reservoir had not been sufficiently demonstrated.

3. In 1989 the three water companies responsible for maintaining supplies in Kent decided to commission a study to review the water resources situation in Kent. They formed a Joint Steering Committee which appointed Binnie & Partners, working in association with Oakwood Environmental, to investigate and promote new resources to ensure a reliable supply of good quality water for the next 30 years.

RESERVOIR PLANNING AND OPERATION

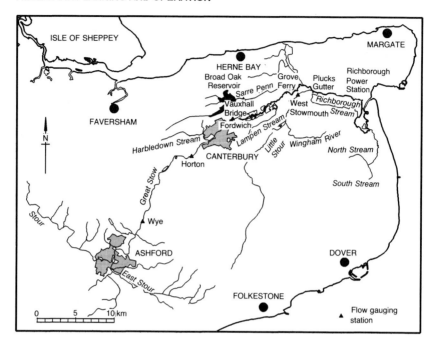

Fig. 1. Location of the proposed reservoir

4. This paper describes the studies that have been carried out to derive yield/storage curves for three basic scheme configurations and to assess the impact of varying operation rules on scheme yields and river flow.

LAYOUT OF THE SCHEME

5. The proposed damsite is at Vale Farm in the Sarre Penn valley (Fig. 1). Following an initial review of economic and physical constraints it was concluded that the practical maximum limit for the Top Water Level (TWL) is 47 mAOD which would provide a gross storage capacity of 24,700 Ml. However a range of smaller reservoirs was also investigated, particularly those that would avoid the inundation of properties in the village of Tyler Hill.

6. The catchment area to the proposed damsite is 18.0 km^2. The main source of inflow to the reservoir will be pumped flows from the river Great Stour. The Great Stour drains the north eastern part of the Weald and flows through Ashford, Canterbury and Sandwich to enter the English Channel at Pegwell Bay. Throughout dry periods, flows in the Great Stour are well sustained by springs which are fed by the groundwater storage of the Chalk aquifer. At Horton gauging station, 4 km upstream of Canterbury, the river has a Base

Flow Index of 0.69.

7. Three basic scheme configurations were considered:

a) a direct supply reservoir with pump intake at Vauxhall Road Bridge, immediately to the east of Canterbury,

b) a direct supply reservoir with pump intake at Plucks Gutter, approximately 12 km east of Canterbury, and immediately downstream of the confluence with the Little Stour,

c) a river regulating scheme which uses reservoir storage to balance natural flows in the Great Stour downstream of Vauxhall Road Bridge to a water supply intake at Plucks Gutter.

8. Tidal influence extends up the Great Stour as far as Fordwich, about 1½ km downstream of Vauxhall Bridge. However, saline intrusion rarely extends beyond Plucks Gutter where there is an existing water supply intake which has a temporary licence to abstract in times of drought.

DESCRIPTION OF THE MODEL

9. Reservoir yield/storage curves were derived and key operating parameters of the system evaluated using a computer simulation model. The model was constructed as a 3-dimensional spreadsheet on a 386 PC computer using version 3.0 of Lotus 1-2-3. Modern spreadsheets have a number of very useful features that make them particularly suitable for models of this type. Their in-built formatting and graphics routines permit the rapid preparation of high quality output. Results of intermediate calculations are easily accessible for perusal/detailed checking. All aspects of program operation are controlled by macros which automate data entry, calculation, printing and plotting procedures. These provide a user-friendly interface which greatly simplifies the use of the model.

10. The model runs as a step by step mass water balance. It assumes that all the water in the Great Stour at the intake site above a prescribed Minimum Residual Flow (MRF) can be pumped to storage, up to a specified maximum pump capacity. No routing effect is applied to inflows when the reservoir is full. Any excess water above total gross storage and drawoff requirements is assumed to be spilled to the stream below the dam the same day. The system is assumed to be 95% efficient in pumping the available river flow to Broad Oak Water when the pumps are operating at less than their maximum rate.

RESERVOIR PLANNING AND OPERATION

11. For the river regulation option, a target flow at the Pluck's Gutter intake is estimated from a knowledge of the flow recorded at Vauxhall Road Bridge, the required drawoff rate and the prescribed MRF. A projected surplus/deficit is then computed from the estimated river flow and a decision is made on whether to fill the reservoir with surplus flow at Vauxhall Bridge or augment river flow by releases from reservoir storage.

MODEL INPUT

Historic flow sequences

12. Although there is no gauging station on the lower reaches of the Great Stour downstream of Canterbury, long term daily flow records are available for the Great Stour at Horton and for the Little Stour at West Stourmouth. These two stations record the runoff from 548 sq km of the 629 sq km catchment which drains to the Great Stour at Pluck's Gutter and their combined flows comprise 91% of the estimated long term mean runoff to Pluck's Gutter (see Table 1).

13. The Horton record begins in October 1964 and the record is continuous to date except for three short gaps totalling 36 days. These gaps were infilled by correlation with the flows recorded at Wye gauging station, 12 km upstream. The Horton record was also extended back to January 1961 by the same method.

14. Records for the Little Stour at West Stourmouth are available from 1966 to date but there are numerous gaps in the record. These were infilled using the HYRROM conceptual rainfall/runoff model. This model was also used to extend the West Stourmouth model back to January 1961.

15. Inflows to the Great Stour from areas between Horton gauging station and the Little Stour confluence were calculated from catchment water balance studies together with the evidence provided by the short term streamflow records available for the Great Stour at Vauxhall Bridge.

16. A record of mean daily flows in the Great Stour at Pluck's Gutter was obtained by combining the flows at Vauxhall Bridge with those at West Stourmouth on the same date and then adding a further 6% of the corresponding Horton flow to represent the inflow from the 42 sq km ungauged area between Vauxhall Bridge and the Little Stour confluence.

17. Broad Oak Water lies in the Valley of the Sarre Penn Stream and has a direct catchment of 18 sq km. Records of mean daily flows in the Sarre Penn are available at

Calcott gauging station, which is close to the damsite, for the period February 1975 to October 1978 and June 1980 to date. These data were correlated with Horton flows and the relationship used to estimate direct inflows to the reservoir from the Horton flow record.

18. There is provision in the model to allow for varying effluent returns from Canterbury sewage treatment works or elsewhere so as to allow investigation of the effect of future increases in effluent returns and of possible options to divert effluent to the Great Stour from elsewhere.

Table 1 Average annual runoff from various subcatchments of the Great Stour basin, 1961-1990

Catchment	Sub-catchment	Catchment area		Mean annual flow m^3/s	% of flow at Horton	% of flow at Pluck's Gutter
		km^2	% of Plucks Gutter			
Great Stour	Upstream of Wye	226.2	36.0	2.230	69.9	-
Great Stour	Upstream of Horton	341.4	54.3	3.186	100.0	70.8
Great Stour	From Horton to Vauxhall Bridge	38.3	6.1	0.212	6.7	4.7
Little Stour	Upstream of West Stourmouth	206.6	32.9	0.914	28.7	20.3
Great Stour	Ungaged remainder	42.4	6.7	0.191	6.0	4.2
Great Stour	Pluck's Gutter	628.7	100.0	4.503	141.3	100.0

Note: Mean annual flow at Pluck's Gutter includes no allowance for effluent Canterbury STW.

Design drought flow sequences

19. The project design criteria stipulated that the reservoir should be capable of supplying the required drawoff throughout a 1-in-50 year drought; i.e. a probability of failure in any one year of 2%.

20. The initial planning of a reservoir scheme involves the evaluation of a large number of possible scheme options, each perhaps having a different critical period. The critical period is defined as the period between the reservoir being full and subsequently reaching maximum

RESERVOIR PLANNING AND OPERATION

drawdown during the design drought. In order to meet the design criteria, reservoir inflow over the critical period must have a return period of 1 in 50 years.

21. To enable a large number of scheme variants to be investigated on a consistent basis a synthetic stacked drought was used for the design flow sequence. In a stacked drought the minimum runoff volume for all durations are nested within each other. Therefore whatever the critical period, the minimum runoff volume within the drought flow sequence for that duration has a return period of 1-in-50 years.

22. 2% drought flow sequences were derived, in conjunction with the Institute of Hydrology (ref.2), from probability analyses of the 30 year flow records at Horton and at Pluck's Gutter. These flow sequences spanned four years starting in February. Fig. 2 shows the drought flow sequence at Pluck's Gutter. A full account of the method used to construct the drought sequence is given by Twort et al (ref. 3). Most of the flow in the Great Stour is derived from groundwater storage which is depleted gradually over an extended dry period. In a four year drought it is desirable to model the scale of recovery in the final year and so minimum flows were arranged in year 3 of the drought with the final year being the wettest of the four.

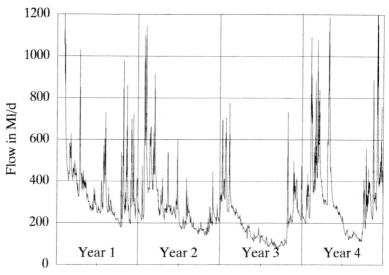

Fig. 2. 2% design drought flow sequences at Pluck's Gutter

YIELD OF THE PRINCIPAL SCHEME VARIANTS

23. The simulation model was run using the design drought flow sequences described above to compute the 2% yield of a large number of possible scheme options. The yield of the three principal scheme variants are given in Table 2 for both a full size and a small reservoir. Although it is difficult to compare a direct abstraction reservoir with a river regulation scheme due to the differences in operating rules, the three variants listed are broadly comparable in terms of pump capacity and minimum residual flow requirements.

Table 2. Design drought yield of the three principal scheme variants

Scheme variant	Intake location	MRF m^3/s	Design drought yield in Ml/d	
			Top Water Level (m AOD)	
			47.0	41.5
Direct abstraction	Vauxhall Bridge	1.3	77	47
River regulation	Pluck's Gutter	1.68	90	48
Direct abstraction	Pluck's Gutter	1.68	97	49

Notes: 1. Pump capacity = 3 times yield

2. The MRF in the river regulation scheme is not maintained

24. Results show that there is much less difference between the yields of alternative schemes involving a small reservoir than with a large reservoir. This is because the yield depends much more on available storage than on inflow. The critical period is short, probably only spanning the 6 month dry season, and river flows for most of this period are close to or below the MRF. Consequently little, if any, flow is available to pump to the reservoir to boost yield. With a large reservoir the critical period will probably span at least one wet season during which significant amounts of pumping can occur to refill the reservoir. The additional runoff available at the downstream intake at Plucks Gutter then becomes significant.

25. The lower yield obtainable from a river regulation scheme compared with a Pluck's Gutter direct abstraction scheme is due to the inability to regulate river flow precisely to the drawoff and MRF requirements at the intake. This results in the wastage of stored water with a consequent reduction in yield.

RESERVOIR PLANNING AND OPERATION

EFFECT ON YIELD OF VARYING MODEL PARAMETERS
26.   It is self-evident that yield increases with the available live storage. Yield/storage curves for a reservoir with intake at Vauxhall Road Bridge are shown on Fig. 3. However it should also be noted how sensitive yield is to the prescribed minimum residual flow (MRF) at the intake. A flow of 0.8 m³/s corresponds to the minimum recorded flow in the Great Stour at Vauxhall Road Bridge. The average daily flow (ADF) at this point is 3.4 m³/s while the 95 percentile flow is about 1.3 m³/s.

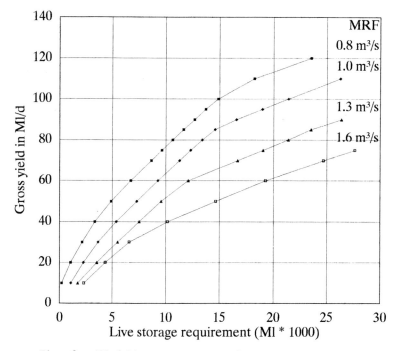

Fig. 3. Yield/storage curves for a reservoir with intake at Vauxhall Road Bridge.

27.   Fig. 4 shows the effect that MRF has on yield and the critical drawdown period for a small reservoir (TWL = 41.5 mAOD) with intake at Pluck's Gutter. As well as reducing yield, high MRFs tend to lengthen the critical period by restricting refill both during and at the end of the period of most rapid drawdown. This effect is also seen on the plots of reservoir drawdown in Figs. 5 and 6.

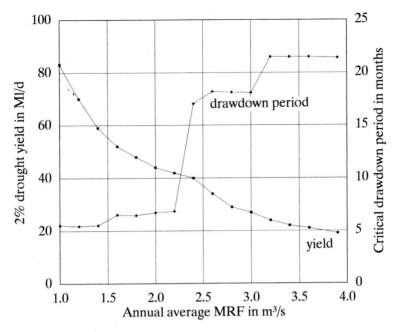

Fig. 4. The relationship between yield, critical period and MRF.

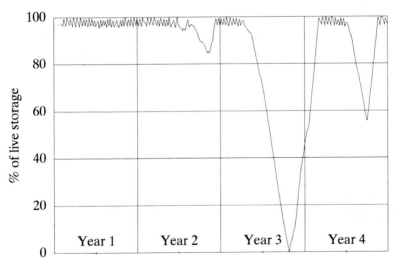

Fig. 5. Reservoir drawdown plot for design yield of a small reservoir with a MRF at Pluck's Gutter of 1.6 m³/s.

RESERVOIR PLANNING AND OPERATION

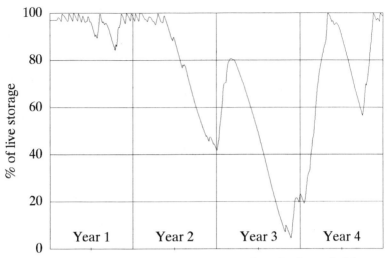

Fig. 6. Reservoir drawdown plot for design yield at a small reservoir with a MRF at Pluck's Gutter of 2.8 m³/s

28. Pump capacity has more influence on refill time than on the yield that is theoretically obtainable from storage. However, the two are linked in that resource managers are uneasy if a storage takes a long time to refill after a prolonged period of drawdown. Table 3 shows the relationship between yield and pump capacity for a large reservoir (TWL = 47.0 mAOD). It can be seen that yield begins to fall off rapidly when pump capacity is less than two times the drawoff rate.

Table 3. Relationship between yield and pump capacity.

Pump capacity Ml/d	Maximum yield sustainable by available storage		Max yield that ensures 100% refill by end of drought sequence
	2% drought yield Ml/d	% of storage remaining at end of drought sequence	2% drought yield Ml/d
300	98	100	98
250	97	100	97
200	96	86	93
150	92	52	80
100	88	20	60

Notes:

1. TWL = 47.00 m AOD

2. Total live storage = 22,800

3. Intake at Plucks Gutter with a constant MRF of 145 Ml/d.

29.   The effect on the catchment water balance of constructing a reservoir is to increase evaporative losses by the difference between open water evaporation and actual evaporation from the pre-reservoir vegetated surface. The increase in evaporation rate that occurs due to the hot, dry conditions of a severe drought is offset by the decreasing surface area of the reservoir as it draws down. Overall losses are obviously greater the larger the reservoir. For all reservoir sizes tested, increased evaporative losses amount to between 1½% and 2% of the net yield to supply in the design drought.

CONCLUDING REMARKS
30.   Simulation modelling has been found to be a very useful technique in screening a large number of possible scheme options. It enables an good insight to be gained into how a water resources system works and which are the key parameters that affect the operation and output of the system. Following the initial screening process, the selected scheme or schemes can then be investigated in more detail and, if necessary, the model refined or alternative design drought flow sequences constructed to test the robustness of the original outline designs.

REFERENCES
1.   HERBERT LAPWORTH & PARTNERS   A Hydro-Geological Survey of Kent. Report for the Advisory Committee on Water Supplies for Kent, 1946.

2.   LAW F.M. and SENE K.J.   A once-in-50 years design drought for Horton on the Great Stour, Kent. Institute of Hydrology Report to Broad Oak Project Managing Consultants, April 1991.

3.   TWORT A.C, LAW F.M. and CROWLEY F.W. Water Supply. Third Edition. 548p. Edward Arnold, London, 1985.

An investigation into the maintenance of high summer water levels at recreationally important reservoirs using a simulation model of the Tees reservoir system

A. I. HILL, BA, MSc, PhD, MRTPI, MIEnvSc, School of the Environment, Sunderland Polytechnic

SYNOPSIS. A model was developed to simulate the operation of the Tees reservoir system under the existing procedures, and a range of alternative operating policies designed to maintain higher summer water levels at recreationally important reservoirs. The results suggest that operation could be successfully modified, with no loss of reliability and at little extra cost, to enhance recreation at the two key reservoirs. It is likely that such possibilities exist in most flexible reservoir systems.

RESERVOIR BASED RECREATION
1. Reservoirs are increasingly being called upon to support additional uses, and water-based recreation has emerged as an important use. Water dependent and water enhanced activities have shown sustained growth; the limited inland water space available in England and Wales has concentrated pressures on water-supply reservoirs. Although only a small proportion of recreation benefits accrues directly to the reservoir operators, the expenditure and travel time associated with reservoir recreation indicate a high social value; the benefits may be substantial and can enhance local economies.
2. The resource requirements of the various activities can be considered in terms of facilities, space, access, water quality and aesthetics. Reservoir drawdown has been found to exert an influence, usually adverse, on most of these requirements. A matrix approach is useful in the analysis of these impacts. Questionnaire surveys (ref 1.) reveal that sailing and other boat-based activities suffer most, but angling problems increase with more drawdown. Sufficient evidence emerged from these surveys to support the contention that more attention should be paid to the impact of drawdown in the operation of reservoirs.

APPROACHES TO ALLOWANCE FOR RECREATION IN RESERVOIR OPERATION

3. Without modifying drawdown regimes, only limited reduction of drawdown impact is possible, through the design of facilities to function over a range of depths, and in some cases through margin revegetation. Such measures should be taken in any case; worthwhile improvements can only be achieved through the manipulation and modification of drawdown regimes in recreationally sympathetic operation. There is scope to do this in flexible, multi-source systems.

4. There are essentially two approaches to allowance for recreation in operation, either through economically derived true multipurpose operation, or through the use of constraints. The former is closely associated with the systems approach, cost-benefit analysis, and optimisation techniques; it is more developed in the USA, but the applications in practice have been limited. There appears to be a reluctance amongst decision-makers to accept the application of the techniques; in practice, recreation has still been sacrificed as a result of powerful interests and entrenched attitudes.

5. In the constraint approach, priority is accorded to one purpose, for which operating rules are formulated. The rules are then modified to respect constraint water levels in the interests of any secondary purpose. This constraint may readily be broken in the interests of the primary purpose, especially if there is no penalty for breaking the constraint. The use of recreation constraint lines in operating rules is well established in the USA; they have been applied in the UK as 'recreation amenity' lines in the Dee and Tees systems. Despite theoretical disadvantages compared with true multipurpose operation, the constraint approach may have more to offer in practice.

6. The application of amenity lines in the Dee and Tees (for Balderhead and Selset Reservoirs) systems has been analysed (ref 2). Under current operating procedures it has proved difficult to respect the constraints, especially at Balderhead and Selset Reservoirs where the constraints were broken in most years for several weeks or more, and for much of the summer in dry years. The Tees system (shown in Fig 1) was selected as a case study to test further allowance for recreation with the constraint approach.

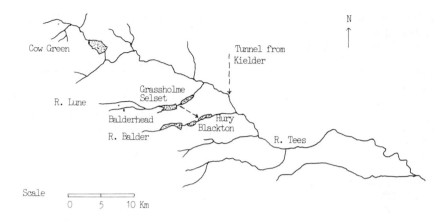

Fig. 1. Map of the River Tees system

THE TEES RESERVOIR OPERATION SIMULATION MODEL

The Tees Reservoir System

7. The system is the main water source for the urban areas and industries of Cleveland. Earlier this century, the five Lune-Balder reservoirs (Selset, Grassholme, Hury, Blackton, and most recently Balderhead in the 1960s) were developed primarily for direct supply, though Selset and Balderhead have a secondary river regulation function. Cow Green reservoir was completed in 1970 as a river regulation reservoir to augment flows in the Tees for increasing abstractions downstream. Lambert (ref 3) and Douglas (ref 4) derived the control rule curves and operating procedures for the six reservoir system, which still form the basis of current operation. Kielder Reservoir (completed 1982) in the North Tyne Valley, was constructed together with 29 kilometres of transfer works to supply the River Tyne, Wear and primarily Tees. Recessions, with industrial decline on Teesside have resulted in most of the Kielder capacity being surplus to current requirements. Modification to operation of the Tees reservoirs has been minimal, and confined to drought situation.

8. Recreational use and facilities at the Tees reservoirs are more fully developed at Selset (including sailing and sailboarding) and Balderhead (including water-skiing). The reservoirs are located in the North Pennines AONB, and nearby are caravan sites, a riding centre, a youth hostel and a good network of public rights of way (including the Pennine Way passing between the reservoirs).

RESERVOIR PLANNING AND OPERATION

These factors, together with accessibility to the North East conurbations, would appear to offer even more scope for promotion of recreational use and provision of a wider range of more extensive recreational facilities, especially if acceptable water levels could be maintained through summer.

The current operating policy incorporates the constraint to maintain levels in Selset and Balderhead above their Amenity Lines throughout the summer; however, this protection is afforded for regulation releases only, but not direct supplies which are maintained even when levels fall below the amenity lines.

The simulation model

9. To compare the operating policy with alternatives that might enhance summer levels, a systems approach was adopted using a model to represent the complex reservoir and river system. An optimisation approach was rejected, as it had proved difficult to quantify the loss of recreational benefits; nor were the other system benefits (water supply, with flood control and hydropower) all readily quantifiable. No appropriate software was found to be readily available. In view of the strong problem orientation and the configuration of the system, it was decided to develop an original simulation.

10. The simulation model TROS (Tees Storage Routing and Reservoir Operation Simulation) was developed to represent the system, with Kielder transfers, as the basis for simulating reservoir operation under a range of operating policies. The computer programme TERESIM (Tees Reservoir Simulation) was prepared to execute the simulation. The simulation was performed with inflows, operation and releases on a daily basis; daily rainfall data were available for the 31 year period 1946-1976, and water travel time through the system is less than 24 hours. The output of the simulation is in the form of resultant daily releases and reservoir storage levels; the number of days for which water level at recreationally important reservoirs is below critical threshold is taken as representing recreation losses. The reservoirs from which daily supplies are taken are determined, thus giving an indication of variations in costs of water.

11. In TROS, the system is represented as five reservoirs and seven sub-catchments (Douglas, ref 4, used six sub-catchments) as shown in Fig 2.

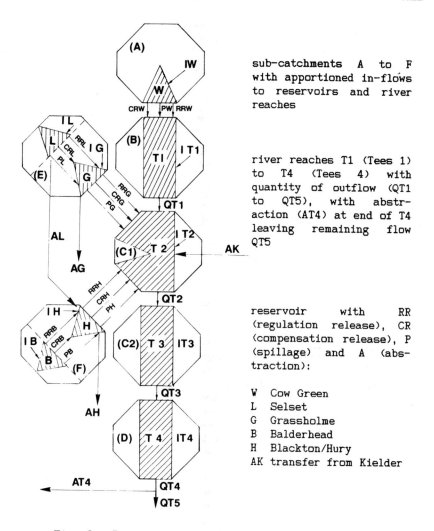

Fig. 2. Representation of Tees system in TROS

Blackton and Hury reservoirs are closely connected and operated on a joint control rule curve, and are combined in the model as one conceptual reservoir. The capacity of Kielder can currently be regarded as infinite in relation to the transfer capacity and limited frequency of transfers. The inflows pass into and through storage in the five reservoirs and four river reaches between Cow Green and Darlington abstractions; TROS thus has nine conceptually separate storage tanks. The physical characteristics of these 'tanks' are indicated in Table 1.

Table 1. Physical characteristics of Tees 'tanks'

catchment for	area (Km2)	maximum altitude (m AOD)	minimum altitude (m AOD)	reservoir size (tcm)	river length (km)
Cow Green	59	893	490	40910	–
Selset	71	788	313	15320	–
Grassholme	8	535	272	6060	–
Balderhead	20	562	337	19670	–
Blackton/Hury	23	484	262	6010	–
Tees 1	183	792	223	–	18.0
Tees 2	79	653	166	–	10.7
Tees 3	65	477	140	–	5.8
Tees 4	308	659	43	–	28.5

TROS combines continuity storage transformation through the reservoirs and channel routing of resulting flows, together with the further sub-catchment inflows through the river reaches. Historical inflows were generated from the rainfall data, using the Tees rainfall-runoff model developed by Douglas (ref 4). Volume of spillage is taken as a function of storage volume in excess of top water level for each reservoir. A simple channel routing submodel representing outflow as a power function of mean daily river reach storage contents was used.

12. The computer programme TERESIM was written in GWBASIC to run TROS. The main part of the programme is a year loop within which the inflows for a year are input, the daily releases and abstractions determined, and daily stage transformations for continuity made. Within each daily release loop, the subroutine (representing the operating policy as control rule curves and operation decision-tree procedures) is called. The procedures require a comparison of refill probabilities under certain circumstances, and this is done in a subroutine. Whilst direct validation was not possible, it is considered that the careful development of TROS and TERESIM, together with extensive checking carried out, produced a sound basis for making comparisons between alternative operating policies.

SYSTEM FLEXIBILITY AND ALTERNATIVE POLICIES

Potential system flexibility

13. Three aspects of system flexibility may be taken advantage of in manipulating reservoir levels in the interests of Selset and/or Balderhead:

(a) reducing the regulation role of Selset/Balderhead (because earlier regulation releases when levels are above the amenity lines may contribute to later breaching of constraints); reduction, or elimination of this role, whereby Cow Green and/or Kielder make up the difference, might be helpful.
(b) redistributing direct supplies within the Lune/Balder sub-system; as the existing high summer control lines for Grassholme and Blackton/Hury are no longer justified in water quality terms, lower summer lines for these reservoirs could enable higher Selset/Balderhead levels to be maintained.
(c) reducing direct supplies from the Lune/Balder reservoirs; flexibility exists in the Teesside supply grid to reduce supplies from the Lartington treatment works (and hence direct supply releases) making up the deficiency by increasing abstractions at Darlington (supported if necessary by increased regulation releases from Cow Green and/or Kielder).

No increase in water supply costs would result from the second option, or the first where Cow Green is used. The third option would involve extra pumping costs from river abstraction, and transfer pumping costs if Kielder water were used, although treatment costs would remain approximately the same.

Alternative policies tested

14. In addition to the above aspects of flexibility, there are further possible changes in policy, including reduced compensation releases (it is possible that the existing requirements may be excessive) and raising the summer amenity lines (in effect, reduces regulation role alone unless used with other changes). To cover varying degrees, and all combinations of the possible approaches would have entailed a vast number of policies. Twenty three policies were selected as representative of the possible options. All were run to satisfy river abstractions of 300 tcmd (somewhat higher than current consumption of around 230 tmcd) and 400 tcmd (to test increased demands); direct supplies were fixed at 120.6 tcmd (standard Lartington throughput) throughout.

The 23 policies comprised 11 separate policies plus 12 combinations of these 11. All 23 policies, at both demand levels, were successfully run through RESOP in TERESIM. After initial evaluation, the number of policies was reduced to six, which were then rerun with more detailed evaluation output. The shortlist six were:

P1. A 'control' or 'baseline' policy: as the current policy, but without summer amenity lines for Selset and Balderhead (their design drought levels were used instead).
P2. The current policy (this is the 1977 policy, with Kielder transfers replacing Lune/Balder regulation releases in the event of Cow Green releases being not maintainable).
P3. No regulation releases from Selset or Balderhead (may through August), together with a lowering of the Grassholme and Blackton/Hury control lines by 10%.
P4. Extra Cow Green regulation releases (abstracted at Darlington) used to replace Selset direct supplies when Selset level is not above its summer amenity line.
P5. Extra Cow Green releases used to replace both Selset and Balderhead direct supplies when their levels are not above the amenity lines.
P6. Policies three and five combined.

EVALUATION OF THE POLICIES

Use of performance indicators

15. Without an optimisation approach against a single parameter it was necessary to examine system performance in the various operating outcomes, with trade-offs between any gains to recreation and other disadvantages or losses. This was done through the use of a series of performance indicators:

(a) Maintenance of summer water levels at Selset and Balderhead:

  (i) number of days from May through September where water level is below the amenity line (out of 153 days total).
  (ii) number of these days is well below the line (1/2m at Selset, with more shallows and 1m at Balderhead).
  (iii) average daily level for this period.

(b) Direct system costs:
  (i) number of days with Kielder transfers (pumping costs).

(ii) volume of Kielder transfers (those to replace direct supplies identified).
(iii) additional Darlington abstractions replacing direct supplies (higher pumping costs).

(c) System reliability and efficiency:
(i) number of days with direct supply reductions.
(ii) number of days with regulation release deficit.
(iii) proportion of water spilled

(d) Mean summer daily levels at the other reservoirs

16. After applying these indicators to the 23 policies, the average performance over the 31 years were compared. It emerged in this analysis that some policies could be identified as clearly inferior to others. It proved possible to identify five representative policies for short-list analysis, without the application of trade-offs. These five, together with a 'control' policy, were those described above.

Analysis of the short-list policies
17. The overall performance of the six policies described is shown in Tables 2 and 3.

Table 2. Summer recreation level outcomes (averages)

Policy Number	Selset Reservoir			Balderhead Reservoir		
	No. of days below		Mean level (% capacity)	No. of days below		Mean level (% capacity)
	Amenity Line	½ m below		Amenity Line	½ m below	
1	93	69	81.3	135	90	82.9
2	74	32	85.1	115	67	86.8
3	43	21	86.4	104	54	88.3
4	52	12	86.7	108	55	88.3
5	46	11	87.3	66	7	92.8
6	12	5	88.0	57	2	93.1

Table 3. Other performance indicators (averages)

Policy Number	Kielder Transfers		No. of days abstraction deficit	Mean level May - Sept. (% total capacity)		
	volumne tcm	Numbers of days		Grass-holme	Blackton /Hury	Cow Green
1	227	3	38	82.1	85.0	82.4
2	396	6	32	82.1	85.1	81.4
3	415	6	13	80.3	77.7	80.8
4	639	6	26	82.2	85.2	79.7
5	865	6	26	82.4	85.4	77.6
6	907	7	12	81.6	78.9	76.6

Averages over the 31 years may be unduly influenced by a few drought years. Therefore, the performance of the short list policies was examined over the years in two ways:

(a) the number of years for which Selset and Balderhead levels were well below (1/2m and 1m respectively) the amenity lines for 5 days (one day per month in summer) or less, and for 22 days (one day per week) or less;

(b) performance through six representative years (two very dry, two average and two very wet); graphical output of weekly levels was produced for all the reservoirs for each policy.

18. Policy 2 produced very worthwhile improvements over Policy 1 at very little extra cost. Although it does not achieve adherence to the amenity lines, the analysis, particularly the graphical output, showed that a noticeable reduction in drawdown results. The existing policy is therefore a good policy, and its use of summer amenity lines is to be commended. Policy 3 produced some further reduction in Selset and Balderhead drawdown and was slightly cheaper. Provided that it could be shown that the ensuing levels at Blackton/Hury do not have an unacceptable impact onangling, this policy is to be preferred. It may be a case of trading off recreation pursuits at Selset and Balderhead against those at Grassholme and Blackton/Hury. The results for Policies 4 and 5 suggest that although replacements of direct supplies may more readily benefit Balderhead levels, it may be more expensive to do so.

There appears to be potential for considerably more recreational development and use of the Lune/Balder reservoirs. If this was the case, the extra expenditure in these policies should be worthwhile. Policy 6 would then offer particularly good value if Blackton/Hury drawdown were acceptable.

CONCLUSIONS

19. The aspects of Tees system flexibility utilised in the alternative policies produced varying gains:
(a) the reduced regulation role for Selset and Balderhead produced only limited improvements (hardly surprising in view of their limited regulation function)
(b) redistributing direct supplies within the sub-system produced worthwhile improvements at Selset and Balderhead, even more so when combined with some changes in regulation releases; this is comparable with the flexibility which might exist in smaller reservoir systems elsewhere, and suggests that even in smaller, simpler reservoir systems, drawdown regimes may be manipulated, perhaps at little extra cost.
(c) reducing total direct supplies from the Lune/Balder sub-system produced the most dramatic improvements in Selset and Balderhead summer levels; in general terms, this can be seen as a product of the flexibility offered through having alternative large reservoirs to draw upon; the surplus backup of Kielder was not found to be significant as Cow Green, could cope in most years, and even when Kielder transfers were used, a large back-up of water was retained in Selset and Balderhead; this suggests that even in systems without the backup of a Kielder level of reserve, drawdown may be considerably manipulated in a large system.

20. The current use of amenity lines is well worthwhile and should not be abandoned. The costs of those short-list policies producing even greater improvements are not great compared with the local income that recreational developments can generate. These policies should be given serious consideration. It has been demonstrated that, in a flexible water resource system, reservoir operation can be modified in the interests of recreation, often at little cost. The costs of substantial modification of drawdown regimes may be higher, but can still be worthwhile. It is sometimes possible to represent a few selected alternatives as fairly simple trade-offs without

resource to formal multiobjective analysis. Other reservoir systems should be examined to explore the scope for further application of these approaches.

REFERENCES

1. Hill, A.I. 'The impact of reservoir drawdown on water-based recreation' IWEM 91 'Water and the Environment', Technical Papers, Birmingham, UK, 30 April - 2 May 1992.

2. Hill, A.I. and MAWDSLEY, J.A. 'The application of recreation amenity lines in UK reservoir operation: case studies of the Dee and Tees schemes', Proc. British Hydrological Society National Symposium, Hull, UK, 1987.

3. Lambert, A.D. Teesdale Reservoirs: derivation of rule curves. Northumbrian Water Authority, Newcastle upon Tyne, UK, 1974.

4. Douglas, J.R. Operating rules for the Teesdale Reservoirs. Northumbrian Water Authority, Newcastle upon Tyne, 1977.

Operational yields

W. L. JACK, Welsh Water, and A. O. LAMBERT, Hydrology and Water Resources, Wallace Evans

SYNOPSIS. The original evaluation of yield for most UK reservoirs was based on the analysis of actual or synthetic drought sequences. It was assumed that the resource is depleted by a uniform demand and that it is just emptied on the day a drought of specified severity (eg. 1 in 50 years) ends. After the 1984 drought Welsh Water decided to base its yield calculations on "Operational Yield". This takes into account the way that sources would actually be operated in severe droughts. It includes effects of hosepipe bans and drought orders, and makes assumptions about the end date of the drought which water resource managers would actually use.

BACKGROUND

1. What drove Welsh Water to derive a new technique, which we called "Operational Yield", was experience in the two droughts 1975/76 and 1984. At the start of the 1975/76 drought the yields had been calculated by the methods then in use. What was wrong was never precisely identified but lack of involvement in the yield estimates by operating staff may have been at the root of it. Certainly yields were derived by others - traditionally by the promoting engineers from Victoria Street, by such eminent engineers as Hawksley, Deacon and Lapworth. More recently specialist hydrologists had invented probabilistic yields, perhaps confusing people with such concepts as statistical synthetic 2% flow sequences.

2. It was not just in Welsh Water that a desire for change arose. Because of widespread feeling, a seminar was held in Birmingham, entitled "Reliability of Water Resources Systems". This was in May 1977 while we were trying to make changes based on the lessons learned in the previous two summers.
Several people were advocating a move towards operational yield but perhaps they were a little too far ahead of their time. Back in Welsh Water it was proposed that we "should cooperate with other Water Authorities and Government bodies in attempts to formulate criteria for the specification of operational yield". In the meantime management resolved to base yields "on a 1 in 50 year return period drought, unless the source is of major strategical importance or where there is no significant back up from other sources, when a 1 in 100 year return period should be used".

RESERVOIR PLANNING AND OPERATION

3. As far as we know the enthusiasm for change then evaporated, only to re-emerge in 1984. The 1984 drought was particularly severe in the North and West of England and Wales. For many sources the drought was as severe as 1976 and we experienced the same problems with resources. This time we resolved to do something about it, and a group was set up by Mike Featherstone to develop a technique for operational yield. The new method was approved by Welsh Water management in May 1985, and has been used through privatisation to the present. One of the strengths of the method is that it was derived by Welsh Water at a time when the water supply and river interests were "under one roof".

BASIC PRINCIPLES

4. The basic principle underlying operational yield techniques is the need to have a single method used both for planning and for operational management.

5. Planners need to know the sufficiency of resources, when demand has risen to meet the available resources and when to bring on the next source. Operational needs focus on how much supply can be drawn from a source during a drought without running too much of a risk of failure of supply. The special feature of operational yield is that it meets both those needs. At the same time it improves upon the traditional yield methods in that the planner now has a more realistic estimate of the amount of water that a source can produce.

6. The key words therefore in this context are:
   a) "unified" - in that all those concerned with the sources are using the same assumptions,
   b) "realistic" - in that the yield derivation incorporates a simulation of how the source would really be used in a severe drought, and
   c) "ownership" - implying a commitment and trust by users in the method, and therefore in the answers that derive from that method.

REALISTIC HYDROLOGY

7. What we use for hydrological input for the calculation of yield is based on the worst recorded sequence for the source in question. This may appear too simplistic but we believe that there are good reasons for it.

8. People might argue that a 2% drought sequence would be a better criterion in that it would allow for the fact that for one source the worst recorded event might be a 20 year event, whilst for another it might correspond to 1 in 100 years. Up to 1984 we used design criteria based on return periods. We might still be using this approach if we had a sufficient number of long hydrological records in our area so that we could have confidence in making statistical estimates for events of this rarity. Unfortunately, individual drought events have widely different return periods depending on the parameter which is used to define the severity.

9. We referred before to the desirability of breaking down barriers between those people working in planning and design, and those who operate sources. The use of statistics in this

context contributed to these barriers. The question the operator used to ask during recent droughts was how would we have got through the 1975/76 drought and other past droughts, if we chose a particular operating decision. The customer during all of this is the operating manager and we found that what the customer doesn't want is a lot of statistical results.

10. The choice of the worst historic drought as the criterion for our sources is strengthened by the fact that without doubt the 1933/34, 1975/76 and 1984 droughts were extremely severe. Many analyses have been carried out and many different return periods have been arrived at, some more believable than others. However we believe that for our sources, if we use the worst of these three droughts, we are using a design criteria not very different to a 1 in 50 year or 1 in 100 year severity.

11. One other very important point concerns the end date of the drought. During a drought the major uncertainty is how long it will be before significant rain arrives. In 1976 and 1984 the drought broke in September, while in 1989 and 1990 the rain arrived in October. The assumptions made for our operational yield calculations are that the drought would continue until mid-October for river regulating systems and river abstractions. For direct supply reservoirs, shallow groundwater sources and springs, the response to rain tends to be slower, and an end date of mid-November is used.

12. The addition of such a dry tail to the drought used in the yield calculation adds to its severity, leading to operational yield being lower than traditional yields. We believe this reduction to be necessary in that it reflects how the sources are actually used in a severe drought. To gamble on an early end to a drought is to put the customer and the river environment at risk.

REALISTIC DEMANDS AND CONTROLS

13. Because the operational yield technique reflects how the source would be used in a severe drought, we need to examine how demand varies. Demand during a drought year has a different pattern to normal, largely, we believe, due to much more external use of water. As the drought develops into a very severe one, measures are implemented to reduce demand. The sequence in the method is as follows:

RESERVOIR PLANNING AND OPERATION

	Control	Demand % of average
Up to end of May	Unrestricted demand	120
June	Hosepipe bans First cuts in compensation water	100
July	First ban on non-essential use	100
August	Further compensation water cuts First relaxations for river abstraction	100
September	Further bans on non-essential use Further relaxations for river abstraction	90
October & November	Rota cuts	75

14. The nature of the controls reflects the type of sources that we operate. A number of our 91 reservoirs have prescribed compensation water requirements and it appears that some can be reduced with little or no damage to the environment. We also have 7 river regulation schemes which, by August in a severe drought, can be expected to contribute extra water, again with little or no damage to the rivers.

15. The above sequence of control measures was derived in 1985 and in some respects has stood the test of time well. For example it brings in restrictions so that the effects on water users alternate with those on the environment. In other ways we are looking to improve the method after the experiences of privatisation and of the 1989, 1990 and 1991 droughts. These include the addition of "Appeals for voluntary restraint" to stand alongside "Hosepipe bans". In recent years the water industry has learned to communicate more with our customers, and the appeal for voluntary restraint has become a very useful tool for the water manager.

16. Over recent years both the quantity and quality of data on water going into supply has improved in Welsh Water. We have found that, as a generalisation, a more suppressed pattern of demand is more appropriate for a drought year. We are building in more flexibility into the demand pattern so that the operational yield for a particular source incorporates the demand pattern appropriate to its supply area.

17. We see these possible adjustments to the method as fine tuning and they in no way detract from the power of the method as a whole.

YIELD OF A SINGLE SOURCE

18. An example of a calculation of operational yield of a single source is given Table 1. The data is for Ystradfellte Reservoir. The calculation has been carried out using inflow data for the major droughts, and 1984 was found to be the most

severe drought year for this source. Thus 1984 becomes the basis for its operational yield. Inflow data of sufficient accuracy was not available for the catchment itself so the flow pattern for that area was used.

19. To arrive at the table shown the amount taken into supply is varied until the reservoir is found to empty in mid-November. In practice this is done by putting in a column for the first of December, and adjusting the supply rate until the negative storage on 1st December is equal to the storage on 1st November.

20. The spreadsheet calculation gives an operational yield of 15.2 Ml/d. However, operations staff need more backup information - in the form of basic control rules - to run the source effectively.

21. Spreadsheets for yields of direct supply reservoirs use a monthly timestep as in the table. For river regulation schemes it is usually necessary to go to a daily timestep. In this form of spreadsheet it is fairly easy to model the transitions, for example when regulation releases start to be made. We also incorporate freshet releases at realistic timings. For example on the Wye we incorporate a freshet for a canoe slalom event in June.

INTERLINKED SOURCES

22. Direct river abstractions have short critical periods, whereas at the other extreme some of the major reservoirs have critical periods of over a year. In practice we have not

Table 1. Yield calculation for Ystradfellte Reservoir

STORAGE (ML) 2840 YEAR: 1984 EXTENDED
CATCHMENT AREA (KM2) 10.14
AAR/O (MM) 1698
COMPENSATION (ML/D) 5.0
YIELD (ML/D) 15.2

MONTH	JAN	FEB	MAR	APR	MAY	JUN	JUL	AUG	SEP	OCT	NOV	DEC
CONTENTS AT START OF MONTH (ML)	2840	2840	2840	2810	2442	1913	1555	1094	712	392	127	-125
INFLOW (%AAR)	20.20	11.40	4.00	1.90	1.10	1.00	.50	.60	.60	.60	.60	.60
INFLOW (ML)	3478	1963	689	327	189	172	86	103	103	103	103	103
SUPPLY FACTOR	1.00	1.00	1.20	1.20	1.20	1.00	1.00	1.00	.90	.75	.75	.75
SUPPLY (ML)	470	424	464	545	564	454	470	470	409	352	341	352
COMPENSATION FACTOR	1.0	1.0	1.0	1.0	1.0	.5	.1	.1	.1	.1	.1	.1
COMPENSATION (ML)	155	155	155	150	155	75	78	16	15	16	15	16
CONTENTS AT END OF MONTH (ML)	2840	2840	2810	2442	1913	1555	1094	712	392	127	-125	-390

experienced great difficulty in deriving the yields of systems which incorporate sources of different critical periods. To a large extent this is due to the extension of the drought (into mid-November for most of our sources) which of itself tends to diminish the importance of critical period.

23. The way operational yields are derived for systems where sources are used together is probably best described by use of examples.

24. In Pembrokeshire, Llysyfran Reservoir is used to regulate the Eastern Cleddau. This is a source with a critical period from 10 March to 15 October, and a yield of 84 Mld. It is used conjunctively with a direct river abstraction from the Western Cleddau at Crowhill. This can pump at a rate of 27 Mld when the water is available, but there is a hands-off flow which effectively prevented abstraction from 10 July in the 1984 drought. On its own, therefore, the Crowhill source has a yield of zero. When used conjunctively with Llysyfran, the yield can be increased by the amount abstractable up to 10 July, divided by the number of days in the critical period of the combined system. The start of drawdown of Llysyfran is delayed until 22 March and the critical period is therefore shorter by 12 days. The combined yield is 96 Mld. The operation of the two sources is modelled on a spreadsheet on a daily basis, and the amount taken into supply is varied until Llysyfran storage reaches zero on 15 October.

25. Another example comes from South East Wales. Here there are several major sources with seven month critical periods, including large direct supply reservoirs and two regulated rivers (the Wye and the Usk). In addition there are some small old reservoirs in the Heads of the Valleys, and these have short critical periods. The large sources are so large in comparison to the small ones that the critical period of

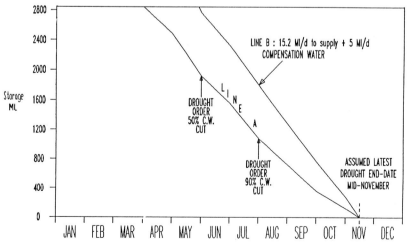

Fig. 1. Ystradfellte direct supply reservoir - simple summer control rules

the conjunctive use system is the same as for the large ones. The yield of the small reservoirs is derived as for single sources and this is then added to the yield of the large ones.

26. The use of interlinked sources can add significantly to yield. However some sources, like unsupported rivers, are only available in the early months of a drought and the enhanced yield is only available provided that this river flow is taken when it is available.

27. This leads to the conclusion that the management of interlinked sources is critical in the spring months, and the decisions on use of sources are made using control rules.

CONTROL RULES

28. It is not enough for the designer/planner/hydrologist merely to inform the operations manager that the operational yield of a certain source, or group of sources, is X Ml/d. In addition to a basic understanding of the assumptions on which the yield calculations were based, operations personnel must have user-friendly control rules to indicate when certain actions are necessary.

29. The Ten Component Method for Control Rules, is used for our sources in conjunction with the concept of Operational Yield. It was specifically designed so that non-hydrologists can understand how control rules for their reservoirs have been derived, and can contribute positively to the process of development. This involvement helps to establish ownership of the rules by operational staff. Many examples of such control rules are given in Reference 1 for those who are interested. Concepts of management can be simply explained, discussed and developed, in particular the need for conservation of upland storage in spring each year.

30. Table 1 showed that the operational yield for Ystradfellte Reservoir is 15.2 Ml/d. Figure 1 shows the drawdown pattern implicit in the calculation (Line A), and compares it to a flat-rate drawoff of 15.2 Ml/d to supply together with 5 Ml/d of compensation water (Line B).

31. Control rules based on these two lines are invaluable to operations personnel. If the storage lies above line B, 15.2 Ml/d can be safely taken to supply for the rest of the year, in the knowledge that no restrictions will be needed (assuming winter refill is not a problem, see Reference 1). If the storage lies between lines A and B, the need for conservation measures or restrictions needs to be regularly assessed, to prevent storage falling below line A.

32. Even if the reservoir is full, storage will always lie to the left of line B in the spring. Management of resources and demands at this time of year therefore offers opportunities for avoidance of possible restrictions on both customers and compensation water later in the year. The objective is to manage so that the storage reaches the zone to the right of line B as soon as possible.

IMPROVED MANAGEMENT OF RESOURCES

33. In our area we have found opportunities for improved management of existing water resources schemes, to the benefit

of water supply, the water environment, customers and recreational interests. For example, many water resources systems have compensation waters or residual flows which are fixed throughout the year. However, in most of the rivers of Wales with migratory fisheries, there is little inward migration before late May, and river temperatures are comparatively low. In such circumstances, compensation discharges in excess of true seasonal requirements increases the possibility of drought orders later in the summer.

34. The concepts of operational yield and control rules, coupled with close examination of local circumstances has already enabled changes to be made on the rivers Dwyfor and Aled in North Wales (References 1 and 2). On both these regulated rivers the pattern of compensation/ regulation releases has been altered. Releases from the upland reservoirs have been reduced in the spring, and more flexibility has been introduced. This has been accomplished with the support of the National Rivers Authority and local angling associations, and they have seen their co-operation rewarded by an avoidance of summer drought orders in the last three years.

35. Another example, described in Reference 3, concerns the River Usk. The new techniques have formed the basis of detailed discussions with the National Rivers Authority. The major improvement so far has been agreed abandonment of the maintained flow condition in the River Usk, for a trial period.

CONCLUSIONS

36. Operational experience in the major droughts since 1974 has highlighted deficiencies with traditional methods for calculating yield. Accordingly Welsh Water has developed the use of operational yield, linked to control rules.

37. The resulting operational yield has significant implications for water resources management. Initially it was identified that in many cases the new yield was significantly less than the previously calculated theoretical yield.

38. However, acceptance of real-world situation has, since 1985, generated control rules based on safeguarding yields up to the assumed latest end-date of a drought. This in turn has lead to changes in operational management. Since 1989 this work is continuing in co-operation with the National Rivers Authority.

REFERENCES

1. LAMBERT A.O. An Introduction to Operational Control Rules using the Ten Component Method. British Hydrological Society Occasional Paper No 1, 1990.
2. WESTON A.E. and HODGSON B.P. The Aled Regulation System - Reconciling the Reconcilable. British Hydrological Society 3rd National Symposium, 1991.
3. GALLIMORE S.A. and MOSEDALE J.C. The South East Wales Conjunctive Use Scheme. Water Resources and Reservoir Engineering Conference, 1992.

The Balquhidder research catchments: development of the results for application to water resources

R. C. JOHNSON, BA, Institute of Hydrology, Balquhidder

SYNOPSIS. The annual water uses of the two Balquhidder catchments are presented for 1984-1991 and relationships derived linking the water uses to a range of weather parameters. The main weather related controls on the individual catchments' water uses are annual and winter precipitation, annual rain days and winter wind speeds. This develops a new way of explaining and applying the results. Further development of these relationships and the Balquhidder processes model should provide valuable tools for water resources management in the uplands.

INTRODUCTION

1. The southern highlands of Scotland are a major source of water for industrial, agricultural and domestic consumption. The Balquhidder catchment experiment was established in 1981, to study the impacts of forest management on water resources in this region. The applicability of the results is essential for a successful experiment and so a range of results, often in model form, are being published which should meet most requirements.

2. The two catchments are situated in the headwater area of the River Forth some 60 km north of Glasgow (ref. 5). Both catchments are steep sided with thin soils and drainage patters comprising networks of small burns. The Monachyle is a moorland (heather and grass) catchment with partial afforestation having taken place in 1986. The Kirkton is a forested catchment (forest and grass) with progressive clear felling having started in 1986.

3. A wide range of monitoring has taken place including detailed precipitation and stream flow measurements. In this paper the difference between the two: the catchment water use, is considered with data from one of the catchment weather stations. A detailed process model has also been developed from site studies on the major vegetation typesp to explain the relative water uses, this is reported in detail in Hall and Harding (ref. 3). This model should also have future applications to land use management, reservoir design, climate change impacts etc.

RESERVOIR PLANNING AND OPERATION

CATCHMENT WATER BALANCE RESULTS 1984-91

4. The Balquhidder catchment water balance results, 1984-89, have been reported before (refs. 4, 6, 1). Table 1 presents these results again and also the 1990 and 1991 data. The additional two years include another extreme year for precipitation with the highest annual total (3683 mm) for these catchments been recorded in the Monachyle. The forested Kirkton catchment still has the lower water use in absolute terms although the water use expressed as a proportion of precipitation shows that in 1990 the Kirkton had a water use higher than the Monachyle. In 1990 and 1991 the Kirkton had the catchment's greatest percentage water uses.

Table 1. Annual water balance results from the Balquhidder catchments

Monachyle

Year	P	Q	dS	WU	%
1984	2648	1929	2	717	27
1985	2612	2956	36	520	20
1986	3280	2522	-21	779	24
1987	2255	1724	-19	550	24
1988	2952	2389	0	563	19
1989	2985	2397	0	588	20
1990	3683	2842	0	841	23
1991	2754	2160	0	594	22

Kirkton

Year	P	Q	dS	WU	%	WUD
1984	2215	1781	1	433	20	284
1985	2285	1960	36	289	13	231
1986	2789	2242	-20	567	20	212
1987	1899	1592	-19	326	17	224
1988	2493	2126	0	367	15	196
1989	2519	2098	0	421	17	167
1990	3092	2354	0	738	24	103
1991	2263	1765	0	498	22	96

P = Annual precipitation, mm
Q = Annual discharge, mm
dS = Net change in snow storage over the year, mm
WU = Water use, (P-Q-dS), mm
% = WU/P*100
WUD = Difference in WU between Monachyle and Kirkton, mm

5. The difference between the water uses of the catchments has been shown previously (ref. 4) to indicate a possible trend from 1984 to 1989 with the difference becoming less. Table 1 shows that this continued during 1990 and 1991 with the difference having reduced at an average rate of 27 mm each year, reaching 96 mm by 1991. The reason for the observed trend is not obvious as clearfelling the Kirkton catchment would be expected to reduce the water use so increasing the difference between the catchments. The steady decrease, starting before the land use changes, suggests a climatic effect.

SITE STUDIES AND A PROCESS MODEL

6. Comparison of the Balquhidder catchments' relative water uses with the Plynlimon results (ref. 8) appears to show an opposite result. The forested catchment at Balquhidder has a lower water use than the moorland catchment. To explain this a composite water use model was developed by mapping the vegetation types and carrying out physically based process studies on evaporation from the main vegetation types. The results from this model are fully explained in Hall and Harding (ref. 3).

7. The process model needs further developments and to be tested on other catchments. If this is successful then there will be many applications in water resources engineering.

CATCHMENT WATER USE AND SEASONAL WEATHER

8. An essential aim of research into the effects of change on water resources is to provide a series of results which can be applied in different situations. Sophisticated models are a product of integrated, high quality research but there is also a need for generalised relationships derived from the catchment monitoring which can be easily applied by water resources engineers to other catchments.

9. In the Balquhidder catchments a network of automatic weather stations (AWS) (ref. 7) has been operating throughout the period 1984-1991. The stations are distributed to record data from a wide range of conditions and one station, the Upper Monachyle, is in a very well exposed location. For this analysis it has been chosen as a reference station to relate both catchments' annual water balance data to.

10. Regressional analyses have been carried out for both catchments with the annual water use as the dependent variable and a series of AWS parameters as the independent variables. The relationship for each regression is here indicated by the coefficient of determination, r^2, and the results presented in table 2.

11. The variables derived from the AWS which indicated the best relationships with annual water use were: WP, WW,

RESERVOIR PLANNING AND OPERATION

P and WRD (see table 2 for explanation of abbreviations). The aerodynamic terms are showing much better relationships than the radiation terms indicating that interception dominates the catchment evaporation. The relevant equations are:

MWU= 265.3 + 0.20 WP (Monachyle)   (1)
KWU= -3.4 + 0.29 WP (Kirkton)   (2)

MWU= 351.4 + 75.8 WW   (3)
KWU= 61.1 + 101.9 WW   (4)

MWU= 39.9 + 0.21 P (Monachyle)   (5)
KWU= -336.8 + 0.32 P (Kirkton)   (6)

MWU= -141.9 + 5.6 WRD   (7)
KWU= -435.0 + 6.3 WRD   (8)

Table 2. Coefficient of determination, r_r^2 for simple regressional analysis between catchment annual water use, Monachyle (MWU) and Kirkton (KWU) and a series of AWS parameters

	Monachyle	Kirkton
Precipitation, mm (P)	.60	.68
Winter P, mm (WP)	.83	.83
Summer P, mm (SP)	.41	.24
No. rain days (RD)	.29	.35
Winter RD (WRD)	.63	.55
Summer RD (SRD)	.04	.01
Temperature, C (T)	.00	.13
Winter T, C (WT)	.08	.26
Summer T, C (ST)	.02	.01
Wind speed, ms^{-1}, (W)	.23	.46
Winter U, ms^{-1}, (WW)	.60	.72
Summer U, ms^{-1}, (SW)	.13	.00
Solar radiation, Wm^{-2} (S)	.02	.00
Winter S, Wm^{-2} (WS)	.02	.01
Summer S, Wm^{-2} (SS)	.26	.22
Net radiation, Wm^{-2} (N)	.15	.20
Winter N, Wm^{-2} (WN)	.24	.13
Summer N, Wm^{-2} (SN)	.04	.05

Winter is January-March and October-December
Summer is April-September

12. Equations 1-8 show consistently steeper gradients for the Kirkton catchment and greater intercepts for the Monachyle catchment. Thiss indicates that the Kirkton is the more weather sensitive catchment, especially to precipitation and wind. Using these equations, values from the AWS parameters can be derived where the water use from both catchments become equal, these are:

Mean winter precipitation = 2986 mm
Winter wind speed = 11.1 ms^{-1}
Mean annual precipitation = 3425 mm
Winter rain days = 419 days

13. In comparison with the observed weather parameters, table 3, the two precipitation values were almost reached in 1990 but the wind and rain days values have never been approached. Thiss analysis suggests that the explanation for the high Kirkton 1990 water use was the very high annual and winter precipitation in 1990. In contrast to this, the 1985 Kirkton proportional water use was exceptionally low compared to the Monachyle; the 1985 winter precipitation and winter wind speed were both the lowest in the 8 year period. These two years illustrate that the Kirkton water use is more sensitive to weather variations than the Monachyle.

Table 3. Annual means/totals of selected observed weather parameters

	MP	KP	MWP	KWP	RD	WW
1984	2648	2215	2022	1692	258	4.3
1985	2612	2285	1193	1012	262	1.6
1986	3280	2789	2264	1870	285	4.3
1987	2255	1899	1329	1075	236	2.8
1988	2952	2493	1755	1442	272	3.6
1989	2985	2519	2036	1737	238	4.6
1990	3683	3092	2918	2415	278	5.6
1991	2754	2263	1735	1373	277	4.1

MP = Monachyle annual precipitation, mm
KP = Kirkton annual precipitation, mm
MWP = Monachyle winter precipitation, mm
KWP = Kirkton winter precipitation, mm
RD = Number of rain days in year
WW = Winter mean wind speed, ms^{-1}

MULTIPLE RELATIONSHIPS FOR CATCHMENT WATER USE

14. A further series of relationships has been derived linking catchment water use to combinations of weather

parameters. Pairs of variables were taken and the relationship again assessed by the value of r^2.

15. For the Monachyle, WP (table 1 explains abbreviations) gave an r^2 value of greater than 0.80 when combined with any of the other variables. The only other combinations which reached this value were P-SP and WW-WT. The combination WP-WT produced the best relationship with an r^2 value of 0.97.

16. For the Kirkton WP again gave consistently high values of r^2. The other combinations with good relationships were P-SP, WW-RD, WW-SS and WW-SN. The combination WW-SN produced the best relationship with an r^2 value of 0.90.

17. These analyses show that excellent relationships can be derived for the individual catchments but the more useful analysis would be to derive a series of equations from the combined catchments which could possibly be applied to other catchments. To do this one variable needs to be retained which will identify the catchment: WP has consistently been shown to relate very well to water use and the combined catchment relationship is shown in equation 9:

$$WU = 73.1 + 0.27\ WP \qquad (r^2 = 0.76) \qquad (9)$$

18. The best variables, selected from the above analyses, which should possibly be combined with WP are P, SP, WW, WT, RD, SS and SN. Only WT made any significant improvement to the r^2 value from equation 9:

$$WU = 110.7 + 0.33\ WP - 53.0\ WT \qquad (r^2 = 0.82) \qquad (10)$$

19. These combined catchment relationships are not as good as the individual catchments. The inclusion of terms for land use, similar to the method employed by Calder and Newson (ref. 2), would seem a sensible further development.

CONCLUSIONS

20. The results from the Balquhidder catchments have consistently shown that the moorland Monachyle catchment has a greater absolute water use than the forested Kirkton catchment. But when expressed as a proportion of the precipitation the 1990 data shows that the Kirkton had a proportionally higher water use than the Monachyle. This is unlikely to be a land use change effect as the difference in water use amounts has declined steadily since the records began.

21. Relationships between the catchment water use data and a series of weather parameters, derived from a catchment AWS, have shown that annual precipitation, winter precipitation, winter wind speed and number of rain days

give the closest relationships. This does not explain the reason for the water use but indicates connections which are quite realistic.

22. The Kirkton catchment has been shown to be the more sensitive to a range of weather parameters which seems to explain the year by year variations in the catchments' relative water uses. This is illustrated by two years: 1985 and 1990.

23. Multiple regressions have been carried out to provide a series of equations for application to other catchments. Relationships were found to be not as close as for the individual catchments and it is suggested that a further term is required in the equations which defines the physical nature of the catchment. A vegetation index is suggested.

24. With further developments of these relationships a series of equations should be available for use in other catchments. This empirical method of estimating water use in ungauged catchments provides a useful contrast to the physically based process model which has also been developed at Balquhidder.

REFERENCES
1. BLACKIE J.R. The water balance of the Balquhidder catchments. J.Hydrol, in press.
2. CALDER I.R. and NEWSON M.D. Land use and upland water resources - a strategic look. Water Resour. Bull., 1979, 16, 1628-1639.
3. HALL R.L. and HARDING R.J. The water use of the Balquhidder catchments: a processess approach. J.Hydrol, in press.
4. JOHNSON R.C. Effects of upland afforestation on water resources: The Balquhidder experiment 1981-1991. Report no. 116, 1991, Institute of Hydrology, Wallingford.
5. JOHNSON R.C. An introduction to the Balquhidder catchments. J. Hydrol., in press.
6. JOHNSON R.C. and LAW J.T. The water balances of two small upland catchments in Highland Scotland with different vegetation covers. In: Hydrological interactions between atmosphere, soil and vegetation (Proc. Vienna Symp., August 1991) IAHS Publ. no. 204, 1991, 377-385.
7. JOHNSON R.C. and SIMPSON T.K.M. The automatic weather station network in the Balquhidder catchments, Scotland. Weather, 1991, 46, 2, 47-50.
8. KIRBY C. and NEWSON M.D. Plynlimon research: the first two decades. Report no. 109, 1991, Institute of Hydrology, Wallingford.

The water resources of Madras

M. C. D. LA TOUCHE, MA, FICE, MIWEM, Binnie & Partners, and
P. SIVAPRAKASAM, ME(PH), MEI, Madras Metropolitan Water Supply and Sewerage Board

SYNOPSIS. Studies carried out during the period 1988-90 for the Madras Metropolitan Water Supply and Sewerage Board, mainly as part of a UNDP technical assistance programme to the Government of Tamil Nadu in southern India, included making estimates of the yield and reliability of the city's surface and groundwater resources. The paper describes the methods of analysis adopted and the results obtained. A combined surface and groundwater model was developed to simulate the conjunctive use of the city's reservoirs and well fields.

INTRODUCTION
1. Madras is the principal port on the Coromandal Coast of southern India and the capital city of the state of Tamil Nadu. The population of the city has grown from 1.8 million in 1961 to about 3.5 million today, and by the year 2000 it is expected to reach 4.5 million.
2. The mean annual temperature is 30°C. Rainfall is seasonal and occurs mainly during the periods of the SW and NE monsoons in July and August, and October and November respectively, although it is only during the latter period that there is appreciable run-off and flow in the rivers. Open water evaporation exceeds 2m per year.
3. The terrain around Madras is flat, being part of the coastal plain that borders the Bay of Bengal, where there is extensive irrigated agriculture. Run-off collected during the monsoon is stored in shallow bunded reservoirs called *tanks*, from which water is released to irrigate adjoining fields during the growing season.
4. Madras obtains its water supply from the Korttalaiyar river which flows to the sea a little to the north of the city and from groundwater, but the potentially available supply is much reduced due to competing agricultural demands on both these sources.
5. Over the last 40 years the development of resources has not kept pace with the growth in demand arising from the growth in population. As a result there have been severe water shortages in the city, particularly in recent years when supplies have had to be limited to a few hours on alternate days. Fortunately relief is at hand, as a project is currently

RESERVOIR PLANNING AND OPERATION

being built to bring water to Madras from the Krishna river in Andhra Pradesh state 450km to the north. The first supplies from the Krishna river are expected to be received in 1993.

SURFACE WATER RESOURCES

6. Surface water supply for Madras is obtained from the Korttalaiyar river and its upstream tributaries (Fig. 1). Recently, water has also been diverted to the Korttalaiyar river from a temporary weir on the Arani river, which flows to the north of the Korttalaiyar.

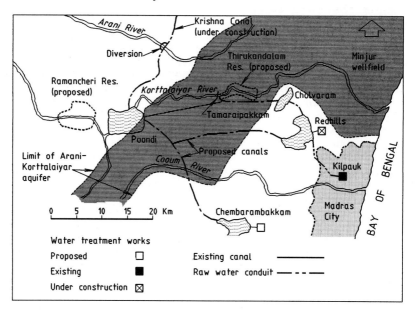

Fig. 1. Madras City Water Supply: Water Sources

7. In most years the rivers flow only during the period of the NE monsoon. During the rest of the year water is drawn from storage and/or groundwater. The inflow to the reservoirs varies considerably from year to year (Fig. 2), and since 1986 the annual inflow has been continuously below the long-term average. In years of low rainfall, the reservoir inflow is more than proportionally reduced, since irrigation *tanks*, reservoirs and diversions upstream have first call on the available flow.

8. Flow from the upstream tributaries of the Korttalaiyar river is collected in Poondi reservoir. Below Poondi, flow in the main Korttalaiyar river is diverted at Tamaraipakkam into a canal leading to Cholavaram and Redhills reservoirs. These reservoirs also have their own direct catchments. Due to the flat terrain, all three reservoirs are shallow and are, in fact, large *tanks* formed by long earth embankments. Details are as follows:

Name	Capacity Mm³	Area at TWL km²	Average depth m
Poondi	75	34	2.2
Cholavaram	25	6	4.2
Redhills	75	20	3.75
Total	175		

9. To accommodate water received from the Krishna river, the capacity of Poondi and Redhills reservoirs is being increased to 95 Mm³ and 88 Mm³ respectively by raising the embankments, bringing the total to 208 Mm³. The construction of two new reservoirs is also proposed for a later stage of development: Ramancheri, 33 Mm³, and Thirukandalam, 28 Mm³ (Fig.1).

10. The canal bringing water to Madras from the Krishna river will discharge into Poondi reservoir. The annual supply to be provided will be up to 340 Mm³, 227 Mm³ being delivered during the months July to October and 113 Mm³ during the months January to April each year. The canal will also carry water diverted from the Arani river, replacing the present temporary arrangement.

11. There is no direct draw-off to supply from Poondi and Cholavaram reservoirs. Water from these reservoirs is transferred first to Redhills reservoir. Water drawn off from Redhills is taken to the main water treatment plant at Kilpauk, which has a capacity of 273 Mld.

12. To handle the additional supplies to be received from the Krishna river, a new 300 Mld water treatment plant is being built at Redhills. As the transfer required from Poondi reservoir will then exceed the capacity of the existing canal, a new link canal is proposed with branches to both Redhills reservoir and Chembarambakkam irrigation tank, where a 540 Mld treatment plant will be built at a later stage.

13. Reservoir inflow data were needed to run a combined surface/groundwater simulation model which was developed to assess the yield and reliability of the resources. These were derived from records of reservoir storage levels, draw-off and overflows, the diverted flow at Tamaraipakkam and evaporation (class A pan). The available data enabled monthly inflows to the reservoirs to be estimated for the period 1969-88. Fig. 2 shows the annual totals of the estimated reservoir inflows.

14. Rainfall records for the reservoir catchments are available from 1945 and an attempt was made to extend the inflow series by correlation with rainfall. However, no reliable correlation could be obtained, since over the period for which inflow data were available, there was a wide variation of inflows for same rainfall. The time available for the study did not permit the reason for the scatter to be investigated in any detail, but the principal reason is likely to be the different incidence of rainfall in different years.

15. The derived reservoir inflow sequences can only be regarded as "best estimates", since the data on which they are based contain many uncertainties. For instance, some of the

RESERVOIR PLANNING AND OPERATION

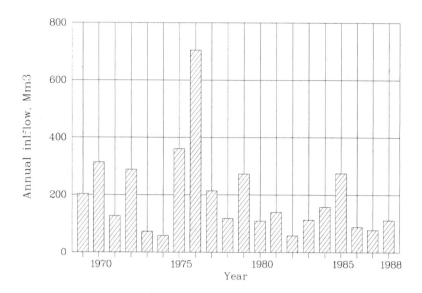

Fig. 2. Mean annual estimated total reservoir inflow

parameters are only measured indirectly, eg evaporation loss, and some not at all, eg seepage loss. Flow measurement records are not necessarily based on levels over weirs or in calibrated channel sections.

16. Inflows to Poondi reservoir have been reduced in recent years due to diversions upstream in Andhra Pradesh state. Estimated annual diversions range up to 36 Mm3 per year. For use in the model, inflows to Poondi before the diversions started were adjusted to give the estimated inflow that would have occurred had the diversion works been in existence since 1969.

17. During an earlier phase of the project, a study was made of the variation of evaporation at the different reservoir sites. It showed that evaporation is higher inland than near the coast, the estimated average annual evaporation at Poondi reservoir being 2400mm compared to 2100mm at Redhills. The highest monthly evaporation occurs in June and the minimum in December, the estimated average figures for Poondi being 278mm and 124mm respectively. The high rate of evaporation combined with the large surface area of the water stored in the reservoirs in comparison to the volume results in an appreciable proportion of the stored water being lost before it can be used.

GROUNDWATER RESOURCES

18. The Arani-Korttalaiyar aquifer comprises an old buried river channel of the river Palar, which now flows to the sea to the south of the city. The aquifer extends from the vicinity of Poondi reservoir to the coast to the north of Madras, and covers

an area of roughly 750 km², which includes the course of the Korttalaiyar river and the lower reaches of the Arani river. The extent of the aquifer is shown on Fig. 1. Water is pumped from the aquifer both by local farmers and for the city supply.

19. In its lower part, roughly downstream of Tamaraipakkam, the aquifer comprises two layers, the lower layer being confined. The alluvium is underlain by an impervious basement of hard Gondwana rocks.

20. Local farmers are the main users of water pumped from the aquifer. Over the four-year period 1987-90 pumping from the aquifer for the city supply has averaged about 35 Mm³/year compared to an estimated total abstraction of between 300 and 400 Mm³/year. There are no records of the quantity pumped by farmers. Estimates are based on electricity consumption.

21. The high rate of pumping from the aquifer has resulted in piezometric levels at the seaward end falling to below sea level, and this is giving rise to the intrusion of saline water from the sea. Since 1969 the saline front has advanced inland by about 5.5 km. To prevent further intrusion, a line of recharge wells has been drilled, into which freshwater is being injected to build up a mound of fresh water with a positive gradient towards the sea.

22. Average recharge is estimated to be roughly equal to abstraction, but varying between 170 Mm3 per year in a dry year to 450 Mm3 per year in a year of high rainfall. Recharge appears to take place mainly from the bed of the rivers and be proportional to the period of flow. An analysis of groundwater quality indicates that recharge is concentrated around the Tamaraipakkam diversion weir and similar weirs elsewhere.

23. The wells pumped for the city supply are concentrated in six separate wellfields, ranging from Poondi to Minjur (Fig. 1). Most of the water pumped is discharged into the conduit carrying draw-off from Redhills reservoir to Kilpauk treatment works. Water from the Minjur wellfield is delivered directly to industries situated to the north of Madras city.

SIMULATION MODEL

24. As part of a UNDP programme of technical assistance to the Madras Metropolitan Water Supply and Sewerage Board, a combined surface and groundwater simulation model was developed to study the conjunctive use of the city's reservoirs and wellfields.

25. The extent of the model is shown schematically in Fig.3. The model program was written in FORTRAN77 for the particular configuration shown and run using the series of estimated reservoir inflows described in #13 above. From given initial conditions and when taking a specified draw-off to supply, the model calculates, month-by-month, the end-of-month conditions over the period of the available inflow data. Months when the specified draw-off cannot be maintained are recorded.

26. The surface water section of the model calculates the change in storage in the different reservoirs during each time

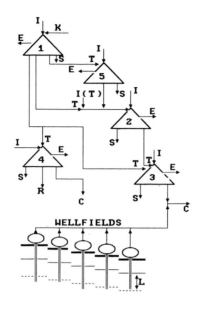

Reservoirs:
1. Poondi (plus Ramancheri (proposed))
2. Cholavaram
3. Redhills
4. Chembarambakkam (future connection)
5. Thirukandalam (proposed)

Key:
C   Supply to city
E   Evaporation loss
I   Inflow to reservoir
I(T)   Diversion at Tamaraipakkam
K   Flow from Krishna river
L   Limits for variation in groundwater levels
S   Surplus (overflow)
T   Transfer

Fig. 3. Simulation model schematic diagram

interval. The increase in storage is given by the inflow (from the direct catchment plus transfer from upstream (if any)) less draw-off (either to supply or as a transfer to a downstream reservoir), less evaporation from the reservoir water surface, less loss by infiltration to groundwater and less surplus (if any). Evaporation loss is estimated by multiplying the average surface area of the volume in storage during the month by the average evaporation rate for the month being considered. Transfers between reservoirs are determined by rules built into the program. The model may restrict actual draw-off to supply to less than the specified figure, either because there is insufficient water remaining in the reservoirs to meet the specified draw-off, or because the storage has dropped below a critical volume of storage specified in the data.

27. The groundwater section of the model calculates the fall in groundwater at each wellfield according to the rate of pumping, using relations between rates of pumping and draw-down, and recovery and recharge, determined from results obtained from a separate detailed finite-difference groundwater model, also prepared under the UNDP programme.

28. The model was used to estimate the reliable yield of the resources under different conditions, to estimate the frequency of supply restrictions if normal draw-off exceeds the reliable yield and to derive a water balance for surface water resources.

SUPPLY STANDARDS AND RESOURCE MANAGEMENT

29. <u>Reliable yield</u>. The reliable yields of surface and groundwater resources estimated using the combined model were:

	Mld
- Korttalaiyar river and tributaries	148
- Arani diversion	23
- Groundwater	29
Total (present)	200
- Contribution of Krishna inflow	796
Total (future)	996

30. For the purpose of the studies "reliable yield" was defined as the greatest constant draw-off that could be taken to supply without failure, assuming a repetition of the inflows for the 20-year period for which data were available.

31. Implicit in the use of the historic inflow sequence to estimate future reliable yield and/or the probability that the supply will need to be restricted is the assumption that the variation of inflows from year to year in the future will be statistically the same as in the past. A corollary of this is that the recent series of dry years is assumed to be part of the natural random variation of inflows from year to year, rather than the result of a long-term change in climate.

32. <u>Restrictions on draw-off</u>. Fig. 4 shows how the draw-off from Redhills reservoir to Kilpauk water treatment works has increased over the years since 1950 in response to the increasing demand for water in the city. For comparison, the estimated reliable yield of surface sources is also shown. It will be seen that since about 1960 draw-off has generally exceeded the reliable yield. The penalty, however, has been that in some years, following a poor monsoon, draw-off had to be reduced to appreciably less than the reliable yield. This has been particularly the case recently as, since 1985, there has been a succession of years with below average rainfall. The effect of drawing more water from the reservoirs than the reliable yield has been to reduce the storage carried over to the next year, which would otherwise have been available to help maintain draw-off at the reliable yield when inflow to the reservoirs was below average.

33. The greatest average annual draw-off shown in Fig. 4 is about 250 Mld. In contrast to this, the estimated unrestricted demand for water in the city is about 1100 Mld. In view of the magnitude of the deficit it would not be practical to restrict draw-off to the reliable yield, if a greater quantity could apparently be taken. The combined model was thus used to estimate the frequency and severity of restrictions that could be expected when the "normal" draw-off exceeds the reliable yield. The most severe situation is to maintain the "normal" draw-off from surface sources until the reservoirs are empty and then rely on groundwater. Although this ensures that the "normal" draw-off is maintained for the longest possible time

RESERVOIR PLANNING AND OPERATION

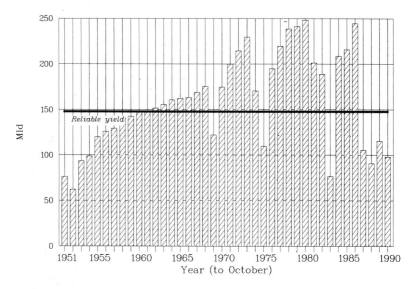

Fig. 4. Surface water draw-off (November to October)

and minimizes evaporation loss, the consequent restrictions arethe most severe. The alternative is to reduce draw-off from surface sources when the remaining storage falls below specified controlling levels depending upon the month of the year. Such an arrangement ensures a greater minimum supply, but at the expense of reducing the time during which the "normal" draw-off can be taken. For instance, for the present supply system and a "normal" total draw-off from surface and groundwater of 295 Mld, model results showed that, if the reservoirs were allowed to empty, the "normal draw-off could be maintained during 80% of months (over the 240 months of the inflow sequence), but draw-off would be reduced to less than 120 Mld for 7% of months. On the other hand, by specifying suitable controlling storages, the minimum draw-off could be increased to about 170 Mld, though at the expense of limiting the period of "normal" draw-off to about 30% of months. Draw-off would be restricted to less than 80% (236 Mld) of "normal" for 40% of months. Based on the model results, a set of controlling storages was recommended for present use.

34. When the new works have been built at Redhills and Chembarambakkam to treat the flow received from the Krishna river, the maximum combined surface and groundwater draw-off capacity will be 1162 Mld. This still exceeds the estimated reliable yield, though by a much smaller margin than at present. It is less, however, than the demand, which by then it is estimated will have increased to about 1500 Mld. Using the model it was estimated that, for a maximum draw-off of 1162 Mld, the draw-off would need to be restricted to less than 80% of the maximum (930 Mld) during 3% of months only. Fig. 5 shows how

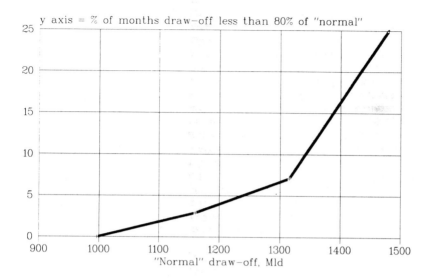

Fig. 5. Reliability of supply with Krishna inflow

the estimated frequency of restrictions would increase, if workswere built to allow an even higher draw-off than at present proposed.

35. <u>Proposed new reservoirs</u>. Since the Krishna inflow will not be received uniformly over the year, storage is needed to balance the inflow and draw-off. Works are already in hand to increase the storage available in Poondi and Redhills reservoirs, but at a later stage the construction of two new reservoirs, Ramancheri and Thirukandalam, has been proposed. The combined model was used to study the likely increase in the reliability of the supply that would result from the extra storage volume. The results showed that the benefit of the new reservoirs would be small and would probably not warrant the cost of construction. In comparison with the results mentioned in the previous paragraph, which allowed for the increased storage in Poondi and Redhills reservoirs but not for the extra storage in the new reservoirs, the model results showed that the frequency of restrictions to less than 80% of the maximum draw-off of 1162 Mld (ie 930 Mld) would be reduced from 3% of months to 2% of months. The reason for the apparently small benefit is that, although as a result of the greater storage overflows are reduced, the increased surface area of the stored water leads to greater evaporation losses, the gain from one being almost exactly balanced by the loss from the other, as can be seen from the following average annual figures:

RESERVOIR PLANNING AND OPERATION

	Without new reservoirs		With new reservoirs	
	Mm³	% of total	Mm³	% of total
Total inflow	680	100	685*	100
Outflows:				
Draw-off to supply	430	63	430	63
Reservoir surplus	120	18	107	15
Evaporation	101	15	115	17
Seepage	29	4	33	5
Total outflow	680	100	685	100

* Increased inflow due to small increase in catchment

36. *Reducing evaporation loss.* The above table shows the high loss due to evaporation from the water surface of the reservoirs. As a proportion of total inflow, the average loss under present conditions is even greater, reaching as much as 25% (55 Mm³/year) of the inflow to the reservoirs. With the object of reducing the loss and thus increasing the amount of water available for supply, strategies were proposed to minimize the surface area of water in store at any particular time. The first recommendation was that, to the extent possible, water should be transferred from Poondi and Cholavaram reservoirs to Redhills reservoir. Secondly, it was recommended that draw-off from Redhills reservoir during the first part of the year should be at rather more than the average rate, the rate of pumping from groundwater being reduced below average, while during the second part of the year the reverse should apply. The model results showed that a reduction in evaporation loss of up to 4 Mm³/year could be achieved in this way.

37. Theoretically, as much as 12 Mm³/year of evaporation loss could be saved by stopping all groundwater pumping at the start of the year and meeting the demand entirely with water drawn from the reservoirs. Then, when the reservoirs were empty, the demand would be met entirely from groundwater. The limited pumping capacity, however, and the need to maintain a minimum reserve storage in case of pump failure precluded this extreme policy being recommended.

RESPONSIBILITY FOR STUDIES

38. The development of the simulation model, derivation of inflow data and analysis of the present situation, was carried out as part of the UNDP technical assistance programme. Analysis of the future situation with the inflow from the Krishna river formed part of a Master Plan study undertaken for the Madras Metropolitan Water Supply and Sewerage Board (MMWSSB) by Tata Consulting Engineers of Bombay, in association with Binnie & Partners of Redhill, UK.

39. The UNDP technical assistance was provided to the Government of Tamil Nadu for the benefit of the MMWSSB and was carried out by the UN Department of Technical Cooperation for

Development, which arranged for consultants from USA, France, Czechoslovakia and Great Britain to work with a team of engineers from the MMWSSB. Apart from the yield and water management studies described in the paper, other aspects covered by the programme included the development of a finite difference groundwater model to provide data for the combine surface/groundwater simulation model and to study measures proposed for reducing saline intrusion, the planning of a series of check-dams on the Korttalaiyar river to increase groundwater recharge, proposals for re-use of sewage effluent to provide non-potable supplies to industries and advice on pump maintenance procedures.

ACKNOWLEDGEMENTS

39. The authors would like to thank the Madras Metropolitan Water supply and Sewerage Board for permission to publish this paper and for the help received from the Board's engineers in its preparation.

Estimation of river flow requirements to meet electricity demands in the Cameroon

P. E. ROBINSON, Power and Water Systems Consultants

SYNOPSIS. Electricity demands in Southern Cameroon are supplied by two hydroelectric stations operated by the Société Nationale d'Electricité du Cameroun on the River Sanaga. Water storage is limited at the first station and negligible at the second, and natural river flows are regulated by releases which take up to 7 days to travel from upstream reservoirs. Downstream flow requirements must accordingly be forecast for up to 7 days ahead. This paper describes application of a computer program to minimise these requirements by optimising operation of the limited downstream storage and the two power stations to meet the demands imposed.

BACKGROUND
1. A contract for the development and supply of computer software for management of flows within the Sanaga River Basin has been placed by the Société Nationale d'Electricité du Cameroun (SONEL) with a consortium consisting of Lahmeyer International, Power & Water Systems Consultants and the Institute of Hydrology. The software has recently been installed and is now being tested.
2. The three software modules developed by the consortium are for Rainfall Forecasting, for Rainfall/Runoff Translation and for River Basin Modelling and Reservoir Release Optimisation. The first two modules are concerned with forecasting inflows to the river system at various locations and the third with optimising reservoir releases to match the balance of downstream requirements. It is the calculation of these requirements that forms the subject of this paper.

The Cameroon Interconnected Southern Electricity Network
3. The Sanaga River Basin and real-time gauging stations installed are shown schematically in Fig. 1. As shown in the figure, the approximate times for water from the Mbakaou, Bamendjin and Mape reservoirs to reach the Songloulou hydroelectric scheme are 6, 4 and 4 days respectively, with a further 12 hours being required for water released from Songloulou to reach the Edea hydroelectric scheme.
4. The capacity of the Edea hydroelectric plant has been increased in stages up to its present value of some 260 MW, with different designs of plant installed at each stage (ref. 1). The available head is normally about 22 m, but may be somewhat less on account of raised tailwater levels during periods of high river flow. The water storage capacity directly upstream of the dam is negligible and thus water must be either utilised as it arrives, or spilled. The plant is located alongside

RESERVOIR PLANNING AND OPERATION

an aluminium smelter and fabrication works, operated by ALUCAM, which normally has a constant supply requirement of 150 MW.

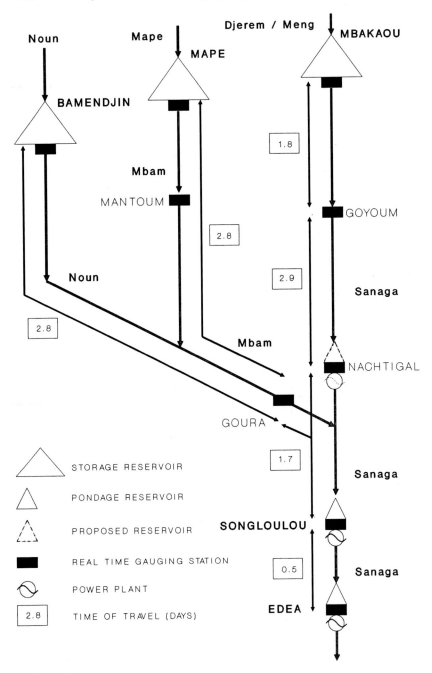

Fig. 1. Schematic of the Sanaga river basin

5. The Songloulou hydroelectric plant, with a total installed capacity of 384 MW, lies some 50 km upstream of Edea. The storage capacity of the Songloulou head pond is limited to about 5 M.m^3, corresponding to a permissible range in upstream water level of some 6 m. The turbines discharge into a canal which parallels the river for just over 1 km before rejoining it, and significant variations in tailwater level arise with change in both turbine discharge and river flow.

6. The two hydroelectric schemes are complemented by a limited amount of diesel plant, generally operated as back-up capacity. As shown in Fig. 1, a third hydroelectric scheme on the River Sanaga, at Nachtigal, is planned for completion at some future date.

7. The maximum electricity demand at present supplied by the existing generating plant described above is about 365 MW, this being made up of the public sector demand with a peak of some 215 MW and the ALUCAM demand of 150 MW.

Time steps for optimisation processes

8. Whilst the telemetry system monitors flows on a 6-hourly basis, it is foreseen that rates of release from the reservoirs will not be changed more frequently than daily. The objective of the reservoir release optimisation process, rather than merely to maintain particular river flows, is to cover the forecast electricity demands. Since these have significant hourly variations, the fulfilment of this objective requires hydroelectric operation to be optimised with a time step of not more than one hour. Such optimisations, making full use of the limited regulating capability of Songloulou head pond, are required either:
- to confirm that a particular set of flows will allow the electricity demands to be covered, or
- to derive target flows at Songloulou which, if met, will allow these demands to be covered.

9. The minimum flow requirement at Songloulou would be obtained with the head pond drawn down completely at least once during the planning period. With this flow, however, the water level at the end of the period may be lower than at the start. Thus, to be sure that at least the same magnitudes of demand could be supplied with the same flow in the following period, the target flow is defined as the minimum which allows not only the full system demands to be supplied but also the end-of-period head pond water level to return to the starting value.

DATA EMPLOYED

10. The data representing operation of hydroelectric schemes and power system in the simulations and optimisations are itemised below.
   a. Electricity demand:
      - daily load curves of public sector demand, summated from transmission feeder outputs for each hour, and
      - total demand from the ALUCAM aluminium plant.
   b. Water flow:
      - river flow at Songloulou,
      - time of water travel from Songloulou to Edea,
      - incremental inflow between Songloulou and Edea.
   c. Songloulou head pond:
      - level/area/volume characteristics,
      - evaporation.

d. Songloulou and Edea power stations (where relevant for each type of unit at Edea separately):
- dam seepage losses,
- spillway level/flow characteristics,
- penstock friction losses,
- turbine rated output and head,
- turbine net head/flow/efficiency characteristics,
- generator rated capacity,
- generator efficiency/output characteristic,
- unit minimum feasible power output,
- tailwater level/canal flow/total flow characteristics, for Songloulou, and
- tailwater level/total flow characteristics for Edea.

e. System operation:
- power station security requirements,
- spinning reserve requirements, and
- minimum unit shut-down times.

METHODOLOGY

11. The work described in this paper has entailed the use of a standard computer program, SYRAP. This employs the established short-term planning concepts of unit commitment and economic dispatch as extended for power supply systems with significant hydroelectric components. It is used to optimise system operation with an hourly or half-hourly time step over a planning horizon of up to a week. An outline of the program and its application to a variety of systems in different countries have been described previously (ref. 2).

12. For the Southern Cameroon system the program is extended by adding a single system-specific routine written to represent the water system topology and control a search for the target flows. This routine could readily be extended to accommodate other hydroelectric schemes, such as that proposed at Nachtigal. The existing small thermal plant or any new thermal plant could be incorporated in the optimisation merely by making appropriate additions to input data.

13. For the current application the program has been used to:
- simulate the Songloulou and Edea hydroelectric schemes operating to supply the public sector and ALUCAM demands,
- optimise operation of this system on an hourly basis for representative days in different seasons and years and with alternative levels of ALUCAM demand, and hence to
- determine the target river flows at Songloulou.

14. The simulations are performed in terms of water flow, storage and spill together with the power and energy capabilities of generating plant units, all as established from data provided by SONEL. Variations in friction losses with flow, efficiency with net head and flow and tailwater level with total flow and power station discharge are all modelled. The resulting input/output characteristics for Songloulou with a range of pond levels and Edea for three plant types are shown in Fig. 2. From the curve shapes it can be seen that, as is common with all types of generating plant, the greatest efficiency is achieved close to maximum output. The number of units operated to supply a particular level of demand should thus be as few as possible. The unit commitment procedure is designed to achieve this while maintaining

the required amounts of spinning reserve, and taking due account of practical operating restrictions such as those arising from:
- water availability,
- storage capacity,
- water flow limits and times of travel,
- minimum unit outputs, and
- minimum unit shut-down periods.

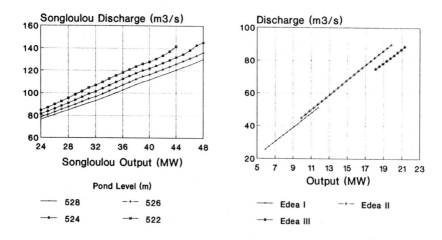

Fig. 2. Hydroelectric unit input/output characteristics

15. Once the numbers of units to be operated have been determined by the unit commitment procedure, load is allocated to the plant by performing an economic dispatch for each time step. This is based on unit characteristics such as in Fig. 2. By ensuring, as far as possible within the feasible output ranges, that the specific water discharge rates are equal for all units the total water discharge is minimised. This solution may be overridden, however, to avoid spill. As can be seen from Fig. 2, a complication at Edea for this application is that the most efficient plant also has the most restricted output range.

16. Schedules produced by the program indicate times for plant to be committed and decommitted, and time step outputs, or set points, such that the forecast demand can be met most economically without operating constraints being violated. It is not necessarily the outputs of specific units that are thus scheduled, however, since decisions as to which of a group of identical units should be run at which times to provide the scheduled output are best made at plant level.

Computer program operating modes

17. The three modes for this application are:
1. Calculate the target river flow as the minimum constant rate allowing the full system demands to be supplied and the end-of-period head pond water level to be the same as at the start.
2. Calculate the hourly flows required to supply the full system demands whilst maintaining specified head pond water levels.

RESERVOIR PLANNING AND OPERATION

3. Given a constant river flow, find an operating schedule allowing system demands to be supplied as nearly as possible.

In Mode 1, plant operation is optimised at each step of a search performed to find the target flow. The starting value for this search is the average of the hourly flows derived as if in Mode 2.

Optimisation procedure

18. The procedure for this application is:
   a) Simulate commitment and economic dispatch of plant units taken in a nominated commitment order to match system load.
   b) Simulate hydroelectric operation to supply as far as possible the outputs calculated in a).
   c) Fix Songloulou outputs for each time step either as initially calculated or reduced to avoid spill at Edea or increased to supply a shortfall at Edea.
   d) Repeat unit commitment and economic dispatch to optimise the different Edea plant types to meet load remaining from c).
   e) Repeat c) and d) until there is no significant change in total Songloulou output from one iteration to the next.
   f) Fix Songloulou outputs for each time step either as previously calculated or increased to supply a shortfall at Edea (i.e. spill at Edea is accepted at this stage of the procedure).
   g) Repeat unit commitment and economic dispatch to optimise Edea operation to supply load remaining from f).
   h) Repeat f) and g) until the total demand is satisfied in every time step or (only in Modes 1 or 3) until there is no significant change from one iteration to the next and it is thus clear that no solution is possible without greater flow being provided. (In Modes 2 and 3 this concludes the procedure).
   i) In Mode 1, adjust the flow and repeat the whole procedure from c) until the minimum constant rate of flow is found.
   j) Increase flow so as to restore the end-of-period pond water level to the starting value and further iterate from a) as necessary until the target is found within a defined tolerance.

19. In consequence of the water travel time from Songloulou to Edea, initial values are required for flows at Edea in the first 12 hours of a planning period. These may be fixed as determined in the previous period or alternatively are found by simulating for the first day of each planning period twice, i.e. assuming that the Songloulou flow and plant operating schedule of the previous day were the same as being optimised for the first day of the current period.

20. In general, the closer the flow is to the target the more iterations are required to find a solution and hence the longer is the calculation time for that flow. On an IBM-compatible 386 computer with co-processor running at 25 MHz, a solution for a given flow (i.e. in Mode 3) for a day thus takes between 20 seconds and about 1 1/2 minutes. The full calculation (i.e. in Mode 1) takes about 10 minutes.

21. The target flow will be correctly estimated only if the:
   - demand forecast is correct,
   - data accurately represent the plant and its operation, and the
   - system is operated exactly as scheduled by the program.

Since all three conditions are unlikely to be satisfied simultaneously, a tolerance is added to the target. This is initially about 5 per cent but is expected to be gradually reduced as experience is gained.

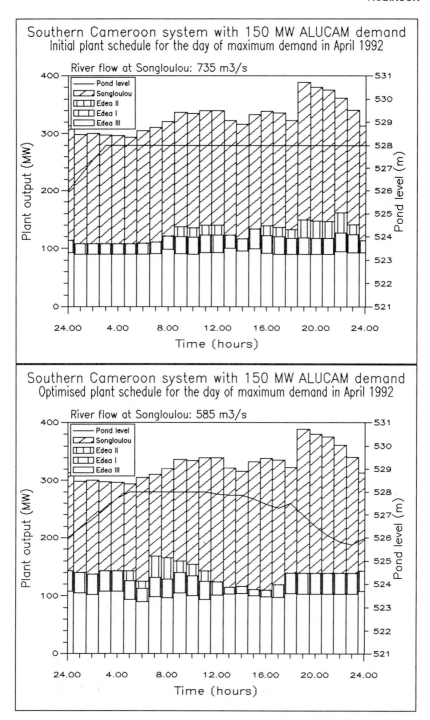

Fig. 3. Comparison of initial and optimised system operation

RESULTS

22. Initial and optimised solutions for a typical day with the same load supplied are illustrated in Fig. 3. The initial solution shows all the plant following the rise and fall of demand. In the optimised solution, however, Songloulou still follows the demand but the Edea outputs are shifted completely out-of-phase so as to utilise as much water as possible 12 hours after it has been discharged from Songloulou.

23. Also shown in Fig. 3 are the required constant river flows at Songloulou for the two solutions, and the associated head pond level trajectories. For the initial solution the level rises rapidly and then remains constant at the spill level of 528.0 m. For the optimised solution the level rises more slowly and then falls again as stored water supplements inflow to supply the fluctuating demand. The reduction in flow requirement of some 20 per cent from 735 to 585 m^3/s is attributable to reduced spill permitted by the changed plant schedule, and also to the utilisation of pond storage. The two effects are separated by performing a further optimisation for the same conditions but with the pond water level held constant as if there were no storage. The required inflow is found to be 678 m^3/s, indicating that the proportion of the total flow reduction attributable to storage utilisation is over 60 per cent. Results for the initial and optimised solutions for the day with and without storage are summarised below.

Solution		Initial	No-storage	Optimised
Songloulou discharge	(M.m^3)	49.02	44.14	48.13
Songloulou spill	(M.m^3)	12.55	14.44	2.36
Storage change	(M.m^3)	1.93	-	-
Inflow	(M.m^3)	63.50	58.58	50.49
Edea discharge	(M.m^3)	47.10	58.58	50.10
Edea spill	(M.m^3)	14.47	0.00	0.34
Songloulou energy	(MWh)	4834	4110	4609
Edea energy	(MWh)	3072	3796	3297
Total demand	(MWh)	7906	7906	7906

CONCLUSIONS

24. It is concluded that:
   a. By optimising unit commitment and economic dispatch and then minimising the spill of water at Edea, there is significant scope for reducing Songloulou flow requirements and hence conserving water in the upstream reservoirs.
   b. Techniques described in this paper enable maximum value to be obtained from limited storage.

REFERENCES

1. SONEL and Electricité de France. Atlas du Potentiel Hydroélectrique du Cameroun, 1983.
2. ROBINSON P.E. The valuation of water for short-term operations planning of hydro-thermal power systems. 8th CEPSI Singapore, 5-9 November 1990, vol 4, paper 4-32.

Evaluation impacts on environment of Narmada Sagar project, India

H. SAHU, E. I. KSP Colony

SYNOPSIS Water Resources Development Projects result in a variety of environmental impacts, both beneficial and adverse. Narmada Sagar is a key project on Narmada River under construction with a reservoir gross capacity of 12.22 Billion Cubic meters. This paper deals with the Evaluation of Environmental impact assessment of Narmada Sagar Project. This paper gives an empirical method for evaluation of Environmental effect of Water Resources Projects.

1.0   INTRODUCTION

The environment is a complex system consisting of physical, biological and social subsystems and the construction of a water resources project results in a variety of impacts both beneficial and adverse, on the environment.
Of late a very high degree of consciousness has been generated among the environmentalists, Social reformers and many others regarding the preservation of our environmental heritage. There has been remarkable increase in public concern about environmental impacts of large Dams.

2.0   ENVIRONMENTAL IMPACTS

Broadly our environment is comprised of many items, aspects and Plant life existing on our planet earth. For a brief resume, these have been listed in the following paragraphs.

2.1   Atmosphere:
i   Clear pollution free
ii.   Pre-ponderance of Gases, fumes/Acid rain, Ozone layer disruption.
iii.   Climate Micro/Macro
iv.   Temperature effects.
v.   Humidity change.

RESERVOIR PLANNING AND OPERATION

2.2	<u>Earth</u>
i.	Land forms.
ii.	Soils, alkalinity, salinity, water logging.
iii.	Minerals resources in mines.
iv.	Rocks types and geological formations.
v.	Wet lands.
vi.	Forest, agriculture, de forestation, plantation, pastures, grass lands.
vii.	Recreational use.
viii.	Commercial uses.
ix.	Industrial use attendent pollution.
x.	Urbanisation.
2.3	<u>Flora</u>
i.	Trees.
ii.	Shrubs and grasses and weeds.
iii.	Crops.
iv.	Rare Plant life / medicinal / ornamental plants.
2.4	<u>Fauna</u>
i.	Birds.
ii.	Reptiles
iii.	Animals on land, trees, water (streams, rivers, sea.)
iv.	Aquatic life, fish life, shells etc.
v.	Food Chain.
vi.	Endangered species.
2.5	<u>Water</u>
i.	Quantity (Supply)
ii.	Quality degradation.
iii.	Natural character viz. Fall, lake.
iv.	Water diversion, changing course of rivers.
v.	Rise in Ground Water Table and decline.
vi.	Seepage, percolation.
2.6	<u>Human Activities & interest and Built Environment</u>
i.	Natural Parks, reservation.
ii.	Tourists and scenic spots.
iii.	Architectural, Archeological monuments (Temples, Mosques, ruins etc.)
iv.	In appropriate uses/slash and burning practice/cutting timber for fuel.
v.	Irrigation, drainage, flood control, power generation, navigation, recreation, pisciculture use.
vi.	Buildings, roads, highways, railways and bridges thereon.
vii.	Water supply and sewage disposal installations.
viii.	Gardens, parks, swimming pools, mela grounds, shopping complex, markets and mandis, schools and hospitals.
ix.	Human resettlement due to displacement from

submergence area.
2.7 Natural resources
i. Air movement, trade and anti-trade winds
ii Rain
iii. Solar radiation, sun shine hours.
iv Erosion/denudation
v. Sedimentation
vi. Flooding regular occasional, sporadic.
vii Emergence and sub mergence of lands due to water level depletion and ponding.
viii. Stability of slopes, slides, subsidence settlement.
ix. Seismicity, reservoir induced, due to dis-aprophic movements in earth-crust, volcanicity.
2.8. Social Factors (Human)
i. Health
ii. Employment
iii. Safety and Security.
iv. Life style and culture.
v. Population density (Dense, sparse)
vi. Socio-economic impact.
vii. Trade and Commerce boost.
viii. Industrialisation.

All these with their detailed sub-heads have to be analysed, without undue bias or prejudice to bring forth the positive and negative impacts with due weighting or importance of the water resources projects.

3. **NARMADA SAGAR PROJECT**

3.1 Narmada Basin
Narmada is the Fifth Largest River of India. Drainage area of Narmada Basin is 98,786 Sq.Km. The Annual Water availability of the river is estimated as 34.54 Billion Cubic Metre to utilise the Narmada Waters. It is proposed to construct 29 Major Projects 135 Medium Project and over 3000 Minor projects in Madhya Pradesh, which is contributing 89.8 per cent of the catchment area. Development of Narmada Basin is envisaged to be completed in the year 2025. Figure 1 gives the major project in Narmada Development Plan.

3.2 The Narmada Sagar Project
Narmad Sagar Project on Narmada River provides an excellent storage site. The proposed Dam will be 92 m. high and is near the village Punasa. The Reservoir created will have gross capacity of 12.22 Billion Cubic Metre, live storage will be 9.75

Billion Cubic Meter. It has an installed capacity of 1000 MW and annual irrigation of 0.265 Million Hectare. The project provides regulated releases of 10.01 Billion Cubic Metre to Gujarat ex.viz.,Omkareshwar and Maheshwar in Madhya Pradesh.

3.3 Environmental Aspects of Narmada Sagar Project

The need for preserving and protecting the environmental and ecological equilibrium has been taken into consideration. All possible measures to compensate for the forest wealth being lost due to the forest being submerged have been taken. Catchment area treatment and command area development programme is also included in the project. Rehabilitation and resettlement of oustees from the Project area and reservoir submergence area have been given highest priority. A settlement plan has been prepared and being implemented to compensate them for better living. A weighting scale from 0 to 10 with positive side for beneficial effects and negative side for adverse effects of environment has been adopted.

Studies have been made and necessary steps taken to minimise the adverse effects of environmental impact on the following:

- Safeguard the monuments of archeological value.
- rehabilitation of micro/macro biology species of flora and fauna.
- Conservation of Wild Life.
- Ensure Control of accelerated erosion of Soil in the catchment area.
- Command Area development programme.
- Ground water and conjunctive use.
- surface and sub-surface drainage.
- All possible measures to compensate for the forest wealth being lost due to reservoir submergence.
- raising of compensatory plantation of degraded forest land.
- lively abverse effects on public health.
- Settlement of people affected due to Reservoir submergence.

4. **ENVIRONMENT IMPACT EVALUATION OF NARMADA SAGAR PROJECT**

Scale from 0 to 10; positive side for beneficial effects and negative side for adverse effects of environment.

A. Catchment Area
1. Change in the Erosion and Deposition pattern -1

2. Some loss of Forest Cover and grasses due to fuel needs and grazing cattle. −2

Total −3 +0

B. Reservoir Area
1. Scenic spots, Archeological sites, temples ruins. −1
2. Mines of semi-precious materials, rocks, boulders, sand would be submerged. −2
3. Flora will be affected. −1
4. Wild life will be affected. −2
5. Sedimentation in Reservoir would occur, causing subsidence of lake bottom. −2
6. Lake Eutrofication. −1
7. Water quality degradation. −1
8. Weeds development in lakes. −1
9. Rehabilitation & resettlement of displaced persons. −8
10. Hazards due to water infestation to human and cattle health. −1
11. Railway and highway dislocation and re-counting. −3
12. Fishery development in lakes. +7
13. Recreation and Tourist resorts. +4
14. Fore-shore bed will be available for cultivation. +2
15. Micro-climatic improvement due to low temperature. +3

Total −23 +16

C. Dam site and Related Area.
1. Communication facility over bridge on spillway. +4
2. Ornamental garden on top of dam, toe of the dam on down stream side. +2
3. Flood control benefits to area situated on river banks of towns and villages. +3
4. Seepage from dam may produce beneficial effects. +1
5. Fish mortality down stream of the spillway. −1
6. Seismic activity may trigger due to impoundment of water in the lake. −1
7. Township and community development in the area. +7
8. Rise of water table in wells thereby reducing pumping lifts. +4

Total −2 +21

D. River regime D/S
1. Bed and bank erosion. −1
2. Structures and habitation on bank affected. −1
3. Navigation hazard. −1

Reassessment of reservoir yields in North-West England in the light of possible climate change

H. SMITHERS, BSc, MSc, PhD, NWW Ltd

SYNOPSIS.
The paper describes methods used in NWW's recent reassessment of yields. Definitions are discussed, and earlier reviews described. Methods applicable to catchments with and without long inflow records are presented, and the provision of suitable data discussed. Evidence for climate change in NW England is briefly reviewed, with some illustrations from NWW's reservoir catchments. It is concluded that recent variations in climate are within the range of the historic record, and that analyses should continue to use the longest possible data set. There is as yet no firm evidence of climate change affecting reservoir yield in NW England.

WATER SUPPLY SOURCES IN NW ENGLAND.
1. North West Water Ltd (NWW) supplies approximately 2400 Ml/d of potable water to around 7 million customers in NW England. 10% of this water comes from boreholes in S.Lancashire, Merseyside, Greater Manchester and Cheshire, and reliable yields of these sources are assessed on the basis of aquifer recharge rates.
2. Regional surface water sources are grouped into the Northern Command Zone (NCZ), which comprises the major Lake District reservoirs and the Lancashire Conjunctive Use Scheme, and the Southern Command Zone (SCZ), made up of Vyrnwy Reservoir and River Dee abstractions. The NCZ and SCZ each provide 30% of total water supplies.
3. Local sources consist of around 150 reservoirs, mainly in the Pennine uplands; together these account for a further 27% of available supplies. The remaining 3% comes from various river abstractions.

DEFINITIONS OF YIELD.
4. The term "yield" has been used with different meanings in different circumstances. Twort et al (Ref.1) propose 5 types of yield:
 a) Yield from experience- the abstraction from a source over a number of years
 b) Minimum historic yield- the steady supply which could just be maintained over the period of record.

RESERVOIR PLANNING AND OPERATION

c) Probability yield- the steady supply which could just be maintained through a drought of specified severity.
d) Operating yield- the supply which could be provided under a fixed set of operating rules.
e) Failure yield- the steady supply which could be maintained for a given percentage of the time.
Options b),c) and d) are currently in use at NWW. Option a) does not provide consistent region-wide estimates and is inappropriate for present-day complex interlinked sources.
Option e) is not applicable to the UK, where interruptions to supply are not expected by the consumer. It is important to distinguish between definitions b) and c), which are based on a steady supply, and d), in which the supplies depend on the choice of operating rules.
5. NWW practice has been to determine a yield of a specified return period. Control rules are then developed to make best use of the water which is not required to protect the yield, which might otherwise spill and be lost to the system. The actual output from the system over a period is termed the average supply. During a dry period the operating yields are also updated to take account of current conditions.
6. However a static long-term yield is still necessary for planning purposes, and provides the basis for assessment of the adequacy of total resources, and comparison of alternative schemes.
7. As water supply systems become more interlinked it also becomes necessary to consider the yields of groups of sources, which are not necessarily equivalent to the sums of the individual values.

RESERVOIR YIELD ESTIMATION IN NWW.

8. Until 1978 assessments of the yield of reservoir sources in the NW had been derived at different times and by a variety of methods. Returns were made to the Department of the Environment using Engineering 100 and W/1 Forms in 1973, and these were updated in 1975 to become the Register of Sources (Ref 2). Methods used included Hawkesley's Rule, the Deacon diagram, and Lapworth charts. McMahon and Mein (Ref 3) provide a historical summary and comprehensive review of methods of reservoir yield calculation. The 1978 Survey of Existing Surface Water Sources (Ref 4) contains estimates of the 2% reliable yields of all NWW's major supply and compensation reservoirs, and essentially the same approach has been followed in subsequent revisions (Ref 5) including the present one.

Yield estimation with long inflow records.

9. The larger sources (ie Thirlmere, Haweswater, Vyrnwy, Stocks, Rivington, and Longdendale) have their own inflow sequences, and these were analysed using programs from NWW's Resource Planning Suite (Ref 6) to produce ranked cumulative inflows for durations of 1-36 months. These were then plotted on probability paper. For durations of up to 12

months the distributions of independent cumulative inflows
are markedly skew, and lognormal probability paper is thus
appropriate. As the duration increases the number of
independent events which can be extracted from the record
decreases dramatically, and with it the number of plotted
points. This problem was overcome by allowing events of
longer than 12 months to overlap, and choosing the same
number of events as the years of record. An event lasting
more than one season extends over a greater part of the flow
range than short period droughts. The distribution is thus
less skewed, and a normal probability plot will produce a
reasonable straight line. A family of straight lines was
drawn by hand for each site, in such a way that flows
increased smoothly and the lines did not cross. Special care
was needed at the changeover between independent and
overlapping events. Inflows were expressed as a percentage of
average annual inflow to allow comparison between sites.
Inflows of durations 1-36 months and appropriate return
period (usually 2%) were then read off the graph, and used in
a further program from the RP suite (Ref 6) which calculates
water balances and hence the minimum steady supply for each
duration. The lowest of these minima is the yield of the
system. The analysis is repeated for a range of storages to
give a yield- storage table. Yields are adjusted to allow for
the fact that a drought may start and finish midmonth.
However if the storage is too large for the catchment, in
practice the reservoir remains unacceptably low for long
periods. Previous analysis of Rivington and Vyrnwy records
indicated that in order to ensure refill within 5 years it
was necessary to limit yields to 76% of average daily flows,
and this criterion has been applied to other reservoirs in
the region.

Estimation of yield for ungauged catchments.

10. The majority of NWW's smaller sources do not have long
inflow records, and yields must therefore be estimated by
comparison with gauged sites. Stocks and Rivington were
located nearest to the majority of the Pennine reservoirs, so
a regional 2% runoff sequence was based on the lower of the
inflows for each duration at the two sites. A dimensionless
regional yield-storage table was then produced, with storage
expressed as a percentage of annual average inflow, and yield
as a percentage of average daily inflow. It was thus
necessary to determine net reservoir capacity and to estimate
average daily inflows. These were estimated using the water
balance equation on an annual basis:

$$\text{Precipitation - Loss = Runoff} \quad (\text{mm}) \quad (1)$$

Mean annual catchment rainfall and losses are thus required

11. Rainfall. The 1941-70 standard rainfall period has been
used in previous assessments. It had been hoped to include
1961-90 rainfalls in the current review, but the data were
not available in time. Catchment rainfalls were therefore

estimated using the 1941-70 annual averages at raingauges in the region, and an isohyetal map of NW England produced in conjunction with the Flood Studies Report (Ref 7) The majority of the reservoired catchments contain raingauges. However in most cases the long-term daily gauges are near the dam itself, in the lowest and therefore driest part of the catchment. The isohyetal map was therefore used as a guide to estimate an adjustment to the gauge total. In the current review these estimates were checked, and amended where necessary.

12. Losses. The 1981 Survey reported that no estimates of 1941-70 catchment losses were available, and instead used Section 14 Survey maps (Refs 8,9). It is not clear where this data originated from. The Lancashire River Authority gives annual potential evapotranspiration (PE), whereas the Mersey and Weaver River Authority map shows estimated actual evapotranspiration (ET). There is thus an inconsistency between parameters in different parts of the region. The water balance equation requires the annual catchment loss, which includes any percolation to groundwater. Fortunately in most Pennine catchments this will not cause significant error. Jack (Ref 10) produced an empirical estimation method, with adjustments to an initial value of 500mm according to latitude, elevation, exposure and afforestation. Loss estimates derived using this procedure were compared with the earlier evaporation estimates, and differences greater than 20mm were adjusted to a mid-value. Wright (Ref 11) suggested a range for mean annual losses of 400-500mm, but very low values may be recorded in mountainous catchments with impermeable bedrock. This agrees with Twort's typical loss rate (Ref 1) of 450mm/year, and that annual rates only exceed 500mm/year where the catchment is afforested.

13. There is an extensive literature discussing the effect of afforestation on yield of reservoired catchments, which is reviewed by Calder (Ref 12), who includes a summary of the work carried out by Law at Stocks Reservoir. From comparative studies of losses in the Severn and Wye headwaters, Kirby et al (Ref 13) estimated that a completely forested catchment would lose an additional 12% of precipitation in comparison with a grassland catchment. If these percentages are applied to a typical Pennine catchment with rainfall 1200mm, loss 450mm, and runoff 750mm, the forested losses would increase to 594mm, with a corresponding decrease in runoff to 606mm. However the proportion of forested land at Stocks is only 23%, so on a pro rata basis losses would rise to around 490mm. In general forested areas are insufficient to affect loss estimates.

14. Within the annual figure there may however be wide variations from year to year. An example is Alt Uaine in the Scottish Highlands, where the mean loss is 242mm (Ref 14), but extremes of 117 and 456 mm have been recorded.

15. The Meteorological Office produces 40km^2 gridded values of Penman PE on a regular basis, but the grid scale is too

coarse for small Pennine catchments. However it is planned to use data from three upland stations as confirmation of the previous estimates. This will also allow investigation of the seasonal and annual variation of evapotranspiration in upland areas typical of water supply catchments.

16. Catchment area. In order to convert runoff in mm to a flow rate the catchment area supplying the source is required. These values have been checked at each review. In some cases where discrepancies were apparent catchment plans were planimetered. A significant number of the Pennine catchments have catchwaters, many of which substantially increase the contributing area. A detailed investigation of the layouts and dimensions of catchwaters was carried out for the 1978 Review, and reductions in effective contributing area were made to allow for loss of storm peak flows. The method used was that described by Twort (Ref 1). Complex catchwaters were analysed in sub-areas, and the results aggregated to give a total effective catchment area. In the current review this information proved particularly difficult to verify, due to limited resources, a large number of organisational changes, and a lack of consistent records. It seems likely that as a result of the general reduction in manning levels and hence field maintenance, catchwater efficiency will have declined as leaks develop and blockages are removed less frequently. A 5% reduction was therefore applied to efficiencies below 60%, and a 10% reduction above this figure.

17. Reservoir storage. The relevant storage parameter for yield estimation is the net "active" storage, ie that between the top water level and the lower limit for abstraction. The original Engineering 100 Forms did not always make a clear distinction between net and gross storage, so in the 1981 Review a 10% allowance was made for dead water and siltation. These estimates were considered likely to be conservative. More recent information has been used where available; for example the Rivington reservoirs were resurveyed in 1989, and substantial drawdowns in recent dry summers have allowed the extent of siltation to be assessed more accurately. At the opposite end of the storage range, top water levels may be reduced, for example due to flood storage provision, or for reservoir safety reasons.

Accuracy of estimates of reservoir yield.

18. There are many potential sources of error in the reservoir yield estimation process. Data measurement errors are the most easily defined. Long reservoir inflow records are notoriously inaccurate over short time periods. However if monthly data is used, some of the associated measurement errors will cancel out. Further inaccuracies arise in determining the 2% inflows at a given site. Curve fitting is fraught with problems, and compromises must be made to provide a consistent family of lines. However these have the effect of smoothing the data sample, and are not necessarily detrimental.

RESERVOIR PLANNING AND OPERATION

19. In the case of ungauged sites further reliance is placed on the accurate estimation of catchment rainfall and evaporation. It is apparent from the description of methods given above that this is unlikely to be achieved. For a typical Pennine catchment with rainfall 1200mm and losses 450mm, if the rainfall estimate is 10% high and the losses 10% too low, the resulting runoff from Eq.(1) is 22% lower. The runoff must then be converted from a depth to a volume; problems with determination of areas and catchwater efficiencies have been mentioned above, but errors are likely to be smaller than those due to rainfall and losses.

20. Yield determination then rests on the assumption that the ungauged catchments behave in a similar fashion to Stocks, in particular that the flow ratio remains constant over the whole range. In the case of a very impermeable catchment with rapid runoff and very low baseflow compared to Stocks, the yield estimated using this method may be too high. This would became apparent during dry summers, when the reservoir would empty faster than expected.

Evidence for climate change in NW England.

21. Wigley and Jones (Ref 15) analysed seasonal rainfall totals for 1873-1987 across different regions of England and Wales, and concluded that the main control on precipitation (and hence runoff) was the number of cyclonic and anticyclonic days. In the NW there was also a significant correlation between precipitation and the strength of the westerly circulation, probably due to the orographic effect of the Pennines. They also showed that the most likely effects of an increase in temperatures would be drier summers and wet autumns. Precipitation variability would also increase.

22. Arnell et al (Ref 16) investigated regional and seasonal variation in runoff, and noted a tendency for clustering of wet and dry years. They calculated values of the Hurst coefficient (Ref 17) for 30 catchments with at least 30 years of data, and found evidence to support clustering, which was backed up by a statistical runs test.

23. Links have also been postulated (Ref 18) between climate in Western Europe and the El Nino circulation in the Southern Hemisphere. This could account for some apparently cyclical behaviour.

24. Several of NWW's longest rainfall and reservoir runoff records were examined for evidence of climate change. 10-year annual running means were calculated for Stocks inflows and areal rainfall, and plotted in Fig. 1. The data suggest two wetter periods with a drier interlude in the 50's and 60's, and some evidence of an upward trend in runoff. Fig. 2 shows running means for three month periods. The seasons are defined following Wigley and Jones (Ref 15), who grouped the months in such a way as to minimise the ratio of within-month variability to between-season variability. The values in Fig.2 are consistent with Wigley and Jones' predictions,

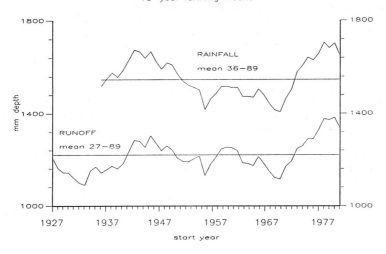

Fig. 1

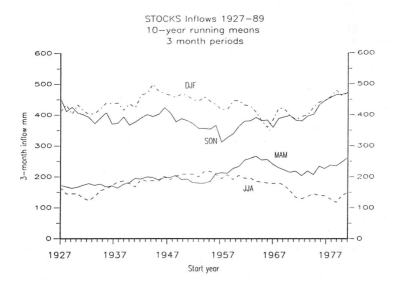

Fig. 2

showing an autumn increase and summer decrease since the late 1950's.

25. Analysis of annual data from Thirlmere and Vyrnwy shows a very similar pattern for the overlap period, and a prolonged dry period at Vyrnwy lasting from the 1880's until around 1910. This has the effect of reducing the mean. Fig. 3 illustrates that over the longer period of record there may well be an increasing trend in inflows. Seasonal data for Thirlmere shows a marked increase in autumn rainfall over the first 20 years of this century, with current values at a historical high, but otherwise little consistent change. The autumn increase is matched at Vyrnwy, which also shows a summer decline since the 1950's, and increases in other seasons.

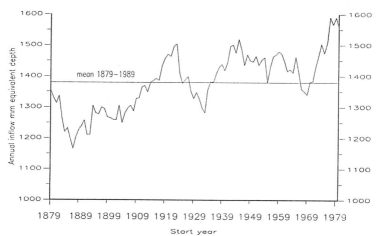

Fig. 3

Implications for yield calculations.

26. The simple approach to time series data described above highlighted the basic variability of hydrological parameters over time. The choice of record period can fundamentally affect the conclusions drawn from the analysis. Statistical evidence of a trend does not help significantly in forecasting next year's flow. All the drought events of recent years are within the range of past experience. Although the analyses discussed above in para. 24 indicate that overall the climate may be getting wetter, seasonal variations may well be more significant in determining yields. This suggests that the longest possible data set should be used in yield calculation, in order to analyse as wide a range of circumstances as possible. It is also necessary to consider the appropriateness of the yield definition, in particular whether it should rest on minimum

historic or return period criteria. While use of minimum historic values may confer a feeling of security it is no more a guarantee against failure than a return period yield. The latter has the advantage of allowing the use of consistent standards and equal risk of failure across the region.

Summary.
27. The method used to calculate a consistent set of reservoir yields for NWW reservoirs has been described and reviewed. Factors affecting the accuracy of these estimates have also been examined. In general adjustments during the present study have been small.
28. The need to alter methods to take account of climate change has been considered. Evidence is presented to suggest that this is premature; recent drought events are well within the range of the historic record, and use of the longest possible record allows consideration of a wide variety of circumstances.

REFERENCES
1. Twort A.C. et al. Water supply 3rd edn. 1985, Arnold
2. NWWA Register of Sources, 1975
3. McMahon T.A. and Mein R.G. Reservoir capacity and yield, 1978, Elsevier.
4. NWWA Survey of existing surface water sources, 1978.
5. NWW Survey of Existing sources 1981, 1986 revision.
6. NWWA An introduction to the Resource Planning suite of programs, 1981.
7. NERC Flood Studies Report, 1975 IH, Wallingford.
8. Mersey and Weaver River Authority First periodical survey under Section 14, Water Resources Act 1963, 1969
9. Lancashire River Authority, First Periodical Survey under Section 14, Water Resources Act 1963, 1971
10. Jack W.L. Annual catchment losses - a practical method. JIWE, 1975
11. Wright C.E. Synthesis of river flows from weather data. Central Water Planning Unit Technical Note no 26, 1978.
12. Calder I.R. Evaporation in the uplands, 1990, Wiley.
13. Kirby C. et al. Plynlimon research: the first 2 decades. IH report no 109, 1991
14. Reynolds G. Rainfall, runoff and evaporation on a catchment in West Scotland, Weather 1969 vol. 24 no3, 90-98.
15. Wigley T.W.L. and Jones P.D. Recent changes in precipitation and precipitation variability in England and Wales: an update to 1987. WRc, Medmenham, PRU 2004-M, 1988
16. Arnell N.W. et al, Impact of climatic variability and change on river flow regimes in the UK. Report to DoE, 1990.
17. Hurst H.E. Long-term storage capacity of reservoirs. Trans. Am. Soc. Civil Eng. 1951, vol. 116, 770-779.
18. Fraedrich K. Grosswetter during the warm and cold extremes of the El Nino/Southern Oscillation, Int. J. Climatol. 1990, vol. 10, 21-31.

Computer software for optimizing the releases from multiple reservoirs operated for flow regulation

T. WYATT, E. V. HINDLEY and T. C. MUIR, Power & Water Systems Consultants Ltd

SYNOPSIS. Electricity demands in Cameroon are primarily met from two hydroelectric plants on the Sanaga river. Flow regulation is provided by three reservoirs, releases from which can take between three and seven days to reach the upstream plant. Computer software has been developed to maximize regulation efficiency and involves rainfall forecasting, rainfall/runoff translation modelling and reservoir release optimization. This paper describes the computer programs developed to optimize reservoir releases, calibration of the flow routing models, concepts underlying the long and short-term release strategies applied, and the results obtained from simulated performance. Conclusions are drawn and opportunities for other applications indicated.

BACKGROUND

1. In June 1990 the consortium of Lahmeyer International GmbH (LI), the Institute of Hydrology (IH) and Power & Water Systems Consultants Ltd (PWSC), was awarded a contract to supply computer software to the Société Nationale d'Electricité du Cameroun (SONEL). This software is designed to improve flow management within the Sanaga river basin.

2. Almost all electricity supplied by SONEL via its interconnected system is generated at the Songloulou and Edea hydroelectric stations located on the Sanaga river. The plants have installed capacities of 384 and 264 MW respectively and very limited local storage.

3. To augment low natural flows, three storage reservoirs have been constructed in the headwaters of the Sanaga basin. SONEL has also installed a telemetry system which polls 33 raingauges, 11 flow gauging stations and 17 humidity sensors. Data is relayed by geostationary satellite (METEOSAT) to a base station in Douala, the major port and commercial centre of Cameroon.

4. Principal functions of the three modules comprising the flow management (FLOWMAN) computer program package, and the consortium members responsible for their development, are as follows :

RESERVOIR PLANNING AND OPERATION

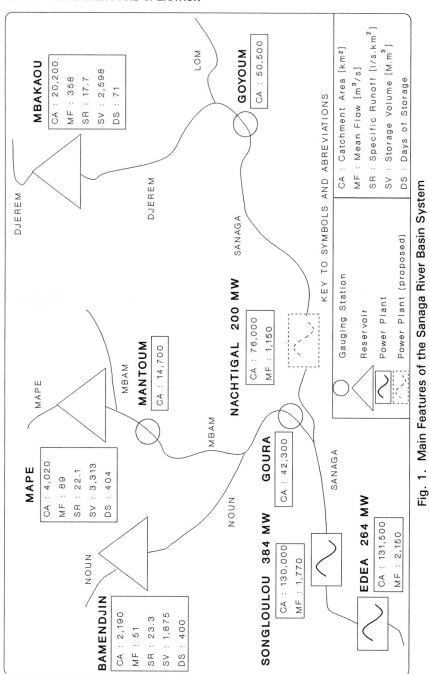

Fig. 1. Main Features of the Sanaga River Basin System

Module 1 : prediction of sub-catchment rainfall, up to 8 days in advance, based on historic rainfall probabilities and the transient behaviour of the Inter-tropical Front (IH)
Module 2 : translation of the predicted areal rainfalls into incremental inflow hydrographs at selected control points on the river network (LI)
Module 3 : routing of predicted flows through the river system and optimization of reservoir releases to satisfy forecast electricity demands (PWSC)

5. In addition to improving reservoir operation efficiency, the software is designed to:
- automate and improve the accuracy of rainfall prediction for the eight designated sub-catchment areas,
- automate and improve the accuracy of runoff prediction for each sub-catchment area and the routing of reservoir release and flow hydrographs down the river basin system,
- provide the system operators with improved information on system behaviour and performance, in the form of graphical and tabular outputs.

6. Software development and calibration began in 1990 and the initial version was installed in November 1991. Real-time testing is now under way and any necessary adjustments will be made following each of the next three dry seasons.

SYSTEM DESCRIPTION

7. Main features of the Sanaga system are illustrated in Fig. 1. The drainage basin covers some 131,500 km^2, approximately 25% of the Republic of Cameroon.

8. The Edea and Songloulou hydro plants entered service in 1953 and 1981 respectively and SONEL has plans to construct a 200 MW plant at Nachtigal. However, due to slowing electricity demand growth, introduction of this station is unlikely before the end of the century.

9. Monthly rainfall patterns differ significantly over the catchment area, and natural streamflows show pronounced seasonal variations. At Songloulou the dry season regime normally lasts between December and May. During this period unregulated flows may decline to some 150 m^3/s, whereas average daily values in excess of 7,000 m^3/s have been recorded in October.

10. As there is minimal thermal reserve capacity within the interconnected system, sufficient water must always be available at Songloulou and Edea if electricity demands are to be fully satisfied. To provide regulatory storage the Mbakaou reservoir was brought into service in 1970, followed by Bamendjin in 1975 and Mape in 1988.

11. Operation of the Songloulou and Edea hydro plants must be scheduled so that their joint power production matches daily load fluctuations. This is complicated by the limited pondage available and a 12-hour time of travel between the two installations (ref.1).

RESERVOIR PLANNING AND OPERATION

PROBLEM DEFINITION

12. To date SONEL have estimated reservoir release requirements using manual computations and graphical routing methods. The efficiency of these procedures is gauged by the proportion of releases used for power generation and an average value of 72% was achieved between 1983 and 1988. Variations in the efficiencies for individual years largely reflect the severity of the dry season, with higher values achieved when lower, and generally more predictable, incremental flows were experienced.

13. For much of the dry season negligible rainfall occurs over the catchment area, and natural river flows decline rapidly. Major scope for improved efficiency lies in improved predictions of rainfall and flows occurring at the onset of the dry season and at the beginning of the wet season, and the reduction of reservoir releases accordingly.

14. Given the need to accurately model the translation of reservoir releases and flows down the river system, the most appropriate method for optimizing reservoir releases was judged to be simulation combined with a search procedure.

RIVER SYSTEM SIMULATION

15. The HEC-5 computer program was developed at the Hydrologic Engineering Center (HEC) of the US Army Corps of Engineers (ref.2), and has been widely used to model the behaviour of multi-reservoir and multi-purpose water resource systems. For each river reach one of six linear routing methods may be specified, and variable time steps can be used. Hydroelectric plant operation can be modelled and sophisticated procedures are available to optimize reservoir releases for flood mitigation, power generation and flow regulation. However, release optimization cannot be undertaken in conjunction with flow routing.

16. All necessary run data, including the definition of system topology, is contained within a single input file. Comprehensive documentation is available as well as inter-active editing programs for input file generation and modification. A large number of optional outputs can be specified by the user and graphical presentations may be obtained via the HEC Data Storage System (DSS).

17. It was originally envisaged that HEC-5 would be exclusively used within the FLOWMAN package to simulate the Sanaga system. However, it was found that even with daily intervals and simulation periods of less than a month, PC execution times preclude use with search procedures requiring a significant number of simulations to reach convergence.

18. Reservoir simulation and flow routing routines were therefore developed which replicate the HEC-5 results but execute much more rapidly. Within FLOWMAN these Reservoir and Routing SIMulation (RRSIM) routines are employed within the release optimization procedure, while HEC-5 is used for final simulations where detailed output is required for tabular and graphical presentation.

19. As described in a companion paper (ref.1), conjunctive operation of the Songloulou and Edea power plants required more detailed modelling than is possible with current versions of HEC-5. PWSC's computer program SYRAP was therefore used to determine the average daily flow required at Songloulou in order to meet a range of electricity demand profiles.

20. Considerable effort was expended in basic data collection and analysis. Many records were incomplete and of dubious validity due to unreliable stage/discharge relationships. Inspection of the 6-hourly telemetry data indicated frequent malfunctions of the water level recorders, while rating curves are not yet available for all stations.

21. The assembled streamflow data were analysed to identify periods for which coincident records were available. These data sets were then used to determine the most appropriate flow routing methods and parameters to be applied to each river reach.

22. As indicated in Table 1, distances between the gauging stations are appreciable, as are the catchment areas and times of travel associated with the corresponding river reaches. This complicated calibration, since incremental inflows are often of similar magnitude to flows entering the reach. The most useful data sets correspond to the start of the regulating seasons when rapid changes have been made to the releases from one or more of the reservoirs (Fig. 2).

Table 1 : River Reach Characteristics

RIVER REACH	LENGTH (km)	TRAVEL TIME (days)	CATCHMENT AREA (km^2)
Mape-Mantoum	54	0.8	20,200
Bamendjin-Goura	257	2.8	
Mantoum-Goura	171	2.0	25,410
Mbakaou-Goyoum	172	1.8	30,300
Goyoum-Nachtigal	272	2.9	25,500
Nachtigal-Songloulou	171	1.7	
Goura-Songloulou	140	1.7	11,700
Songloulou-Edea	55	0.5	1,500

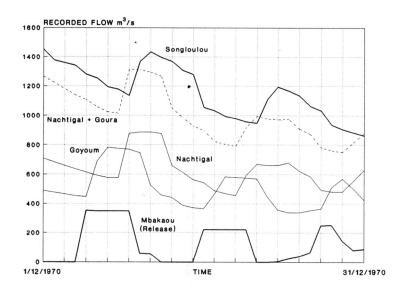

Fig. 2 : Example of Data Set Used for Flow Routing Calibration

23. Initial calibrations were made using the Muskingum routing method, but it was found that the observed attenuation of reservoir releases within the Sanaga river system could better be matched by specifying up to six routing coefficients and applying the formula :

$$O_n = C_1 I_n + C_2 I_{n-1} + C_3 I_{n-2} + C_4 I_{n-3} + C_5 I_{n-4} + C_6 I_{n-5} \cdots$$

24. In this equation O_n and I_n are the outflows and inflows in routing interval n, and C_N are the routing coefficients which sum to unity. Selection of the appropriate coefficients can involve subjective judgements, and spreadsheet derived graphics of inflow and outflow hydrographs proved valuable in their determination.

RESERVOIR RELEASE OPTIMIZATION

25. The reservoir release optimization logic and the RRSIM simulation routines are incorporated within Program RESRO. This program also assembles all information required for the HEC-5 and RRSIM simulations, including the recorded and predicted flow data.

26. Trial reservoir releases specified by the search procedure, together with the recorded and forecast incremental flows, are routed through the river basin to determine the corresponding regulated flow at Songloulou. To account for flow lags in the river reaches, the simulations are carried out for a period starting 15 days before the current day (Day 0) and ending 7-days after (Day +7).

27. The optimization logic takes account of conflicts between short and long-term objectives. In the short term, maximum efficiency would be achieved by favouring releases from the nearer reservoirs viz. Bamendjin and Mape. However, the contents of all three reservoirs must also be kept in balance if long-term security of the regulated flow at Songloulou is to be maximized.

Long Term Strategy

28. Using reconstructed flow series for the 25-year period between July 1965 and June 1990, daily operation of the three reservoirs was simulated to assess which balancing strategy would lead to the highest sustainable regulated flow at Songloulou. In the absence of coincident flow data at the intervening gauging stations, routing effects were not considered.

29. Allocating releases so as to maintain equal percentage fullness in all three reservoirs, subject to maximum and minimum release constraints, was found to be superior to use of a 'space' rule whereby the remaining storage space in each reservoir is held proportional to the expected inflow over a specified future period.

30. As indicated in Fig.1, the three reservoirs have disparate characteristics in terms of storage capacity, mean inflow and hence the degree of regulation provided. These parameters are reflected in markedly different refill patterns for Mbakaou on the one hand and for Bamendjin and Mape on the other, the relatively smaller Mbakaou reservoir refilling and spilling every year.

31. Simulations were therefore made with a modified percentage fullness rule in which Mbakaou reservoir is 'overdrawn' and its contents are maintained a certain proportion below the other two reservoirs. Fig. 3 shows the performance of the three reservoirs over the critical period of complete drawdown and refill, with releases made from Mbakaou which hold it 20% below Bamendjin and Mape. It was found that an increase of around 3-4%, in the regulated flow at Songloulou could be expected by following this strategy.

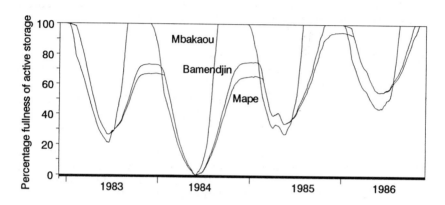

Fig 3 : Application of Reservoir Release Allocation Strategy

Short Term Strategy

32. The short-term strategy determines, for Day 0 through Day +7, the minimum total reservoir release consistent with future flows at Songloulou being not being less than required. Based on this total, the individual releases are allocated according to the long-term strategy described above, taking into account any restrictions imposed by maximum and minimum release rates or rates of increase. On moving from day to day it is assumed that future releases will be the same as those made on the current day.

33. In real-time application the optimization is repeated each day, with updated forecasts of flow requirements and incremental inflows being used for the rolling 7-day horizon; within the simulation previously predicted flows for Day -1 are replaced with recorded values. In this way releases from Bamendjin and Mape can be increased if rain predicted to fall towards the end of the period fails to materialize, and reduced if contributions from the intermediate catchments proves to be greater than expected.

34. Performance of the optimization procedure has been simulated over several historic dry seasons assuming perfect forecasting of the incremental flows. Results for the 1966 dry season are shown in Fig. 4, the associated reservoir release efficiency being 95%.

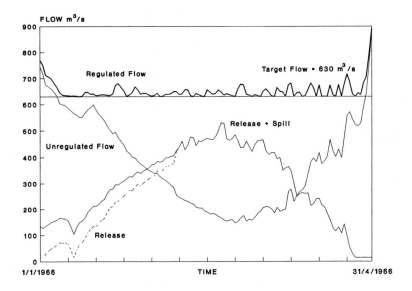

Fig. 4 : Simulated Performance of Reservoir Release Optimization Assuming Perfect Forecasting

CONCLUSIONS

35. The FLOWMAN computer program package has been developed to improve the efficiency with which multi-reservoir river regulation systems are operated.

36. Operation of the Sanaga river system poses unusual, but not unique, problems due to the times of travel between the storage reservoirs and the flow regulation point.

37. Sophisticated modelling and optimization techniques require reliable input data if they are to fulfil their potential for improving operational efficiency.

38. Especially in developing countries, the value of telemetry systems can be seriously diminished unless adequate resources are made available for their maintenance and calibration.

39. Performance of the reservoir release optimization procedures outlined in this paper can only be quantified as part of the FLOWMAN package. In particular, real-time benefits will depend on the accuracy with which rainfall and incremental inflows can be forecast over the optimization horizon.

REFERENCES

1. ROBINSON P.E. Estimation of River Flow Requirements to meet Electricity Demands in Cameroon. Conference on Water Resources and Reservoir Engineering. Stirling, 24-27 June 1992.
2. 'HEC-5 Users Manual', US Army Corps of Engineers Hydrologic Engineering Center, 609 Second Street, Davis, California, April 1982.

Modelling the uncertainty of sediment deposition upstream of flood control dams

G. W. ANNANDALE, PrEng, BSc(Eng), MSc(Eng), DEng, FSAICE, MAWRA, MUSCOLD, MIAHS, MIAHR, Surface Water Engineering, Steffen Robertson and Kirsten (US), Inc.

SYNOPSIS. Flood control dams are usually operated in a fashion which causes the reservoir basin to be empty or close to empty between flood events. Sediment which is deposited during flood events is exposed and the cause of various environmental concerns. This paper describes a methodology which was used to model the uncertainty in sediment distribution patterns as a function of space and time. The findings of this investigation was used in the environmental assessment of the adequacy of the design of the flood control facility.

INTRODUCTION

1. The practice to keep flood control reservoir water levels drawn down between flood events exposes sediment which was deposited during such events. An example of such a facility is a flood control dam which is planned in South Africa. The Department of Water Affairs in South Africa plans to construct a flood control dam upstream of the town of Ladysmith, which is designed to remain empty between flood events. The dam is situated in a high sediment yield area, giving rise to environmental concerns regarding the unsightliness of large volumes of exposed deposited sediment between flood events. The findings of a procedure which modeled the uncertainty of sediment deposition as a function of time and space was utilized in the environmental evaluation. This paper describes the stochastic modeling procedure, and presents some of the results.

MODELING PROCEDURE

2. The outline of the procedure which was used to model the uncertainty of sediment deposition as a function of time and space is presented in Figure 1. The procedure consists of five phases, viz

(a) Phase I: Data preparation and modeling of sediment deposition under unsteady, non-uniform flow conditions for selected single events
(b) Phase II: Stochastic modeling of sediment deposition
(c) Phase III: Empirical estimates of the volume of sediment which is expected to deposit in the reservoir basin

RESERVOIR DESIGN AND CONSTRUCTION

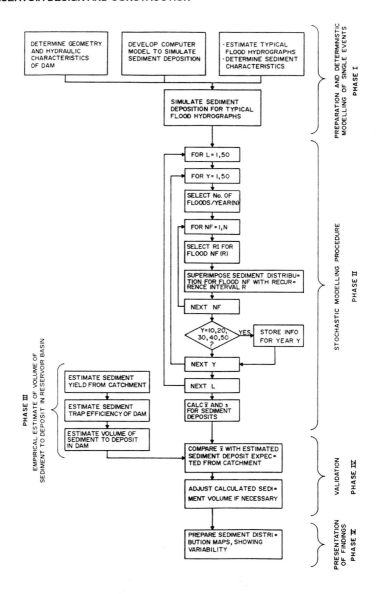

Fig. 1.   Stochastic analysis procedure to model sediment deposition under unsteady, non-uniform flow conditions

(d) Phase IV: Validation of modeling results
(e) Phase V: Presentation of findings.

Phase I: Data preparation and single event modeling

3. Data collection required collation of relevant geometric data, sediment characteristics, and hydrologic data. The hydrologic data included precipitation data; catchment characteristics such as size, slope, stream patterns, vegetation, soils, and geology; and flood data, including flood hydrograph shapes and flood peak data (obtained from the gauging station at the dam site). The geometric data which was used consisted of a description of the topography of the reservoir basin, and the height and design of the dam. The sediment characteristics which were required consisted mainly of grading distributions of bed and suspended material, and records of sediment loads (which were obtained from a resurvey of an upstream reservoir, and generalized sediment yield maps).

4. The hydrologic data was prepared by deriving unit hydrographs and preparing typical hydrograph shapes for a range of recurrence intervals. These typical hydrograph shapes were used to model single event sediment distribution patterns. In addition, the flood data was analyzed by determining the probability distributions of the number of flood which are likely to occur in any one year, and of the flood peaks of <u>all</u> floods on record. The latter two probability distributions were used in the stochastic modeling procedure.

5. Single event sediment distribution patterns were modeled by making use of a computational hydraulics model which utilizes the full St Venant equation and an option to choose between the use of three sediment transport equations. A computer model was developed specifically for the project by adding the facility to model sediment transport to a computational hydraulics model which was previously developed to simulate unsteady, non-uniform flow (ref. 1). The sediment transport equations which were added to this computer model are the equations developed by Yang, Engelund and Hansen, and Ackers and White (refs 2, 3 and 5).

6. Phase I was concluded by modeling the expected distribution deposited sediment for a range of flood recurrence intervals. The results of this simulation was stored in digital format for use in the stochastic modeling procedure (Phase II).

Phase II: Stochastic modeling procedure

7. The stochastic modeling procedure entails modeling likely sediment distribution patterns over 50 project lifetimes (L in Figure 1) of 50 years (Y in Figure 1) each. This is accomplished as follows: the number of floods (NF in Figure 1) which is likely to occur in any particular year is first determined by random generation from the probability distribution representing the number of floods which could occur in any one year at the dam site. Once this number is known, the recurrence intervals (RI in Figure 1) of floods associated with each of these likely events are determined from the probability

RESERVOIR DESIGN AND CONSTRUCTION

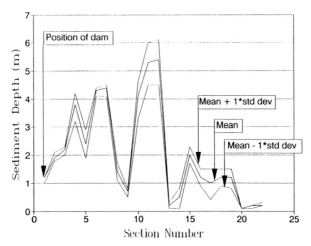

Fig. 2. Mean and variability of distribution of deposited sediment after ten years as a function of space

distribution representing the likelihood of occurrence of <u>all</u> floods at the dam site. These recurrence intervals are then used to select the associated sediment distribution patterns, which were modeled during the course of Phase I, for superposition. The stochastic procedure is designed to record the sediment distribution patterns after 10, 20, 30, 40 and 50 years for each simulated project lifetime. Summary statistics are calculated for the sediment depths at various positions along the long section of the reservoir basin. This process is repeated for 50 project lifetimes.

Phase III: Empirical estimate of the volume of sediment expected to deposit in the reservoir

8. An empirical estimate of the volume of sediment which is expected to deposit in the reservoir basin is required to validate the modeled sediment volumes. This is done by first estimating the likely sediment yield from the catchment upstream of the reservoir, whereafter an estimate of the trap efficiency is made, and the volume of sediment calculated.

9. The sediment yield was estimated by making use of a generalized sediment yield map (ref. 4) and resurvey data of an upstream reservoir.

10. The trap efficiency of the reservoir was estimated by making use of the Brune curve (ref. 6) and by interpreting the results of the sediment modeling procedure in Phase I.

11. Once this information was known, an independent estimate of the volume of sediment which can be expected to deposit in the reservoir over a lifetime of 50 years was made for comparison with the results of the computer simulation.

Phase IV: Validation

12. The empirically estimated volume of sediment which is expected to deposit in the reservoir was compared with the mean

volume of deposited sediment determined by computer modeling. The assumption was made that the sediment distribution pattern would not be significantly influenced if the volume is adjusted within limits of accuracy of the sediment transport equations used during the computer modeling procedure.

Phase V: Presentation of findings
13. The results of the investigation were presented in tables and on graphs by providing the mean and standard deviation of the depth of deposited sediment after 10, 20, 30, 40 and 50 years of operation. The finding was that the expected variability of depth of deposited sediment decreased with increasing lifetime of the project, ie the uncertainty of depth of deposited sediment is larger during the early years of the project. An example of the mean and one standard deviation variance of depth of deposited sediment after a project lifetime of 10 years is presented in Figure 2.

CONCLUSION
14. It is possible to model the uncertainty in distribution of deposited sediment in flood control reservoirs where unsteady, non-uniform flow conditions dominate. A stochastic modeling procedure by means of which the probability of sediment depth can be calculated as a function of space and time is described in this paper. By determining the probability distribution of the simulated sediment deposition depths, it is possible to calculate the probability that various depths could be exceeded. The findings of the investigation were used to assess whether the expected annual increase in depth of deposited sediment would allow establishment of vegetation. Such establishment will reduce the environmental impact significantly, as it would improve both aesthetics and the possibility of physical impacts such as the effects of windblown sand. The procedure described herein can therefore be used to evaluate optional designs and operating procedures.

REFERENCES
1. Benade N. et al. Optimization of the management of irrigation canal systems (in Afrikaans). Laboratory for Systems, Rand Afrikaans University, Johannesburg, 1990.
2. Engelund F. and Hansen E. A monograph of sediment transport in alluvial streams. Teknisk Verlag, Copenhagen, 1967.
3. Ackers P. and White W.R. Sediment transport in channels. Hydraulics Research Station, Wallingford, England, 1972, Report INT 104.
4. Rooseboom A. Sediment yield map of South Africa. Department of Water Affairs, Pretoria, South Africa, 1975, Technical Report no 61.
5. Yang C.T. Unit stream power and sediment transport. Jnl of the Hydr Div, ASCE, vol 98, HY10, 1972.
6. Brune G.M. Trap efficiency of reservoirs. Trans Am Geophys Union, vol 34, no 3, 1953.

The construction of a cut-off in a volcanic residual soil using jet grouting

L. J. S. ATTEWILL, MA, MSc, FICE, MIWEM, J. D. GOSDEN, MA, MSc, MICE, MIWEM, and D. A. BRUGGEMANN, BSc(Eng), MSc, DIC, MICE, Howard Humphreys & Partners Ltd, and G. C. EUINTON, BSc(Eng), MICE, Howard Humphreys (Kenya) Ltd

SYNOPSIS. The foundation cut-off for Thika dam in Kenya was constructed using jet grouting. The factors leading to the choice of this technique, the evolution of design parameters and the results of the technique are discussed.

INTRODUCTION.
1. Thika dam is part of the Third Nairobi Water Supply Project which is presently under construction. This project, which comprises some 50 km of raw and treated water transmission, new treatment facilities and extensions to the urban distribution system, will provide an additional 305 Ml/d of potable water when it is completed in 1994.
2. The Thika dam is situated on the Thika river which flows off the Aberdare hills eastwards to the Tana river. The dam site is located at the confluence of three tributaries and thus offers the best ratio of storage volume to dam fill volume of all the sites considered.
3. The dam location is illustrated in Figure 1.

GEOLOGY
4. The rocks underlying the area are of Pleistocene age and are of volcanic origin being predominantly pyroclastic - that is they were formed by violent explosive events involving the expulsion of much fine debris into the air. The resulting deposits are tuffs, which can be fine or coarse grained and more or less welded depending on their origin. Two origins can be recognised;

    (a) Pyroclastic flows consisting of fragments of rock dispersed in a medium of fluidised fine material. The resultant rock typically grades from coarse deposits at the base to finer material at the top.

    (b) Pyroclastic falls from material thrown into the air by the volcanic explosion. These particles cool as they fall and may thus be less welded than the flow

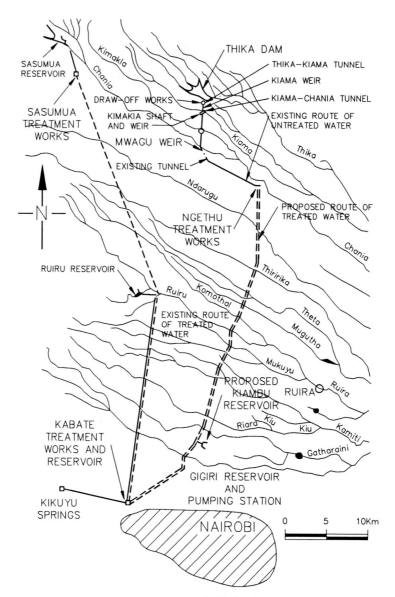

Figure 1: Location of Thika Dam

deposits. Falls tend to become finer as the distance from their source increases, and they may be deposited over pyroclastic flows originating from the same explosion.

5. The remainder of the volcanic sequence comprises flows of phonolite lava. Lavas represent the height of volcanic activity with eruptions occurring from localised vents, flowing downslope to collect in hollows or depressions. Their deposition was sometimes accompanied by air fall activity and thus the phonolite may be found either as massive units or interbedded with the tuffs.

6. Six periods of volcanic activity can be recognised. The end of each period of deposition is marked by a weathered horizon at the top of the sequence, and the presence of a residual soil horizon (latasol) indicates an ancient erosion surface which was subsequently covered by later volcanic deposits. Earlier units in the sequence of volcanic deposition appear to have been laid on a relatively flat topography. Later deposits, especially flows, tended to fill the depressions left in the underlying units.

7. The modern drainage pattern has deeply dissected this volcanic sequence and a mantle of highly to completely weathered material covers the slopes. Outcrops of rock are restricted to small areas of very steep slope in the valley sides and to the waterfalls formed where streams flow over the more resistant lava flows.

DAM DESIGN

8. The soft nature of the foundations and the absence of rock at reasonable depths dictate the choice of an embankment dam at the site. The embankment is of earthfill, won from a borrow area within the reservoir area, with an upstream sloping clay core of the same material as the shoulders but placed at a higher moisture content.

9. The embankment is founded on grade IV/V residual soil with only the grade VI material, generally less than 2 m deep, being stripped. The embankment is shown in cross section, plan and longitudinal section in Figures 2 3 and 4.

10. Although permeability values of the dam foundations were only moderately high it was considered prudent to provide both a foundation cut-off and drainage in order to limit seepage and control pore pressures. However, the design of the cut off was problematic due to the presence of the succession, at depth, of weathering zones, so that soft weathered layers are, in places, overlain by fresh rock. It is well known that highly weathered tropical rock cannot be successfully treated by conventional injection grouting, and therefore an alternative method must be used in this material. Three options were considered:-

    i)   bulk excavation
    ii)   diaphragm wall
    iii) jet grouting

RESERVOIR DESIGN AND CONSTRUCTION

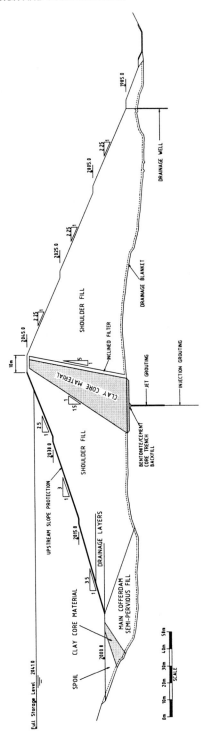

Fig. 2. Thika Dam cross section

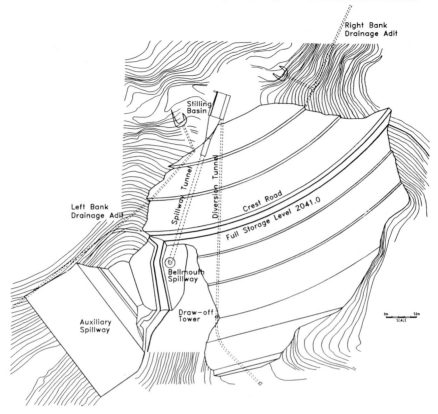

Figure 3: Thika Dam Plan View

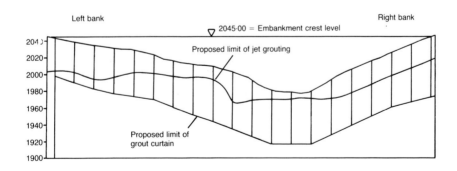

Figure 4: Thika Dam Longitudinal Section

RESERVOIR DESIGN AND CONSTRUCTION

11. Diaphragm walling was rejected because of the steepness of the abutments and it was doubted that it would be possible to penetrate the layers of phonolite without recourse to explosives. At the design stage jet grouting was adopted in preference to bulk excavation on the grounds of cost. A further comparison, repeated early in the construction programme with the benefit of actual construction rates, confirmed the original selection and moreover, prompted doubts over the stability of the very high excavated slopes that would be involved. Thus the cut-off comprises a series of interlocking columns constructed by the jet grouting technique to a depth, where, it was judged, the rock would respond to injection grouting.

THE JET GROUTED CUT-OFF
The Technique

12. The jet grouting technique is well known, but has not been widely applied to dam construction. It comprises the following operations:

    i) the drilling of 150 mm diameter pilot holes at the required spacing.
    ii) the introduction to the bottom of the hole of the tool which houses two nozzles, one for water and air, the second lower nozzle for cement grout.
    iii) the tool is rotated and raised up the hole, while water and air is pumped through the upper nozzle at high pressure, and grout through the lower nozzle.

13. As the tool is raised, the water/air mixture erodes a column, the diameter of which depends on the energy applied, which in turn is a function of the speed of withdrawal and the pressure. Typical values for this job were 100mm/minute and 490 bar. Grout from the lower nozzle mixes with, and partly displaces, the eroded material to form a mixture of soil, cement and water. Displaced material, essentially the same mixture as that which remains in the columns is displaced, and thus the process can, at low withdrawal speeds, be considered wasteful of cement.

Trial Panel

14. It was considered necessary to construct a trial panel both to confirm the practicability of the technique and to select the main parameters. The trial panel consisted of 18 columns constructed to a depth of 19 m. The following parameters were raised:

- column spacing
- water pressure
- nozzle diameter
- speed of advance/withdrawal
- water flow rate

15. Following the completion of the trial, the columns were investigated by means of three cored drill holes, and the soil surrounding the panel was excavated to a depth of 12 m, enabling a physical inspection of the columns to be made.

16. Data collected during the trial enabled a loose correlation to be established between erosive energy and column diameter, and the inspection enabled column diameters calculated from these data to be checked by visual inspection.

17. The trial confirmed that the basic jet grout technique was appropriate to the ground conditions, and enabled parameters to be selected for the grout curtain.

Modified Procedure

18. Arising from the experienced gained in the early part of the trial, the jet grouting procedure detailed above was modified in order to reduce the wastage of cement. Instead of eroding and grouting in one operation, the cohesiveness of the material enabled the column to be eroded in the descending stage, and to be grouted during the ascent. This modification, which was tried in two trial panel columns, greatly reduced the amount of grout wasted, but also resulted in a column material compound of pure grout, rather than a grout/eroded soil mixture. The modified technique was subsequently adopted, but with a changed grout mix which included added clay.

Site Investigation

19. The area of the Trial panel was investigated prior to the start of the trial by four holes. The ground compressed completely to highly weathered residual soil, grade VI to V. For the purposes of the trial three strata were identified with the following mean characteristics.

Depth	0-5m	5-13m	13-20m
Water Content %	60	60	50
Bulk density	1.50	1.63	1.73
Porosity %	64.6	61.3	55.7
Void ratio	1.83	1.58	1.28
Saturation %	0.87	1.0	1.0

Predrilling

20. The predrill holes were bored with a rotary rig in 150 mm diameter. In 13 of the 18 holes the deviation from the vertical was less than 0.5%, and 2 holes were marginally outside the specification limit of 1%. During predrilling the torque, thrust and rate of advance were recorded and plotted automatically, although it was found difficult to correlate these data with geotechnical characteristics.

Prejetting

21. In the modified technique, prejetting is the term given to the process of eroding the ground using a high pressure jet

RESERVOIR DESIGN AND CONSTRUCTION

of air and water. The intention of the trial was to establish a relationship between erosive energy and column diameter in various ground conditions to enable the supervisory staff to vary erosive energy in order to ensure the column diameter remains close to 1 m. In order to achieve this, it is necessary to calculate the theoretical column diameter from measurements of the rate of flow and densities of the air/water mixture and of the spoil density. Erosive energy has been plotted against theoretical column diameter for each of the strata identified in the site investigation, as shown in Figure 5. That the correlation is not very good is probably due to the fact that the resistance of the ground to erosion is not perfectly correlated to depth; but until a routine method of determining erodibility to any of the characteristics measured during predrilling is available, the depth/diameter/energy correlation is the best guide available for the determination of the appropriate erosive energy. From this plot it was decided to proceed to the main curtain with a jetting energy of 30 MJ/m and 100 MJ/m at depths less than and greater than 6m respectively.

Investigation of the Panel

22. The panel was investigated by drilling 3 inclined investigation holes in the plane of the panel to confirm the continuity of the grout and to obtain samples for testing. The surrounding soil was subsequently excavated to a depth of 12m to expose the columns. Inspection of the columns also confirmed continuity of grout with column diameters in broad agreement with those calculated theoretically. A section of the panel in place at a depth of 5m is shown in Figure 6.

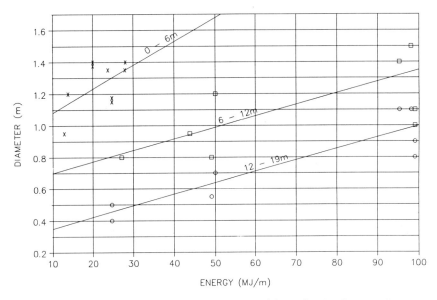

Figure 5: Variation of Column Diameter with Erosive Jet Energy

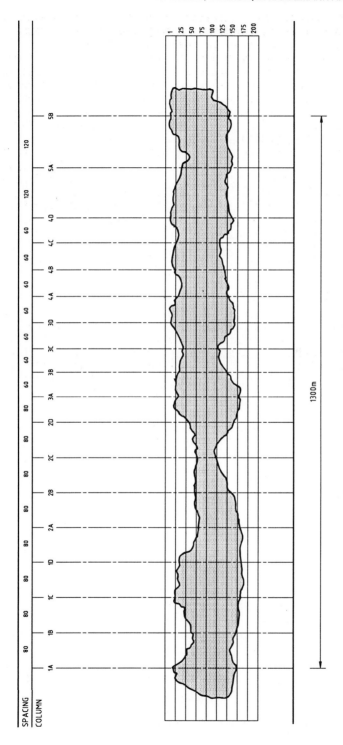

Figure 6: Thika Dam Foundation Treatment Trial Panel No. 1
Plan Section at 5m Depth

Grout Material

23. A series of grout mixes was prepared in the laboratory with water: cement ratios varying from 1.8:1 to 3:1. To each mix was added soil, varying in content from 0 to 20% by weight of cement, giving 12 mixes in all. The laboratory tests showed that a water cement ratio of 2.6:1 resulted in a grout of the specified strength and tangent modulus, the results being relatively insensitive to the clay content.

24. The water cement ratio of the grout pumped into the columns must allow for the dilution and mixing with the water from the erosion jet and with soil slurry present in the hole. The degree of dilution is also dependent on the method, there being less dilution with the pre-jetting method. The grout mix actually pumped in the early stages of the main curtain from the results of the trial panel had a water cement ratio of 2 with 10% added clay in the form of bentonite with a liquid limit of 100. Consolidated-Drained triaxial tests carried out on cores taken from the completed columns in the main curtain indicated average deviator stressess at 5% axial strain of 1640 kPa and 2680 kPa at confing pressures of 400 kPa and 800 kPa respectively. The average tangent moduli at 1% axial strain were 39 350 kPa and 61 900 kPa. Permeability of the mix was less than 1×10^{-9} m/s.

Conclusion

25. At the time of writing, the jet grouting is some 25% complete. Preliminary results indicate that the techniques is viable. Further results will be presented at the conference and will be subsequently published.

ACKNOWLEDGEMENTS
The authors would like to thank Mr W.J. Odhiambo, General Manager, Water and Sewerage Department, Nairobi City Commission, for permission to publish this paper.

Two embankment dams on alluvial foundation: Durlassboden and Eberlaste

H. CZERNY, Dipl-Ing, Federal Ministry of Agriculture, Forestry and Water Management, Austria

SYNOPSIS: The following paper will give an example of the progress of dam technology, 25 years before the sites were judged to be unsuitable for dams because of the deep alluvial foundation.

1. DURLASSBODEN DAM

It is a 83 m high earthfill dam with central earth core and a grout curtain in the alluvial foundation. Its situation is almost on the "Gerlos-pass" between the Ziller-Valley and the Upper Salzach Valley. It was constructed in the years 1965 und 1966 by the TKW, an Electricity Supplying Company, first full-filling was reached in September 1968.

The run off from the natural catchment of the Gerlosstream is angmented by the diversion of the Salzach and Nadernach streams. The Durlaßboden is an annual storage reservoir and it is exklusive used for power production.

1.1 Geology and foundation treatment

The bedrock of pale-grun sericite plyllites with gypsum intercalations and block plyllites is filled with a neterogeneous mass of crystalline gravels up to 136 m deep in the middle of the valley. A sill layer varying between 6 and 30 m in thickness is present at a depth of 30 to 50 m.

The right valley flank is formed by an early rockslide mass of the laminated limestone series, the left flank being composed of green schist.

Due to the low elevation of the bedrock, the embankment type was the only feasible solution. The poor-quality fill material available dictated a flat-sloped earthfill dam. The grout curtain with an area of 10.600 m^2 extends 60 m deep into the foundation not reaching the lower gravel layers and ends in a horizontal silt-layer with an natural imperviousness of about 10^{-6} to 10^{-7} m/s.

RESERVOIR DESIGN AND CONSTRUCTION

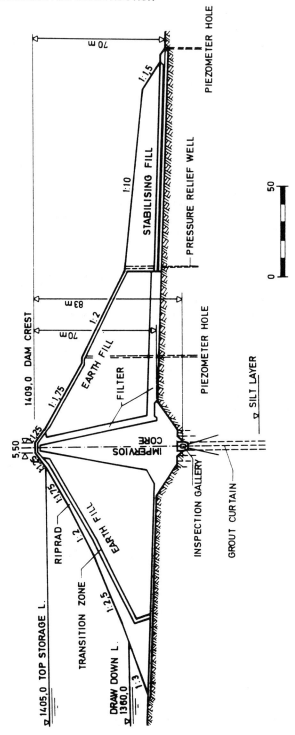

Fig. 1. Durlassboden: cross section

The grout consisted of a clay-cement of bentonite suspension
with Algonit gel admixture to preserve soil-deformability.
The grout holes were arranged in 8 lines, of which only
three in the centre extended to full depth. Injection was in
5 m sections, grout acceptance averaging 3,5 m^3 per linear
metre. The grout curtain has reduced the permeability of the
soil to 10^{-6} m/s.

1.2 The embankment structure

For the impervious core and the 2 m thick impervious apron,
beneath the upstream shell a mixed-grained talus material
with a maximum grain size of 80 mm was used.
The core material was dried to a water content of 8 percent.
In order to increase plasticity, 1 to 2 percent bentonite
was added. Placement was done by bottom dumptors in 30 cm
layers, compaction by pneumatic rollers.

The upstream shell consists of sandy gravel to El. 1380 m,
and of talus gravel above this level. Rip-rap of blocks not
less than 350 kg in weight protect the dam face against wave
action. The rip-rap rests an a transition layer of mixed
stone material.

Downstream, a 3 m thick filter zone follows the core. The
shell consists of talus material to a particle size of 1 m.
A stabilising fill with a filter layer beneath and 7 relief
wells, protect the dam toe from hydraulic ground failures.
The shells were placed in 60 cm layers upstream and in 2,5 m
layers downstream, compacted by regular traffic of heavy
equipment.

A site laboratory was responsible for all checks and measurements.

1.3 Instrumentation and surveillance

The effectivness of the grout curtain is monitored by 7 relief wells, 25 piezometers rand 22 pore pressure pick ups.
For the measurement of hydrostatic pressures upstream and
downstream of the grout curtain, 4 drill holes on each side
were sunk from the inspection gallery, which was located at
the base of the impervious core. It was made of bending-re-
sistant concrete rings varying between 2 and 6 m in length
to resist the anticipated settlement of the fill material.

Measurements results have shown that the headwater pressure
is reduced by abaout 70 percent in the grout curtain. Seepage emerging in the relief wells is somewhat smaller than forecast and amounts to 25 - 28 l/s.
Deformations of the dam are monitored by observation in
three settlement profiles comprising 7 settlement gauges

RESERVOIR DESIGN AND CONSTRUCTION

each, and by geodetic measurements taken in the inspection gallery and in the north exploration gallery.

The foundation settled 0,95 m during construction, than by another 20 cm. Total settlement of dam body (core) and foundation has been measured to be 1,9 m at the crest, of which some 80 percent during construction.

Downstream displacement of the inspection gallery under full headwater pressure is 2,5 cm.
The drainage facilities located downstream of the core suggest no permeability.

Further deformation measurements are made by means of water levels and extensometers. An extensive geodetic surveillance system has been provided for the dam and the reservoir slopes.

In the last 5 years some springs were observed in a little pond downstrem of the stabilising fill. For the pressure which was met in drilling down 2 new piezometers was very high, the stabilising fill has been made longer for ablout 40 m and some new wells were drilled down on the downstream toe of the lengthened stabilising fill to reduce the pressure in the upstream area.

This dam is an example of the progress of dam technology: In 1940 the site was judged to be unsuitable for a dam (plan, crossection, longitudinal section).

2. STILLUP-DAM

It is a 28 m high earthfill dam with asphaltic-concrete core and slurry-trench foundation cut off. Its situation is in the north of Mayrhofen, which is the main town of the Ziller-Valley.
It was constructed in the years 1966-1968 by the TKW, an Electricity Supplying Company, the top storage level has been reached first in 1971.

The runoff from the natural catchment of the Stillup-stream is augmented by the downstream water of the power stations of Roßhag and Häusling (the great power stations of the Schlegeis and the Zillergründl arch dams).

The Stillup-reservoir has different economic functions, it is the tailwater basin for the two power stations, lower basin for superimposed pumped storage operation and weekly-storage reservoir for the Mayrhofen power station.

The reservoir has an important key position between the upper station and the lower station of the power scheme,

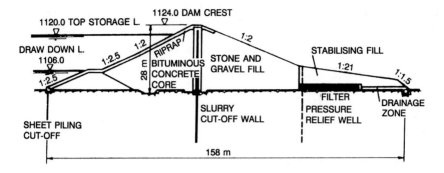

Fig. 2. Eberlaste: Stillup: cross section

which supplies high-valne peack load energy.

2.1 Geology and foundation treamtment

The flanks are composed of biotite and muscovite granitic gneiss and mignatite strata dipping at about $60°$ towards the middle of the valley, where drill holes attaining a maxium depth of 125 m did not encounter bedrock. The rock channel is filled with very heterogeneous sandy/gravelly river - deposited material which is interlocked with permeable talus material and bouldevs at the foot of the vally flanks.

In the middle of the valley, there is a 20 m deep surface layer of silts of sufficient density but subject to erosion. The cut off is 22 m deep in this area. In the permeable zones interspersed with boulders prevailing at the foot of the slopes, the cut off extends to a depth of 52 m. The slurry-trench cut off was preferred over a grouted one because of the great heterogeneity of soil types and anticipated settlement. Only the contact between slurry-trench cut off and bedrock on the valley slopes was grouted in places.

2.2 The embankment structure

The site material with less than 80 mm in grain size, with bentonite admixture, would have been suitable even for an earth core, but the high amount of water contained in the material would have involved prohibitive drying cost.

Therefore, for the first time in Austria, an asphaltic-concrete impervious core 40 to 50 cm thick, continuous with a slurry trench foundation cut off was provided. The impervious core was subjected to substantial stresses due to differential settlement.

RESERVOIR DESIGN AND CONSTRUCTION

The asphaltic concrete was made of 0 to 25 mm gravel with 8 percent limestone addition and 7 percent type B 300 bitumen, and was placed in 25 cm layers by means of a crawler-type vehicle especially designed for this purpose.

The slurry used for the foundation cut off consisted of soil-cement with 0 to 40 mm aggregate, cement, bentonite and a chemical admixture. Quality tests and material checks during construction were carried out in a special site laboratory.

2.3 Instrumentation and surveillance

Measurements mainly relate to underseepage, which is monitored by 15 relief wells and 14 piezometers. Both the seepage rate-approximately 150 l/s in time off full reservoir - and the contend of solid matter are measured. As a progress of self- sealing seepage decreased till nowadays to 110-120 l/s.

Vertical and horizontal dam deformations are monitored by geodetic means, with crest alginement, ·traversing and precision levelling.

During construction, the foundation settled about 2,20 m in the middle of the valley. Since the end of construction work, secondary settlement of 20 cm been measured. Settlement of the dam body itself has not exceeded 10 cm; i. e. 0,4 % of the dam height, since the completion of the structure.

REFERENCES

1. KROPATSCHEK H. and RIENÖSSL K.: Effectivness of the grout curtain of Durlaßboden Dam, No. 18 of the series "Die Talsperren Österreichs" 1970.

2. RIENÖSSL K. and SCHNELLE P.: Durlaßboden and Eberlaste Embankments - large Settlements and Underseepage in the Overburden, Qu. 45, R. 14, 12. Congress on Large Dams, Mexico 1976.

3. RIENÖSSL K.: Embankment Dams with Asphaltic Concrete Cores. Experience and Recent Test Results, Qu. 42., R. 45, 11. Congress on Large Dams, Madrid 1973.

Reservoir competency study made in respect of limestone formations found upstream of Kodasalli Dam, India

H. V. ESWARAIAH, V. S. UPADHYAYA and C. R. RAMESH,
Karnataka Power Corporation Ltd, Bangalore

1. INTRODUCTION

The occurrence of limestone formations across a reservoir raises the apprehensions of a transbasin leakage which is due to the likely formations of solution channels within the limestone mass. This phenomenon may become more pronounced when a valley with higher floor level is dammed and the reservoir formed creates an additional driving head because of the raised hydraulic depth. Many instances have been reported wherein dams constructed under such conditions could not be of utility as reservoir impoundment was not possible. Hence under such situations it becomes essential to conduct a thorough investigation of the entire reservoir area and plan for remedial measures to be taken up before a reservoir is formed by the construction of a dam. Apart from the loss of reservoir water the other associated problems may be (i) sudden slumping of the abutment hillmass due to the formation of sink holes, (ii) reduction in reservoir capacity due to repeated landslides and rapid silting of the reservoir.

2. PROJECT DETAILS

Kodasalli dam is situated in Uttara Kannada District of Karntaka State. It is one of the series of the dams built across the Kali river and has been taken up for execution under second stage development works under which the balance Power Potential to the extent of 270 MW is proposed to be tapped. The project has the financial aid of the World Bank. The first stage development with an installed capacity of 910 MW is complete and under operation. An index map of the Kali river project is shown in fig. 1.

Kodasalli dam is a concrete dam (with an embankment dam for a short length in the right flank) of maximum height of about 46 m above the deepest river bed level. A power house with an installed capacity of 120 MW (3 units of 40 MW each) is located at the toe of the dam on the right flank. It is about 502 m in length at top RL of 79.00 m, which includes a spillway with 9 gates of 15 m x 9 m size.

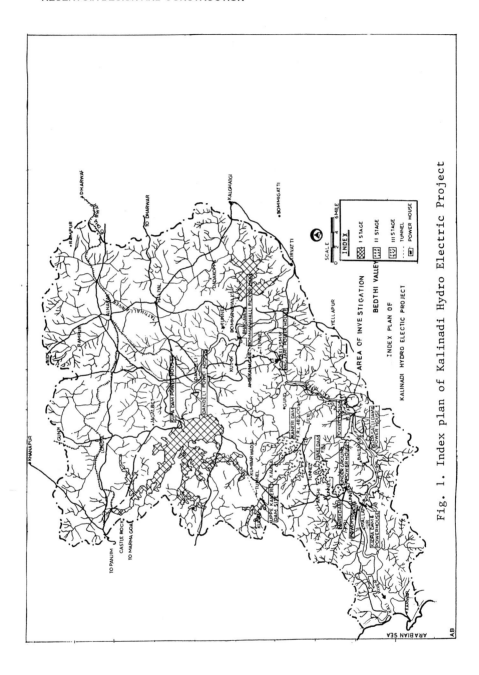

Fig. 1. Index plan of Kalinadi Hydro Electric Project

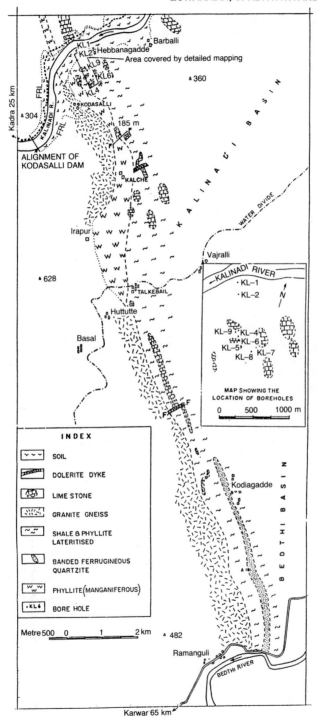

Fig. 2

3. TOPOGRAPHY AND DRAINAGE

The area forms a part of the Western Ghats made up of steep ridges. The highest level is 628.0 m. The Kalinadi river bed level is about 34 m near Kodasalli. A NE-SW trending water divide at a height of 500 m separates the Kalinadi basin from the adjacent Bedthi river basin (see fig. 2). Northerly flowing streams draining into Kalinadi exhibit a dendritic drainage pattern. Towards the south of the water divide, the drainage pattern is parallel to subdendritic. Ten springs are noticed discharging water into the Kalinadi valley.

4. GEOLOGICAL DETAILS

Kodasalli dam is located in the Granite Gneiss horizon. The reservoir area encompasses the rock types of phyllite, shale, chart and limestone, banded ferruginous quartzite belonging to the Dharwar series. The demarcation line between the schistose group of rocks and the granite gneiss lies at about 2.5 km u/s of the dam site. Among the schistose group of rock, limestone bands are found exposed at the river bed from about 3 km u/s of the dam site up to 6 km u/s of the dam site as disconnected parallel bands. The general trend of the schistose rock is NNW-SSE and this trend is almost perpendicular to the course of the river at this location. The schistose rocks associated with limestone bands are found extending over a long distance in the south-south-easterly extension, manifested in the form of disconnected outdrops. The width of individual bands of limestone varies between 250 m and 2000 m. A thin linear band of limestone is occurring beyond the water divide in the adjacent Bedthi valley which is about 6 m below the Kali river bed along the strike direction of the schist band. When the Kodasalli reservoir is formed, the difference between the FRL of the Kodasalli reservoir and the Bedthi river bed will be about 48 m. The distance between Kali and Bedthi basin is about 17 km along the strike direction of the limestone bands.

The reconnaissance geological survey made on the left bank of Kalinadi valley indicated outcrops of limestone standing to heights of 20 m to 30 m from the ground. The moderate karstification exhibited in the body of limestones raised apprehensions regarding the transbasin leakage possibility.

5. INVESTIGATIONS

(i) Geological mapping

An area 15 sq.km was geologically mapped in the region between Kalinadi and Bedthi (to the scale of 1:50,000) along the strike of formation with a view of delineate limestone exposures. The limestone occurs as isolated discontinuous bands in the NNW-SSE direction, while minor discontinuities of the order of a few tens of metres may be suggestive of a sink hole or a doline; the long gaps of the order of a few kilometres may indicate the discontinuity in the limestone bed due to non-depositional conditions.

Detailed geological mapping was carried out in an area of 300,000 sq.m to a scale of 1:1000 on the left bank of Kalinadi to bring out the structural features of the limestone bands and to locate the boreholes precisely. The detailed mapping depicted limestone outcrops occupying certain definite contour elevations on the left bank, rising to heights of 10 m to 15 m above the surrounding ground. The detailed outcrop pattern of limestone is along a regional synformal axis which has undergone folding and faulting. Dolerite dykes have been found emplaced along the fault zones. The existing outcrop pattern of the limestone can be attributed to two reasons.

(a) At the time of limestone formation paleogeographic features and physico-chemical conditions were favourable for the precipitation of calcium carbonate as lensoid bodies, in the shales/phyllites.

(b) Well bedded limestone undergoing intense folding and subsequent chemical decomposition. The interference pattern of successive folds produced domes and basins.

The geological examination of the right hill slope (i.e. on the other side of the water divide) of the Bedthi river valley indicated that there is no limestone along the Bedthi river in the strike continuity of those exposed along either flanks of Kalinadi in the Kodasalli reservoir area. The nearest limestone band from Bedthi river towards Kalinadi is at elevation 120 m which is much above the FRL 75.50 m of Kodasalli reservoir.

(ii) Borehole details

As the left bank of Kali basin in the said reach is covered by laterite and soil with dense growth of vegetation, it was necessary to drill a few boreholes along the strike continuity of the limestone bands and recover the cores to identify whether the limestone occurring underneath the soil cover is cavernous in nature.

In the first stage, diamond drill holes were restricted within FRL of the reservoir. Boreholes KL-1 at RL 50.62 m and KL-2 at RL 12.41 m (see fig. 2) were intended to establish the occurrence of limestone below the soil cover along the strike continuity of a band exposed on the right bank. They were drilled to depths of 35.65 m and 47.64 m respectively to reveal the strata below the river bed level. They intersected massive limestone with 100% recovery without indicating the presence of any caves and solution channels. The soil cover over the limestone strata was of the order of 13.0 m to 20 m. The permeability tests carried out in the holes reveal that the top soil is impervious to semi-pervious.

In the second stage, diamond drill holes were located close to the limestone outcrops at higher elevations of RL 100 m and RL 120 m. While borehole KL-5 intercepted limestones below a soil cover of 9 m, the other three boreholes KL-4, KL-6 and KL-7 did not encounter limestone even after drilling to a depth of 80 m. These boreholes might have passed through either Dolines infilled with soil or intervening part of two limestone bands.

In the third stage, one borehole (KL-8) was located at elevation RL 131.10 m across the strike continuity of the limestone band to verify the discontinuity of limestone below the laterite cap. The depth required to reach the river bed level was about 90 m. No core was recovered from this hole and the sludge samples recovered revealed that this hole has passed completely through weathered manganiferous/ferruginous phyllite. A definite stratification of ferrous and manganese bearing silty horizons was identified in sludge samples which show that the hole might not have passed through a sink hole infilled with soil.

(iii) Study of natural springs occurring on the left slope of Kalinadi up to the water divide

Periodic observations were kept over several years on the ten natural springs occurring along the left slope of Kalinadi up to the Kali-Bedthi Water Divide. This was done to find out whether there is any reversal of the ground water gradient when the ground water table rises

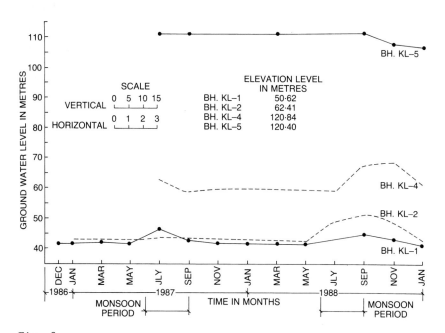

Fig. 3

Table 1

SL no	Sample no	Location	Depth in metres	Loss on ignition in %	Silica insoluble residue in %	$Al_2O_3 + Fe_2O_3$	Cao in %	Mgo in %
1	L-1	Left bank	–	44.58	1.08	2.84	39.90 (62.122)	15.175 (31.86)
2	L-2	BH KL-2	21.50	44.18	1.00	3.22	35.08 (62.44)	15.052 (31.60)
3	L-4	BH KL-1	16.30	44.04	1.42	3.18	36.36 (64.72)	14.719 (30.90)
4	L-5	BH KL-1	35.40	44.54	1.04	3.58	34.94 (62.19)	14.827 (31.13)
5	L-6	L/B exposure	–	43.80	3.68	3.76	31.58 (56.21)	17.094 (35.89)
6	L-7	L/B exposure	–	43.20	1.38	3.52	31.06 (62.40)	16.956 (35.60)
7	L-9	L/B exposure	–	41.87	5.82	4.12	31.98 (56.92)	16.04 (33.68)
8	L-10	R/B exposure	–	42.24	2.46	3.34	32.64 (58.09)	17.63 (37.02)
9	L-11	R/B exposure	–	43.01	6.94	3.32	32.24 (57.38)	14.056 (29.51)
10	L-12	L/B exposure	–	46.41	0.70	3.48	31.78 (56.56)	17.053 (35.81)
11	L-13	L/B exposure	–	44.00	2.00	5.70	36.42 (54.82)	11.837 (24.85)

N.B. The figures within brackets are the $CaCO_3$ + $MgCO_3$ percentages corresponding to the Cao + Mgo values.
L/B – left bank; R/B – right bank.

during the monsoon period due to the occurrence of any subterranean channels. The study revealed that the groundwater gradient is always towards the Kali valley and there is no ground water leakage towards the Bedthi valley.

(iv) Ground water fluctuations in the boreholes

The ground water levels in the boreholes KL-1 and KL-2, drilled close to the river, were kept under observation for nearly 3 years. The water level fluctuation, seasonwise, is indicated in fig. 3. The water level used to rise during the monsoon and used to fall during the dry season. The water levels in these boreholes were always higher than the river water level, thereby indicating a hydraulic gradient always towards the Kali river.

(v) Chemical analysis of the limestone and its solubility

Samples of limestone collected from different outcrops have been chemically analysed. Table 1 gives the details of the analysis.

It can be seen that the percentages of Cao and Mgo varied from 31.06 to 39.90 and 11.84 to 17.90 respectively. It could be classified hence as Dolomitic variety. Nine samples were subjected to solubility tests for a period of one year, by immersing them in water of pH values ranging from 4 to 6. The percentage loss in weight was as given below.

pH value	Loss in weight: %
4	0.055 to 0.190
5	0.053 to 0.156
6	0.076 to 0.160

However, Kali water samples collected and tested are found to be non-acidic.

6. CONCLUSIONS

1. Large-scale geological mapping of the area traversed by limestone bands has indicated that the limestone occurrence is in the form of lenticular bodies without any linear continuity to extend from Kali valley to Bedthi valley. Along the shortest distance between the two valleys, manganiferous phyllite rocks exist along the strike continuity of limestone bands.

2. The area has undergone intense folding due to which limestone is found in the form of domes with intervening basins filled with soil derived from the decomposition of manganiferous phyllite rocks.

3. In several places, the limestone bands are found faulted and Dolerite Dyke intrusions are found across the bands which act as a barrier for the movement of ground water, if any, subsequent to the

storage of water in the limestone bands.

4. The nearest limestone band observed in the right hill slope of Bedthi valley is at elevation 120 m which is much above the FRL of Kodasalli reservoir (75.50 m).

5. The groundwater gradient on the left slope of the Kali valley up to the water divide is towards Kali valley even during the monsoon period when the ground water level rises considerably and hence there is no posssibility of any subterranean channels to promote transbasin leakage.

6. The limestone occurring in the area is of Dolomitic type and the solubility factor is low to cause any anxiety of dissolution.

7. The hill slope up to FRL on the left bank is covered by soil of more than 10 m which is impervious to semi-pervious. The likelihood of reservoir water permeating through this soil cover and reacting with limestone is very remote.

ACKNOWLEDGEMENTS

The authors wish to thank Sri. G. Ashwathanarayana, Managing Director, Karnataka Power Corporation Ltd, for all the encouragement given and also permitting publication of this paper.

SUMMARY

The occurrence of the limestone formations in the reservoir area formed by the Kodasalli dam in the Kalinadi river valley raised the apprehensions of a possible transbasin leakage into the adjacent Bedthi river valley. The detailed investigations which included geological investigations, ground water observations, chemical analysis of limestone etc., revealed that this possibility is very remote.

The instrumentation, monitoring and performance of Roadford Dam during construction and first filling

J. D. EVANS, BScTech, FICE, FIWEM, South West Water Services Ltd, and **A. C. WILSON,** MICE, MIWEM, Babtie Shaw & Morton

SYNOPSIS. Roadford Dam is constructed of low grade sandstone/mudstone rockfill and has an asphaltic concrete membrane on its upstream face. The paper describes the instrumentation installed and relates the significance of the results to the performance of the dam.

INTRODUCTION
1. The first filling of South West Water's Roadford Reservoir commenced in the Autumn of 1989. By Spring 1991 the water level had risen to within 150mm of the spillweir at 126.4m AOD, the highest level reached to date. The reservoir now forms a significant feature in the landscape of West Devon and the environmental management of the planning and construction has been described in ref. 1.

DESCRIPTION OF DAM
2. The geology of the site consists of sedimentary rocks of the Crackington Series of Upper Carboniferous age. The rocks comprise rapidly alternating mudstones and sandstones which have been extensively folded. The uppermost part of the rock mass is considerably weathered particularly where mudstone strata predominate.
3. The 41m high embankment is formed from a 50:50 mixture of sandstone and mudstone quarried on site. The construction, using low grade rockfill, has been described in ref. 2.
4. An asphaltic concrete membrane on the upstream face provides the waterproofing element. The lower periphery of this is sealed to the top of a toe structure, which incorporates an inspection gallery throughout most of its length.
5. A graded stone drainage layer is provided beneath the membrane to collect any seepage passing through the asphalt. Any flow discharges through pipes into the inspection gallery where the location and rate can be monitored.
6. To minimise the strain on the membrane from differential settlement downstream of the toe structure a transition zone of sandwaste is used. This material, which has a compressibility about one third that of the rockfill, is a by-product of the china-clay industry.

RESERVOIR DESIGN AND CONSTRUCTION

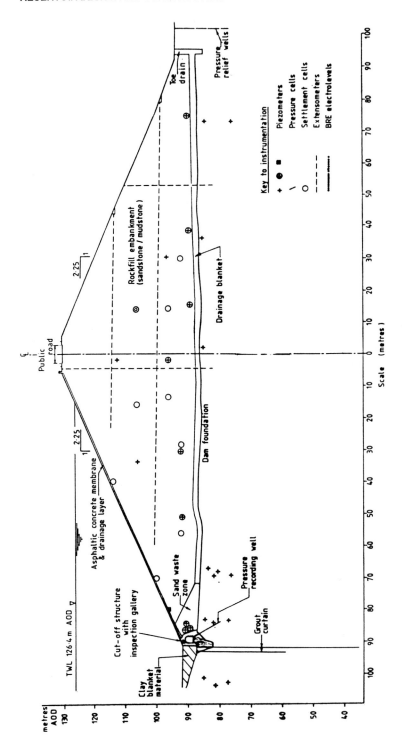

Fig. 1. Cross section of dam at instrumentation zone B

7. A granular drainage blanket is installed between the dam foundation and the embankment fill to collect groundwater from beneath the dam and any rainwater percolating through the fill. Underdrainage flows can be monitored before discharge downstream.

8. Seepage from the reservoir is reduced by a concrete filled cut-off trench and a double row grout curtain. Further sealing is provided by a blanket of low permeability weathered material placed upstream of the toe structure over the disturbed working area to join with the natural clayey mantle in the valley.

INSTRUMENTATION

9. The embankment has been extensively instrumented with piezometers, vertical and horizontal extensometers and inclinometers. Earth pressure cells are provided at the interfaces of the fill and the access and inspection gallery structures. Survey stations and provisions to measure drainage water flows complete the arrangements which will allow the dam's behaviour to be monitored.

10. The instruments are located mainly at three cross sections of the dam; Zone B is the highest section, shown in figure 1. Zones A & C are on the western and eastern flanks respectively. Details of the instruments are given in Table 1 together with the minimum frequency of reading required by the Construction Engineer under the Reservoirs Act 1975.

11. Installation of the instruments presented no particular problems and caused little disruption to rockfill placing despite numerous trenches being required to carry about 130km of hydraulic tubing from the instruments to the read-out locations. With the exception of one piezometer, all the instruments were commissioned successfully although even with semi-automatic equipment, de-airing was particularly tedious. Initial readings were taken from temporary gaugehouses before transfer to the permanent location in the pumphouse.

12. Most instruments were read manually with the data later entered into a computerised storage system. Readings from the inclinometer were directly downloaded to the computer. Programmes prepared by the instrumentation contractor enabled the data to be presented in tabular or graphical form but difficulties were encountered in the use of these when plotting graphs because there is no operator control over the choice of scales. More flexible programmes that are now available would undoubtedly result in considerable time saving.

13. The technician who was to be ultimately responsible for the reading of the instruments was seconded on a full time basis from SWW from the commencement of installation. He was thus able to build up the experience and knowledge of the instruments which has proved invaluable during the subsequent monitoring. This commitment of staff time is essential for the production of reliable information.

RESERVOIR DESIGN AND CONSTRUCTION

TABLE I ROADFORD DAM MONITORING INSTRUMENTS

Instrument	Location	Zone A	Zone B	Zone C
Hydraulic Piezometers (low air entry)	Reservoir basin	3(F)	3(W)	3(F)
	Dam foundation	7(F)	11(W)	10(F)
	Embankment rockfill	2(M)	3(M)	3(M)
Hydraulic Piezometers (high air entry)	Sandwaste zone	3(F)	3(W)	3(F)
	Embankment rockfill	2(M)	6(M)	4(M)
	Associated with pressure cells	-	-	11(Q)
Disc piezometers	Under membrane	-	1(M)	1(M)
Hydraulic Pressure Cells	Inspection gallery	2(M)	2(M)	2(M)
	Access culvert	-	-	33(Q)
Overflow Settlement Cells	Under membrane	1(M)	2(M)	1(M)
	Embankment rockfill	-	7(M)	1(M)
Vertical Extensometers & Inclinometers	Centreline of dam	1(F)	1(W)	1(F)
	Downstream face	-	1(F)	-
Horizontal Extensometers	Embankment rockfill	1(Q)	2(Q)	1(Q)
BRE Electrolevels	Under membrane	1(W)	1(W)	1(W)
Survey stations	Downstream face	25 monuments (Q)		
	Crest Wall	115 locations (Q)		

Note: Frequency of reading shown.
(W) Weekly (F) Fortnightly (M) Monthly (Q) Quarterly

PIEZOMETRIC PRESSURES

14. Piezometers in the reservoir basin upstream of the cut off have recorded pressures about 25m below reservoir level in Zones B & C indicating the effectiveness of the upstream blanket. In Zone A the reservoir level has been reflected much more closely probably because it was difficult to place blanket material on the steep hillside.

15. Since impounding, pressures have increased in the foundation downstream of the cut-off. Only occasionally have these exceeded the top of the drainage blanket, indicating its success in controlling foundation pore pressures.

16. Piezometers installed in the sand waste zone have shown pressures close to or just above the floor level of the inspection gallery. In Zone B these correspond to those measured in the pressure recording wells. (See para.20).

17. Taken together, the piezometric information shows a significant pressure difference between the upstream and downstream sides of the grout curtain and confirms that the cut-off provisions appear to be operating effectively.

DRAINAGE FLOWS

18. Prior to impounding, the flows from the foundation drainage system were estimated to be in the range 2.3 l/s to 4.6 l/s. As the reservoir level approached T.W.L. the underdrainage flow increased to about 16.2 l/s. This is well

within the estimate of 43 l/s, made during design, for the capacity of the drainage systems.

19. As there is significant seasonal variation in the flow, it is not yet possible to distinguish the relative effects of rainfall and reservoir level. However, consideration of the flow patterns, together with chemical analysis, seems to suggest that most of the flow is diverted groundwater and that direct seepage from the reservoir is small.

PRESSURE RECORDING WELLS

20. To provide a positive indication of groundwater levels, pressure recording wells were drilled from the inspection gallery in the lowest part of the valley into the rock immediately downstream of the cut off trench. When impounding commenced water flowed from some of these holes with the number and flow increasing as the reservoir level rose. The area with the greatest flow was at the toe of the western hillside where the cut off trench is deepest due to fractured rock.

21. The wells are now normally kept closed to minimise possible erosion of material from rock fissures. They are opened for a period of 24 hours each week so that the 'pressure', 'flow on opening' and 'steady state flow' can be measured. Pressures up to 14m above the gallery floor level have been observed with the valves closed; these reduce by about half when steady valve open flow conditions are achieved. Flows show a similar pattern with the average steady state flow of about 2.3 l/s being about one third of that when the well is first opened. Both pressures and flow are now reasonably constant and do not seem to be affected by changes in reservoir level.

22. The reason for the flow from the wells has not been fully explained but it appears that groundwater is diverted and concentrated at the toe of the western hillside where the cut off structure turns sharply along the valley floor.

MEMBRANE DRAINAGE

23. Shortly after impounding, small flows were observed from some of the pipes connecting with the under-membrane drainage layer. Three months later there was a dramatic increase in flow to a maximum of about 1.4 l/s particularly at the toe of the western hillside.

24. Chemical analysis suggested that the water originated from the reservoir but inspection of the membrane by divers revealed no imperfection. A bacteriophage survey similar to that previously used at Colliford Lake (ref. 3) was also carried out. At Roadford the tracer was detected in small amounts in the membrane drain water some six days after having been introduced into the reservoir. Conversely, tests on the water failed to reveal any traces of poly-aromatic hydrocarbons which were experimentally proved to be present in water which had been passed through the asphalt.

RESERVOIR DESIGN AND CONSTRUCTION

25. By Summer 1990 the flows from the membrane drains virtually ceased but reappeared with the onset of the next Winter to reach a new maximum of about 3.1 l/s but at a location nearer to the centre of the dam. The flows again became negligible during the Summer but recent indications are that the seasonal pattern will be repeated during the Winter of 1991/92.

26. Various theories attempting to explain the source of the water have included small cracks in the asphalt and rainwater passing through the fill but none is completely convincing.

GROUNDWATER REGIME

27. Observation holes installed around the area of the reservoir during the site investigation period established that the basin would be watertight. Subsequent monitoring of the water levels has confirmed that the reservoir has had no effect on the groundwater regime along its margins.

28. On the west side of the valley 40m downstream of the dam a drainage adit was driven 100m into the hillside to drain an area of potential instability. Before impounding, flow from the adit averaged about 0.35 l/s which was probably artificially high due to the new construction drawing down the existing water table. Post impounding flows have averaged 0.5 l/s which suggests that the adit is effective in reducing seepage pressures in the area.

PRESSURES ON STRUCTURES

29. The pressures of the embankment fill against the downstream face of the toe structure have been highest in Zone B. These were approximately 265 kN/m^2 before impounding and 363kN/m^2 when the water level reached 85% of its total depth.

30. The pressures of the fill acting vertically and horizontally both along and perpendicular to the direction of the access culvert are measured by groups of cells installed at five positions on the crown and at the springing points of the arched roof. Piezometers are installed adjacent to the cells so that a correction can be made for pore pressure.

31. At the end of construction the pressures against the culvert immediately beneath the crest of the dam reached values of 776 kN/m^2 vertically and 638 kN/m^2 horizontally. When the reservoir was near T.W.L. the vertical pressure at the crown had reached 825 kN/m^2 equivalent to about 1.2 times the height of fill at that point and about two thirds of the pressure used in the design.

SETTLEMENTS AND MOVEMENTS

32. Settlements within the fill have been analysed from the vertical extensometer data. Hydraulic overflow settlement cells were also installed, but have not produced satisfactory results because of difficulties with the overflow drain tubes and insufficiently accurate pressure measurement transducers.

33. During construction, as the height of the embankment increased, the lower layers of fill were compressed but this settlement was 'built out' with the crest being brought to its correct level shortly before impounding. The measurements indicated a maximum settlement of about 320mm which correlated closely with the predictions of the finite element analysis.

34. Laboratory tests prior to construction confirmed that the mudstone rockfill was susceptible to collapse settlement. This occurs when fill that is placed comparatively dry 'wets up' to its natural moisture content causing the point contact between individual particles to weaken and crush. Heavy compaction will reduce the effect although it may not be sufficient to eliminate it entirely (ref. 4).

35. Since impounding, settlements have continued at a reasonably constant rate apparently not related to water load. The steady reduction in settlement that has occurred in other dams has not been evident. The reason may be that the fill has taken longer than expected to reach its natural moisture content hence delaying the collapse settlement effect.

36. The total compression to date is consistent with theoretically derived values and recent measurements have indicated a reducing rate of settlement. It is expected that this trend will continue until the moisture content of the fill reaches equilibrium.

37. Levels taken on the crest wall have shown settlements to be proportional to the height of fill. However there are local depressions coincident with instrument Zones B & C. These may be related to the number of trenches required for the instrumentation tubes and the less effective compaction. The depressions could also reflect an easier access for water resulting in quicker collapse settlement effects.

38. Lateral movements within the fill have been assessed from the positions of the magnetic plates of the horizontal extensometers. Although no accurate survey of the ends of the tubes was made during construction it was deduced that both upstream and downstream shoulders moved outwards from the centreline to a maximum of about 120mm or 30% of the settlement. After impounding the upstream movement of the embankment ceased but there were continuing movements downstream. These were confirmed by survey of surface monuments which also showed that there had been lateral convergence of the embankment face towards the valley bottom.

39. Deformations within the fill have also been monitored by a biaxial probe inclinometer which detected only small movements. Towards the end of construction, difficulties were experienced which resulted in the probe becoming jammed at the bottom of Zone A and eventually abandoned. Recent results undertaken by an instrument contractor are obviously anomalous which has reduced confidence in the reliability of the equipment.

RESERVOIR DESIGN AND CONSTRUCTION

MEMBRANE DEFORMATIONS

40. The deformation of the asphalt concrete membrane was measured by a special instrument designed and installed by the Building Research Establishment (ref. 5). This consists of electro-levels mounted within articulated stainless steel boxes laid within the top 100mm of the drainage material immediately under the asphalt binder layer. The instruments extend 11m up the face in Zones A & C and 14m in Zone B.

41. The electro levels have been very successful in operation and measurements have shown the deformations to be in line with those predicted by the finite element analysis. The greatest angular deflections occurred immediately downstream of the cut off structure and about 5m up the slope at the sand waste/rockfill interface. These deflections represent settlements of about 10mm and 5mm respectively. This indicates that the sand waste zone has been effective in reducing the settlements at the joint and the deflections that have occurred can easily be accommodated by the flexibility of the asphalt.

CHEMICAL DEGRADATION

42. The mudstone in the embankment rockfill contains about 1% of pyrite (iron sulphide). When pyrite is exposed to air and water it oxidises to produce sulphuric acid which in turn attacks other clay minerals in the fill. This acid also reacts with carbonate present in the drainage blanket material to produce calcium and bicarbonate.

43. Laboratory tests have confirmed that the strength of fully degraded fill material is not less than that assumed in the embankment design. The degradation processes occur on the surface of rock particles and result in a minor loss of density which will not significantly affect the total volume of the embankment.

44. Regular analysis of drainage waters indicate that degradation products are flushed from the dam following periods of high rainfall. It has been assessed that 13kg/day of sulphur is leached from the dam resulting in an annual reduction in volume of $1.85m^3$. Similarly the loss of calcite from the drainage blanket has been assessed at $14.3m^3$/year. These losses will be offset by some depositions within the embankment such as iron from pyrite oxidation and calcium sulphate (gypsum) neither of which will be carried away in the drainage water.

45. Monitoring of the drainage waters has confirmed the expected chemical reactions and more details are given in ref. 6.

CONCLUSIONS

46. Installation and commissioning of the extensive instrumentation was generally more successful than is sometimes the case. Substantial manpower resources are required for the regular reading and recording of data from the instruments. The electrolevels measuring deflections of the membrane have been particularly successful.

47. The impressive array of data for analysis and interpretation has confirmed that the performance of the dam has been satisfactory and in accordance with the design predictions. However settlements have continued over a longer time scale than expected.

48. The time span of post impounding readings is quite short and further monitoring is still required to help to resolve the outstanding queries and to give a better understanding of the behaviour of the dam during its service life.

ACKNOWLEDGEMENTS

The authors would like to thank all those who have been involved with the installation and reading of the instruments and providing the data on which this paper is based. The permission of South West Water Services Ltd. to publish this information is gratefully acknowledged.

REFERENCES

1. GILKES P.W., MILLMORE J.P. and BELL J.E. The Roadford Scheme : Planning, Reservoir Construction and the Environment. Journal of the Institution of Water & Environmental Management 1991, vol. 5, no.6, Dec., 659-670.
2. WILSON A.C. and EVANS J.D. The use of Low Grade Rockfill at Roadford Dam. Conference proceedings, British Dam Society, The Embankment Dam, Nottingham 1990, 21-27.
3. MARTIN C. The application of Bacteriophage Tracer Techniques in South West Water. Journal of the Institution of Water and Environmental Management 1988, vol.2, no.6, Dec. 638-642.
4. CHARLES J.A. Laboratory Compression Tests and the deformation of rockfill structures. NATO Advanced Study Institute Conference, Advances in Rockfill Structures, LNEC, Lisbon, June 1990.
5. TEDD P., PRICE G., EVANS J.D. and WILSON A.C. Use of the B.R.E. electro-level system to measure deflections of the upstream asphaltic membrane of Roadford Dam. Proceedings 3rd International Symposium on Field Measurements in Geomechanics, Oslo 1991.
6. REID J.M. and DAVIES S.E., Roadford Dam : Geochemical aspects of construction of an embankment of low grade rockfill. Proceedings International Conference on Ground Chemistry/Microbiology for Construction, Bristol 1992.

Recent examples of reinforced grass spillways on embankment dams based on CIRIA report 116

R. FREER, BSc(Eng), FICE, Construction Industry Research and Information Association

SYNOPSIS The experimental work carried out between 1984 and 1987 by the Construction Industry Research and Information Association on the practical design of reinforced grass spillways resulted in the publication in 1987 of CIRIA Report 116 *Design of reinforced grass waterways*. Since then many waterway channels have been built using the design procedures recommended in this report in circumstances where previously a concrete spillway would have been considered. This paper illustrates some recent practical examples of this design and draws attention to the technical and financial benefits of this research project.

BACKGROUND
1. In 1984 the Department of Environment (on behalf of small reservoir owners) together with a number of major reservoir owners and suppliers and manufacturers of materials suitable for reinforcing grass slopes agreed to a proposal by the Construction Industry Research and Information Association for a broadly based research programme with the objective of producing a guidance document and practical recommendations on the use of reinforced grass as the surface of a waterway subject to occasional high velocity flow.
2. Research work on the project was carried out by several different research organisations over the period 1984 to 1987 and resulted in the production of the CIRIA Report 116 *Design of reinforced grass waterways*[Ref.1]. This Report has been prepared in the absence of any previous comprehensive published information on the design of reinforced grass waterways. It is a guide to good practice for design engineers, and is intended to provide a framework within which an appropriate site specific design can be developed.
3. This design is particularly appropriate for auxiliary spillways for dams which are subject to intermittent flows of short duration. In many instances auxiliary spillways are additional structures which have had to be added after the main structures were completed and the designer has had to consider the effect on both the stability and the appearance of the main structure of adding an auxiliary spillway. Reinforced grass

spillways are particularly appropriate where the appearance of the structure is an important consideration.

4. This paper illustrates a number of recent examples where the designers have used reinforced grass spillways using the recommendations given in CIRIA Report 116. The designers reported that these designs showed a financial saving compared with the alternatives using a concrete spillway, and the total saving reported is many times greater than the original cost of the research work.

5. The implementation in 1985 of the Reservoir Act 1975 required local authorities to enforce safety provisions on any large raised reservoir (capacity of 25,000 m^3 or more) in Great Britain where the panel engineer had recommended an increase in spillway capacity in his inspection report. The ICE Guide on Floods and Reservoir Safety draws attention to the possible use of low-cost auxiliary grass spillways in extreme flood events, and to the requirement for the panel engineer to use his judgement in allowing rare overtopping of embankments. Reinforced grass spillways are a cheaper alternative to spillways of mass concrete but their use had been inhibited by a lack of confidence both in their performance and in the existing ad hoc design practices.

6. In order to remedy this situation CIRIA started a research project on reinforced grassed waterways in 1984 to try to establish the design criteria for these waterways, although the design cannot be considered in a wholly analytical manner because the performance of reinforced grass is determined by a complex interaction of the constituent elements.

FULL SCALE TRIALS

7. In 1986 CIRIA conducted trials on a number of specially constructed sloping channels to test at full scale different types of reinforced grass waterways. Supporting laboratory scale tests were carried out at Salford University and desk studies were made of the use of vegetation as an engineering material. The result of this work was drawn together in a comprehensive CIRIA Report 116 *Design of reinforced grass waterways* which describes the procedures and principles for the planning and design of reinforced grass waterways.

8. The report was produced by Rofe, Kennard & Lapworth under the direction of Mr M F Kennard in association with and under contract to CIRIA. The Report was written by Mr H W M Hewlett (Rofe, Kennard & Lapworth), Dr L A Boorman (Institute of Terrestrial Ecology) and Mr M E Bramley (CIRIA) and in association with Hydraulics Research Ltd.

9. The materials tested as a base for the grass cover in the different channels were geotextiles and concrete blocks, and various combinations of the two:
- Two dimensional woven geotextile fabric 0.75 mm aperture size
- Two dimensional geotextile mesh 27 mm aperture size laid on biodegradable establishment aid

- Open three dimensional geotextile mat, 18 mm thick
- Bitumen bound three dimensional geotextile mat, 20 mm thick, mass 20 kg/m^2
- Non-tied interlocking concrete blocks 100 mm thick, mass 110 kg/m^2
- Cable tied concrete blocks (3 no. products) 85 to 100 mm thick mass 135 to 160 kg/m^2 on geotextile underlayers with aperture size 0.145 mm to 0.28 mm
- In situ concrete 150 mm mass 200 kg/m^2 on geotextile underlayer with aperture size 0.1 mm.

10. One channel was lined with unreinforced grass as a control channel. The same grass mixture, containing 40% Perennial Ryegrass, 30% Creeping Red Fescue, 20% Smooth-stalked Meadow Grass, 10% Creeping Bent was used throughout.

11. The channels were built on the slopes of a disused reservoir embankment 10 m high at Jackhouse near Bolton, and had an mean longitudinal gradient of 1 in 2.5 down the 25 m long slope. The channels were each 1 m wide with 1.1 slide slopes 0.7 m high. Intermittent flows of increasing discharge up to 1.1 m^3/s and increasing velocity were directed down each of the channels in turn. Failure was recognised when an area of the grass cover was eroded or the concrete units were lifted out of position.

DESIGN RECOMMENDATIONS

12. As a result of these tests a number of design recommendations were made to assist designers to prepare a rational design. It should be emphasised that the CIRIA report provided the essential basis of confidence without which this innovative form of construction examples of which are described later in the paper would not have been used.

13. The design criteria recommended included:
- Consideration should first be given as to the principles for design, specification and construction and whether the circumstances are appropriate for the use of reinforced grass. For instance reinforced grass would not be suitable where the flow was continuous for more than 2 or 3 days at a time or where regular maintenance of the grass cover was not possible.
- Consideration of the risk associated with failure of the reservoir, i.e. Classification A, B, C, D as listed in ICE publication "Floods and Reservoir Safety".
- Frequency and duration of the anticipated flow and hydraulic loading. Limiting maximum velocities are recommended between 1m/sec and 8m/sec depending on the foundation materials and duration of flow, as shown in Figure 1.
- Properties of subsoil. Seepage of water from the spillway into the subsoil may increase the pore water pressure and reduce the soil

RESERVOIR DESIGN AND CONSTRUCTION

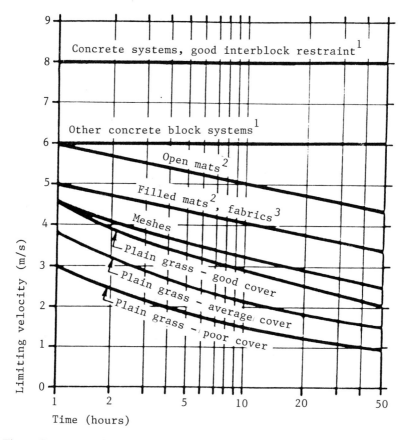

Fig. 1. Recommended limiting values for erosion resistance of plain and reinforced grass

strength and increase the possibility of a deep failure of the soil mass. A shallow surface slip is considered to be less likely because of the development of root growth, which increases as the grass matures. The bulk of root growth is within 200 mm of the surface.
- Appearance. In situations where the visual appearance of the waterway is important the grassed waterway is eminently suitable and may make it possible to provide a spillway in circumstances where a solid concrete spillway might not be acceptable.
- Capital and maintenance costs and strategy for future maintenance. A grassed waterway will be significantly cheaper than the alternative concrete spillway, but will require maintenance.

COSTS OF TRIALS

14. The cost of the Jackhouse trials, approximately £250,000, was supported in cash and kind by the DoE, PSA, the then NOSHEB, BWB, the then water authorities and a number of industries.

15. A number of reinforced grass spillways have been subsequently designed and built using the information in CIRIA Report 116, and the designers have reported that they have shown a total capital saving, compared with the conventional designs in concrete, in excess of £1,000,000. The research has therefore paid for itself many times over.

PRACTICAL EXAMPLES OF REINFORCED GRASS WATERWAYS

16. The following are some recent examples of the reinforced grassed waterways designed using the CIRIA Report 116.

a) **Welsh Water Authority**
   Cantref reservoir, Taf Fawr Valley
   Designed by Rofe, Kennard & Lapworth
   Welsh Water required to increase the spillway capacity of two consecutive reservoirs in the Taf Fawr Valley, upstream of the town of Merthyr Tydfil. At the lower reservoir the restricted site conditions made it necessary to build a reinforced concrete air regulated siphon whereas at the upper reservoir, Cantref, there was sufficient space to build a reinforced grass spillway using concrete blocks on a geotextile base and a small air regulated siphon. A comparison of costs at Cantref showed that the reinforced grass spillway with a small air regulated siphon was approximately £250,000 cheaper than the extra spillway capacity being catered for entirely by a larger air regulated siphon. Also the reinforced grass spillway with a small siphon was more environmentally appropriate at this particular site in the Brecon Beacons National Park and adjacent to a main road than a larger siphon.

b) **Esso Petroleum UK Ltd in conjunction with Tamworth Borough Council**
   Stoneydelph Retention Dam
   Designed by Ove Arup & Partners
   The design of this new earth dam for flood protection classified as Category B incorporates the facility for the full length of the embankment to be overtopped in a controlled manner in the event of a Probable Maximum Flood. The dam is 7 m high by 56 m long and the downstream face of the embankment has been designed as a reinforced grass spillway using tied and interlocking cellular concrete blocks laid on a sand filter bed. This design is environmentally more acceptable in this location and also has been estimated to cost £500,000 less than the alternative concrete

spillway.
c) **Melton Mowbray Borough Council**
Scalford Brook Reservoir, Melton Country Park
Designed by Severn Trent Water Authority
A flood alleviation reservoir with an earth embankment designed for overtopping in a 100 year flood. The downstream slope at 1:4 was designed as a reinforced grass spillway using tied pre-cast armorflex 140 and 220 cellular concrete blocks laid on geotextile filter membrane Nicolon HD 625 or HD 430 over 50 mm of compacted topsoil. Blocks above downstream waterlevel were filled with topsoil and those below downstream water level were filled with gravel. This design was adopted for environmental and economic reasons.

d) **British Waterways Board**
Tringford Reservoir and Marsworth Reservoir
Designed by British Waterways Board
For the auxiliary spillways at both reservoirs a reinforced grass spillway has been designed using Dycell tied pre-cast concrete units laid on a sand bed and Terram 35C textile fabric. BWB selected reinforced grass spillways for environmental reasons both because conventional concrete weirs were considered to be incongruous at both these particular sites.

e) **Carr Bottom Reservoir**
Burley Moor, Near Bradford
Yorkshire Water plc
Designed by Rofe, Kennard & Lapworth
The discharge under p.m.f. conditions is 5.5 cumecs. The main spillway can discharge 1.5 cumecs and above this discharge the auxiliary spillway comes into operation. Reinforced grass was used for the auxiliary spillway because the reservoir is located in an environmentally sensitive area and the designers wished to ensure that new works would blend into the landscape. The auxiliary spillway will be used only occasionally and it was considered preferable to avoid using a concrete structure. The spillway construction consists of 120 mm of topsoil overlain by one layer of Enkamat 7220. The surface was finished with topsoil and hydroseeded. The spillway is approximately 8 m wide. The cost of placing the reinforced grass spillway was £3,200 which was half the cost of the equivalent reinforced concrete construction.

f) **NRA Welsh Region**
Pont-y-Cerbyd flood storage, Solva, Dyfed
Designed by Wallace Evans Ltd
Reinforced grass was used on the crest and downstream face of

this flood storage dam to protect it from annual overtopping. The dam is 2 m high and has a downstream face at a slope of 1 in 5. Armour flex concrete units were used to protect part of the downstream face of this dam to protect it from damage by floods in excess of 0.3 PMF. This scheme is reported to have cost £70,000 compared with the alternative proposed for flood alleviation with reinforced concrete walls which would have cost £270,000.

g) **Welsh Water (Dwy Cymru)**
   Lower Lliedel, Llanelli
   Designed by Binnie & Partners
   The dam is 20 m high and has a downstream face at a slope of 1 in 2.5. Cellular concrete units provided by Tarmac Masonry Products were used to cover an area of 850 m^2. This scheme was costed at £32,000 whereas the alternative would have been to replace the masonry channel wall with higher reinforced concrete walls at an estimated cost of £100,000.

ACKNOWLEDGEMENTS

17. The research project, including the field trials and production of RP116, was funded by CIRIA and the Department of the Environment, and was supported with funds or contributions in kind by:

Anglian Water	Property Services Agency
Ardon International Ltd	Brooklyns Ltd
British Waterways Board	Schlegel Technology Ltd
ComTec (UK) Ltd	Severn Trent Water
Don & Low Ltd	SLD Pumps Ltd
Grass Concrete Ltd	Southern Water
MMG Civil Engineering Systems	Welsh Water
Netlon Ltd	Wright Rain Ltd
North West Water	Yorkshire Water
North of Scotland Hydro-Electric Board	

18. The full scale trials were carried out under contract to CIRIA by:
   - Tarmac Cubitts, construction and operation of the trials facility
   - Salford University Civil Engineering Ltd, hydraulic field trials (Mr M Smith) with assistance from Institute of Terrestrial Ecology (Dr L Boorman)
   - Geostructures Consulting - in situ shear tests (Mr D H Baker).
19. Supporting activities were carried out by:
   - Mr R Baker - study of performance of concrete block systems in high velocity flow, carried out at University of Salford under grant from the Science and Engineering Research Council (to be published).

- Richards, Moorehead and Laing - information project to produce a report giving guidance on the use of vegetation in civil engineering as an engineering material, under contract to CIRIA (reported separately).
- Mr R Baker, University of Salford
- Richards, Moorehouse and Laing

20. CIRIA gratefully acknowledges permission to use in this paper information supplied by:

Ove Arup & Partners
Binnie & Partners
British Waterways Board
Rofe, Kennard & Lapworth
Severn Trent Water Authority
Wallace Evans Ltd
MMG Civil Engineering Systems Ltd
Ardon International Ltd
Tarmac Masonry Products Ltd

REFERENCES

1. Hewlett, H.W.M., Boorman, L.A. and Bramley, M.E. (1987)
Design of Reinforced Grass Waterways
CIRIA Report 116, CIRIA, London

Reservoir construction development for irrigation in the United Kingdom 1960-92

S. M. HAWES, BScEng, ACGI, FICE, FIAgrE, FIWEM, FBIAC, Consulting Engineer

SYNOPSIS. This paper deals with developments relating to irrigation and reservoir construction in the United Kingdom during the period from 1960 to the present. This period of modernisation in British agriculture saw the creation of a large and efficient industry where public money was injected to encourage private sector development, in which that private sector always had to provide 50% or more of the capital involved. Thus cost effective techniques were adopted which may have applications for the more effective use of resources in the public sector.

THE AGRICULTURAL REVOLUTION

1. Many years of Commonwealth and Empire preference to the importation of food had left British agriculture in a derelict condition in 1939. The war years saw an enormous effort to resurrect the industry by opening up old drainage systems, installing new ones, and generally no effort was spared to produce as much food as possible.

2. The post war era started with the emphasis on mechanisation, partly on machinery imported from the USA, and partly on a new industry of tractor production in the UK. Imported combine harvesters led to a requirement for bulk grain storage, knowledge of animal health requirements led to the application of air conditioning principles to stock housing, which in turn produced improved live weight gain for less feed.

3. A pre-war Czechoslovakian immigrant named Sigmund (later Sigmund Pulsometer Pumps, later SPP) who had been manufacturing fire fighting equipment in Gateshead for the National Fire Service saw the potential for irrigation and started production of lightweight galvanised steel pipes and sprinkler systems, as had been in pre war use in eastern Europe. And so the irrigation industry in the UK was born.

4. Subsequently equipment was imported from Italy and the USA and copied for manufacture in UK, the pumping units being shaft driven by tractors, and where rivers were adjacent to suitable soils the systems thrived. That is until the effect of abstracting substantial quantities of water from the rivers in summer began to show us that rivers could indeed be pumped dry.

5. Over zealous salesmen sold equipment to farmers who

had little more than a stream running through their farms, with the advice that a few sandbags in the stream would form a small dam and then they could pump to their hearts content. The supply barely primed the pump!

6. Co-incidently the east coast floods of 1953 had shown up the inadequacy of our river walls, and progress was being made in the study of soil mechanics. Thus while marshes were being protected by higher river walls, the fields required greater freeboard for tile drainage, so land drainage pumps were required.

7. Government grants were available for most agricultural activities, except when frequent credit squeezes held up development.

8. The author had been on a school visit to the then new King George V graving dock at Southampton. The engineer showing the party round remarked that here was a highly expensive structure with massive foundations to take the weight of the Queen Mary, but there she was floating in the muddy estuary which cost nothing. Twenty years later when asked to design a land drainage pump it became apparent that frequently the only reliable foundation material on the marshlands was the water. The only problem to floating the pumping station lay in designing a simple but effective pipe joint which would allow the movement caused by water level variations. Complete construction apart from minor site works was then possible in a workshop rather than on a remote river wall.

9. Between 1959 and 1963 some eighteen floating land drainage pumps from twelve to twenty eight inch diameter were accomodated in pontoons at less than half the cost of conventional stations, thus enabling River Boards and Internal Drainage Boards to provide pumps during credit squeeze periods at the same price which they would have had to pay with a 50% grant. Simplified designs for smaller pumps are used on many reservoirs both as filling pumps and as the main irrigation pump delivering at pressures up to 12 bar and with up to 100 KW motors.

10. These floaters are still afloat and pumping some thirty years on. Several have been modernised with large bore polythene flexible delivery pipes to replace the old leaky barrel joints. Cathodic protection has kept the steel hulls as new except for some corrosion at the water/air interfaces. Many engineers who considered that resources were more important than resourcefullness, looked upon such hydrids with contempt as they contaminated civil engineering with marine engineering. Eventually in the 1963 reorganisation of River Boards the traditionalists got the upper hand and the "sound engineering solution produced in-house" (i.e. design without thought for cost) became the norm.

11. Meanwhile the private sector forged ahead with cost effective systems, particularly in the irrigation sector, albeit frequently inefficient in running costs due to a predominance of trade rather than professional engineering input. Such input was later imposed by the Ministry of Agriculture requirement for

proper design when grant aid was involved. Nevertheless the farmer had to provide sixty percent of the cost of his installations, and when water storage became necessary he had to compete with those who had free water from an adjacent river.

DEVELOPMENTS AFFECTING RESERVOIRING.

12. Boreholes. These were becoming increasingly common in chalk and limestone areas, and subtantial improvements in yield from small diameter (12" downward) were enabled by the use of improved acidisation techniques and the ability of the small bore local well sinker to cope with diameters above six inches. However low yields were frequently overcome by providing storage for peak usage periods.

13. Sand beds. Initial development for irrigation water supplies by was by seepage reservoirs, but these were unreliable. Whilst well points for civil engineering purposes were commonplace, their prime requirement was that the substrata should not be affected and fine screening meant that they required re-jetting when clogged. For water supply however it was essential to develop the aquifer, sothe fines needed to be removed. Initially clay tile drains were jetted in to form a small bore well, but as PVC pipes became available they enabled efficient cheap well tubes to be constructed. Essentially the slot size, which was adjusted by the width of saw blade used to slot the tubes, was arrived at by considering the sieve analysis. Unfortunately the design was copied by those who had no idea of the design principles so there are many which have become inefficient or failed completely now that water tables are low. To avoid damage to pumps, sand traps were designed to intercept the sand drawn out of the aquifer during development, before it entered the pump. Substantial improvement in yield is now made possible by improved techniques of sinking the well tubes with sand or geotextile filters when the sieve analysis shows self development is likely to be ineffective. The main design parameter is water velocity at the screen.

14. Weed screening. Abstraction from drainage ditches required systems of screening which were not liable to frequent blockage. Floating pumps on two metres square floatation units enabled bottom entry screens to have four square metres of screen area, which allowed the blocking material to drop off when the pump was stopped. Again the chief design parameter was water velocity. Frequently this cleaning operation was improved by allowing water from the system to automatically backflush. Pressure switches which detect screen blockage and then start a cycle of system close down, automatic backflush, followed by automatic re-start enabled night time operation to continue unhindered by screen blockage. This feature was invariably used on pontoon mounted land drainage pumps which operate mainly during off-peak reduced night rate tariff.

15. Application to reservoir design. These developments have had a bearing on subsequent reservoir design, particularly in relation to safe filling and emptying methods.

RESERVOIR CONSTRUCTION TECHNIQUES.

16. <u>Siting</u>. The farmer's idea of his ideal reservoir site is usually astride a stream in a valley. Thus the first requirement is to point out to him that the essence of low cost construction is with mechanical equipment which does not usually like bad ground conditions, and that his site may well be mainly alluvium. Frequently good sites are on the top of a hill, since the hill is there because it is capped by impermeable material. Wooded areas are also clues to impermeable soils which were difficult to cultivate. Distance from the water source is frequently immaterial since distribution mains can have a dual use, that of distributing the water to the fields, as well as connecting the water source to the reservoir.

17. <u>Design and Specification</u>. Small scale contracts could not be effectively designed or managed by traditional civil engineering methods. This could not have been better illustrated than by the major consultancy who produced a Specification and Bill of Quantities weighing over two kilograms for a reservoir of 60,000 cubic metres capacity. The only price received was ten times the then going rate per million gallons stored.

18. <u>Basic faults</u>. These were expecting a small job to carry the weight of site offices, phones, site transport, survey equipment and full time supervision. Soil mechanics technology was then used in an academic and totally unpractical manner to change the moisture content of the indigenous materials to meet optimum strength characteristics, rather than modifying the design to accomodate those characteristics.

19. In the early days of farm building design, unimaginative architects did great damage to the cause of advisory work, and in the end the RICS have taken over that role. Engineering consultancies with experience of large dam design seldom had an approach to small dam work which was not readily acceptable to individuals spending their own money. The DOE in refusing to adopt a suitable Panel under the new Act, are causing farmers to go direct to local work hungry plant hire firms which results in their throwing up nests of badly designed reservoirs which are outside the Act, and which will be a constant cause for maintenance, and could well become a safety hazard.

20. <u>Full time site supervision</u>. An engineer costing £20 to £30 per hour cannot be cost effective when the sole plant required may be but one tractor and scraper hired at £27 per hour. Thus the criteria of design and specification becomes one whereby the contractor is instructed in the methods of contruction which are known to produce the right results, as well as an easily understood specification as to the end product. For example specifying the route which scrapers must take and where on that route they may discharge, enables all compaction to be more than adequately carried out purely by construction traffic. The owner can be appraised of these methods and call in the engineer between his normal visits for compaction testing if they are not being obeyed.

21. <u>Impounding or non-impounding</u>. Very seldom do

impounding reservoirs as defined by the 1930 Act, (a dam across a valley) have an application in agriculture. Their disadvantages are:- a. Problems in arranging residual flow to meet licensing requirement.

b. Overflow works to suit catchment area rather than reservoir capacity.

c. Siltation from annual catchment flow rather than from annual usage.

d. Field drain outlets are submerged causing internal siltation.

Where frost protection is involved it is often essential that melt water can be recovered, and here the 1975 Act definition that a reservoir is impounding if it obstructs a watercourse can become a problem since it requires a Panel AR member endeavouring to apply the minutiae of the Flood Studies Report to the very small catchments (less than a square kilometre or two) which apply in agricultural situations.

The most cost effective solution is usually to have a filling bay alongside a watercourse with dual weir control to meet licensing requirements. The bay is then easily de-silted when necessary and a filling pump takes settled water from it to the main reservoir.

22. Bank section. The majority of water retaining earth embankments in the UK prior to 1960 were river banks. These were seldom wider than 0.6 metre at the top, having been built by hand, but the 1953 flood showed the main fault to be that they were two steep and thin and had dried out and cracked. Since the scraper was the most cost effective reservoir construction tool, to bring in a dragline to complete a narrow top to an embankment was more costly than moving additional material to complete with the same slopes to a three metre wide top. This had the added advantage of ensuring that any scour had to be very severe before drying out of the core arose, and also meant that intermediate scour faces could be accepted without affecting the stability or strength of the main bank.

23. Impermeable site conditions. Initially most sites were in clay land because the permeable areas had been served with bores and wellpoints, and there was ample winter flow to fill a reservoir, but often zero flows in summer. Additionally competition, particularly with potatoes, meant that whilst the water retentive heavy land farmers had had the edge on the light land producers, the reverse now became true. In the 80's quality superseded quantity as the criteria, and irrigation has become essential to prevent drought conditions making a crop unsaleable. As had been the case with blackcurrants in the 60's, when Beechams would not contract with farmers who did not have frost protection, so nowadays quality potatoes cannot be economically grown in East Anglia without irrigation.

24. Permeable site conditions. Subsequently sites became more difficult and even small deposits of boulder clay in predominantly chalk areas were developed with the use of full or partial linings. In the 1960's the thickest PVC or polythene

lining material was 250 microns, and the author still acts as Supervising Engineer for several such reservoirs which, provided the soil cover has been well maintained, are still in good order after twenty years or more.. Other reservoirs where Butyl rubber was used but left exposed, and which cost four times as much, have had high maintenance cost due to animal damage and ultra violet deterioration. Whilst still having sufficient strength to form a watertight seal, some are having to be covered with a protective soil layer to prevent having to replace them in the near future.

25. Lining thickness. Present trends in lining material, largely caused by local authorities demanding unnecessary thickness for waste tip lining, have lead to engineers adopting thicker and thicker sheets welded together to form an impenetrable barrier. Such practices are environmentally self defeating in that the extra plastic produced itself causes environmental pollution by additional power requirement and past experience shows that thin sheets with unwelded joints are equally effective. It is pointless to have a specification for total watertightness when losses through evaporation far exceed anything which may seep through a small defect when the law of diminishing returns applies.

26. Appropriate technology. Not always are the techniques for large dam construction applicable to small embankments, in fact they are sometimes damaging. For example the provision of drainage blankets in the outer shoulder, not only very costly when there is no suitable material on the site, but cases have been known where they have caused excessive drying out of the back of the core causing cracking of the embankment. Just as large banks require care to avoid pore pressure build up, so small bank design requires care to prevent loss of moisture.

ANCILLARY STRUCTURES

27. A reservoir is of no use without a means of filling, abstraction and overflow. There was a view in the 60's that it was also essential to have a scour pipe to empty it. Research into the causes of reservoir failure soon brought to light the problem caused by inflexible structures within, or adjacent to, the loading caused by an embankment upon the original ground beneath, and the failure of pipe joints under such conditions.

28. Scour pipes. If necessary this is the first thing to be constructed. The author considers that the only situation requiring such a pipe is an impounding reservoir, simply because it is necessary to have a route by which the impounded stream may discharge though the embankment whilst it is under construction. If that stream can be safely diverted then it can be dispensed with. The size of the pipe will depend upon the summer storm flow which must be safely dissipated without overtopping. This pipe can then be converted into an overflow pipe and a scour pipe by forming an overflow tower from concrete chamber rings upstream of the core.

29. Piped overflows. Agricultural reservoirs are seldom filled by pipes in excess of 150mm dia. The exceptions are large

capacity reservoirs with restrictive licences which must be filled from short lived storm flows when a maximum of 300 mm. might be used. Thus when non-impounding the overflow is related to the filling pump capacity. The predominant requirement should be that all pipes are accessible with minimal excavation and without lowering the water level more than a metre or so to make the repair. The pipe should have an entry which cannot easily be blocked, and in order to ensure the maximum weir length it is usual to have an upturned easy bend ending at top water level. By placing pipes obliquely down the back of the embankment just below the surface, settlement movement can be taken up in the compression joints. Using short lengths of pipe with frequent compression joints, and leaving an ample gap between the pipes within the joint, gives further safeguard against fractured pipes.

30. **Emergency spillways.** These are usually a simple low section in the embankment top where the height is minimal, and if the outside slope is steep then geotextiles, or a solid clay outer shoulder are used to prevent scour.

31. **Filling pipes.** The simplest system has been to design the pipe to be just below the surface of the embankment face similar to that described above for the overflow. Once through the core of the embankment close to top water level the pipe goes obliquely down the embankment to discharge at base level such than any erosion cannot affect the stability of the bank.

32. **Emptying pipes.** Previous paragraphs have mentioned floating pumps. The irrigation pump, usually a conventional end suction motorpump with a flat head characteristic, can be vertically mounted such that it is self priming and frost protected. The delivery pipe in early days was constructed from alloy irrigation pipes with a double knuckle joint made from allow 90deg. field bends. Thus the pipe weight was minimal as it was normally below water level other than where it lay on the bank face. With the advent of high pressure lay flat hose and latterly welded high density polythene it is no longer necessary to pipe to the bottom of the reservoir, but merely to let the pipe lay just beneath the water surface.

33. **Safety aspects.** The use of floating pumps enabled water to be pumped from deep reservoirs where suction heads would have been excessive for a pump mounted above top water level. This contributes to reservoir safety by eliminating the need for pipes deep within high embankments

WAVE PROTECTION

34. Due to the small fetch of agricultural reservoirs scour below top water level is rare, and paragraph 22 mentions the added protection given by the practical construction techniques. Top water level scour is usually localised by the prevailing wind, the less affected sides becoming protected by reed growth. Methods to prevent scour at top water level were legion, but few were effective. Boards fixed to posts were about the most expensive and lasted least time. Old electricity poles

floating on the surface were cheap but totally ineffective. Scrap tyres, provided they were filled with stone which usually has to be imported, have been reasonably successful, but the least costly and most used system is to wait for a scarp to form and then fill it with hardcore. Because grant-aid was only payable on initial work some owners put stone around the top edge before filling for the first time, but experience shows that seldom is full armouring required. Most damage occurs if the scarp face comes too close to the core which then cracks, dries out, and erodes even more rapidly.

EFFLUENT STORAGE

35. Agricultural irrigation reservoirs are frequently combined with effluent settlement ditches. This allows the slurry content to settle, whilst the excess water is conserved for irrigation use. Dilution, wind action and sunlight all combine to provide free purification, and fertiliser value is returned to the land. The least expensive way to deal with effluent is to combine the natural biological cycles with the production cycle.

36. Presently, because the producers are fearful of high fees for inappropriate technology, they frequently opt for the trade solution, in the form of mechanised slurry separators and glass lined steel storage tanks, all of which are very costly.

SUMMARY.

Agriculture is the largest industry in the country, but perhaps has the least site input from engineers. Its advice comes largely from trade sources which is far from independant.

Britains standard of living relies on competing in world markets. Water plays a vital part in British food production, without which we will loose out to Continental suppliers with more amenable climatic conditions. Cost effective engineering advice is vital to that industry.

Technology, if used to produce cost effective rather than academic specifications, aimed at enabling local contractors to carry out work economically, can assist in our maintaining our competitive position in world markets.

The use of stepped blocks for dam spillways

H. W. M. HEWLETT, BSc, MICE, MIWEM, Rofe, Kennard and Lapworth, and R. BAKER, BSc, MSc, MICE, MIWEM, University of Salford

SYNOPSIS. A study has been carried out for CIRIA to evaluate the effectiveness of stepped blocks developed in Russia to protect embankments from high velocity flow. The appraisal study has included a literature review, hydraulic model testing and preparation of a design guide.

INTRODUCTION
1. Stepped-block spillways were developed in Russia as the result of extensive model studies and prototype trials. A number of articles about the Russian work had been published in technical journals (Refs 1, 4 and 5) which indicated that stepped-blocks could provide an effective and economic solution to erosion problems. An interest subsequently grew in the UK and USA into their use as a low-cost method for the construction of chute spillways and the protection of embankments from erosion by overtopping. The lack of first-hand understanding and experience was, however, a constraint on their use in the West where there is a natural reluctance to use unfamiliar techniques. CIRIA therefore set up an appraisal study with the objective of producing a design guide.

2. Stepped-blocks can provide a flexible but relatively robust protection system suitable for high intensity discharge on service spillways or diversions subject to long-duration flow. Typical applications include (a) main or auxiliary spillways on earth dams, (b) on the downstream side of embankments subject to overtopping during extreme flood events and (c) as temporary spillways to convey floods greater than the design diversion discharge across the downstream face of a partly completed dam embankment.

3. Much of the Russian research was directed towards the design of spillways to withstand high prototype discharges of up to 100 $m^3/s/m$, which is equivalent to an overtopping head of 15m. The highest discharge successfully withstood in Russia was 60 $m^3/s/m$ at Dneiper Power Station (ref.1). It is likely that typical applications in the UK would be much

smaller with overtopping heads of 2-3m and discharges up to about 10 $m^3/s/m$.

4. Figure 1 shows a typical section through a stepped-block spillway. The blocks are made of precast concrete which may contain a nominal amount of reinforcement. Various different shapes of block have been investigated, some of which are illustrated in Figure 1. The blocks must contain holes or grooves to reduce excess pore pressures in the subsoil and permit drainage of the underlayer. They are normally laid in stretcher bond with the side joints between blocks staggered in successive rows.

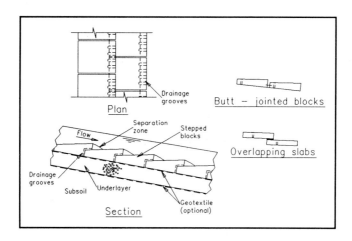

Fig. 1. General features of a stepped-block spillway.

5. The underlayer may comprise two or more layers of granular material, or a combination of granular material and a geotextile. The underlayer assists the relief through drainage vents of excess hydraulic pressure at any location below the block, acts as a filter to restrain soil particles on the subsoil formation against movement due to seepage flow, and provides an even foundation for the stepped-blocks.

6. The subsoil may comprise either original ground or fill material. Seepage flow during operation of the spillway could affect subsoil stability. Provision for seepage flow may be either nominal in the case of low-permeability clayey soils or must be purpose designed for higher permeability soils such as gravel.

7. Stepped blocks have a number of basic advantages: their upstream edges are protected from flow stagnation pressures which could lead to extreme lift and drag forces. The flow pattern causes a low-pressure **separation zone** downstream of each step which helps to evacuate seepage flows

beneath the blocks. The block shape is inherently stable: if any block moves off the slope, the sloping upper surface experiences a stabilising downthrust. In addition, the stepped upper surface has a high roughness which helps to dissipate flow energy.

8. Experience in Russia has shown that the cost of stepped-block spillways is potentially competitive in comparison with other forms of spillway construction such as in-situ reinforced concrete, flat concrete slabs, rip-rap or roller-compacted concrete. The main economic advantages are (a) the on-site works can be reduced to relatively straightforward and low-skill activities, (b) a reduction in the time taken to construct the spillway and associated works, (c) the installed cost of stepped blocks per unit discharge is substantially lower than for other materials where a much greater thickness is required.

MODEL TESTING
9. The model studies were carried out in a high-velocity open-channel flow test facility constructed at the University of Salford. The basic facility was constructed in 1986 as part of a SERC project to investigate the stability of flat concrete blocks in high velocity flows of water (Ref 2). This was modified using a CIRIA grant to allow a hydraulic jump to form on the blocks.

10. The test facility consisted of two tanks, with a vertical difference in level of just over 5m, joined together by a 600mm wide timber acceleration ramp and two by-pass pipelines each containing a valve. Water could be supplied to the upper tank at rates up to 300 l/s (0.5 m^3/s/m prototype), and the balance of flow between the pipes and ramp, regulated by adjusting the valves. The angle of the ramp for these tests was set at a fixed slope of 1 vertical to 2.5 horizontal. From the bottom of the ramp a 6m long horizontal perspex channel extended to the second tank. The lower 2m of the ramp had perspex sides and a 300mm drop in invert to allow model blocks to be bedded flush with the acceleration ramp on either impermeable or permeable bedding material. Within this watertight structure a false floor was constructed so that the gradient of the channel changed gradually from the 1 in 2.5 slope of the ramp to a gradient of 1 in 6, finishing above the floor of the perspex lined channel.

11. Throughout the programme of research a longitudinal fin drain, 4mm thick extruded polyethylene core of Trammel was laid along the entire length of the existing timber bed of the ramp. This provided underdrainage with a capacity of about 1.1 l/s/m width. The effect of wedge-shaped blocks on the ramp were simulated by fixing triangular strips of timber with the same dimensions as the blocks under test onto the

RESERVOIR DESIGN AND CONSTRUCTION

drain. This ensured the correct boundary layer, turbulence, air distribution and sub-block flow at the start of the test area. All of the tests were carried out with the model blocks laid tightly packed in stretcher bond with a physical restraint at the toe, unless otherwise stated.

12. Model blocks of different sizes were used in the tests. Overlapping blocks with an average thickness of 50mm, step height of 30mm and exposed plan area of 150 x 150mm were cast in concrete, to carry out detailed flow measurements, Fig. 2. These blocks were sufficiently large for the flow pattern to be clearly visualised and were stable at all flow rates. A test block with fourteen pressure tappings that could be connected individually to a single pressure transducer was fabricated in brass. Values of maximum, minimum and mean pressures along with statistical parameters and spectral analysis are presented in Baker (Ref 3).

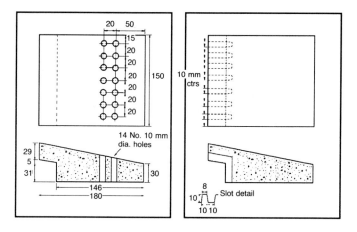

Fig. 2. Details of test blocks

13. The large size blocks were stable in the test facility at all flow rates and hence empirical failure tests were carried out with two smaller block sizes. Overlapping and non-overlapping blocks of average thickness 17.5mm were cast in mortar and overlapping blocks with average thickness 8.5mm were formed in epoxy-resin. All of these blocks were stable at the maximum flow in the test facility under free flow conditions, the smallest blocks representing a 1:12 scale of a 100mm thick prototype block subjected to a Froude scaled flow of $20m^3/s/m$.

Tests affecting block design
14. <u>Holes through the block</u>. The standard 50mm average thickness block had fourteen 10mm diameter holes linking the low pressure zone to the underdrain. These holes were sealed in pairs with plasticine and the pressure in the underdrain

measured using three U-tube standpipes. The data showed a reduction in underpressure once the low pressure roller formed in the lee of the step (Refs 2 and 3). 14 holes gives 5% open area as recommended by Grinchuk and Pravdivets (Ref 4), although the Salford data suggests that there is very little benefit in more than (say) 2.5% open area. The disadvantage of the holes is that under low flow conditions they allow a considerable increase in the quantity of water entering the underdrain.

15. <u>Step height to length ratio</u>. Pressures were measured on the top of blocks with different step height to length ratios. The results showed that with too small a step, e.g. a ratio of 1:15, the low pressure zone is too small to be effective. At a ratio of 1:5 the low pressure zone is at its optimum with the maximum holding down force. Below about 1:3.5 the blocks become unstable as the flow trajectory ceases to land on the next block downstream and hence the low pressure zone covers the whole of the block top surface. A step height to length ratio of at least 1:4 is recommended for design.

16. <u>Longitudinal joints between blocks</u>. The effect of joint width between blocks was investigated and it was found that a large joint had four detrimental effects: (a) it broke up the roller in the low pressure zone, reducing the suction on the underside of the block, (b) it allowed large quantities of water to enter the underdrain, (c) the benefit of inter-block friction was lost and (d) there was the possibility of loss of material from the underdrain.

17. <u>Sliding tests</u>. Sliding tests were carried out by laying the 17.5mm average thickness blocks in a loose packing and it was observed that the blocks slid across the surface of the underlying geotextile at very low flow. Prototype blocks should be laid tightly packed since otherwise they would be expected to slide into this condition after the first flood, potentially exposing the embankment at the top of the slope.

18. <u>Hydraulic jump tests</u>. These tests were carried out on the overlapping blocks with average thickness 17.5mm. The blocks entered a failure mode as soon as the high velocity flow ceased to generate a low pressure zone in the lee of each block step. This was not at the start of a rolling hydraulic jump but somewhere under the jump as shown on Fig. 3. Two failure modes were observed, individual blocks vibrating normal to the embankment surface and groups of blocks waving up and down as a panel. With both types of failure if the blocks were laid in stretcher bond the panel of blocks kept its integrity and a sudden progressive failure did not occur.

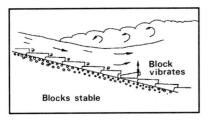 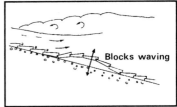

Fig. 3. Failure under hydraulic jump

19. A toe block of the type recommended by Pravdivets and Bramley (Ref 5) was installed at the end of a panel of blocks. The investigation confirmed that the toe block created a surface jump in the channel downstream of the spillway and since the blocks are always subjected to high velocity turbulent flow there is no risk of failure provided that the toe block is not drowned (Ref 3).

DESIGN
20. The detailed design of a stepped-block spillway needs to be considered in terms of (a) the overall spillway and the stability of its foundations and (b) the individual elements of the spillway, in particular the channel cross-section, the block, crest and toe details and the underlayer design. The alignment of the spillway in plan should be straight, since any change of direction would cause an uneven distribution of flow and wave patterns which perpetuate downstream. The width of the spillway will be determined by the design discharge intensity. The spillway cross-section is normally trapezoidal.

21. The general slope of the spillway will depend on the stability of the underlying embankment fill (or original ground if the spillway is constructed on the abutment). The major consideration affecting slope stability is the effect that water movement will have on the subsoil. A ground investigation will normally be undertaken to assess the degree to which the subsoil will become saturated during spillway operation. Stepped block spillways can be constructed on granular soils of medium permeability, where the soil becomes fully saturated during spillway operation. It is, however, important to ensure that adequate drainage is provided to evacuate the seepage flows and prevent unacceptable uplift forces.

22. Different shapes of block have been investigated and shown to be stable. The choice of block shape is likely to depend on a number of factors including economics, flexibility, ease of replacing damaged blocks, potential clogging of drainage holes and resistance to vandalism. One of the most important design considerations is the determination of block size. A minimum average block

thickness of 0.1m is recommended, principally on the grounds of handling the blocks and concrete durability. Design curves have been derived from the laboratory tests relating average block thickness to discharge intensity for various spillway slopes and subsoil materials.

23. One of the advantages of stepped blocks is that their high surface roughness helps to dissipate flow energy. The results of laboratory tests in the UK and Russia have been used to develop design curves to estimate the unaerated depth of flow at any point on the spillway, which is needed for (a) design of the sides and (b) determination of the flow conditions entering the tailwater. Laboratory tests have shown the air-bulked flow depth to be between 130% and 160% of the unaerated depth obtained from the design curve and this together with an extra 0.5m for freeboard should be allowed for when determining the depth of the channel. Careful detailing of the channel invert/side joint is needed to avoid gaps which could lead to erosion of the underlayer and subsoil.

24. The crest should be formed of in-situ concrete as a slightly curved overflow. The upstream end should be terminated in a low velocity area so as to avoid the risk of erosion by the approaching flow from upstream. The downstream end of the crest should be adequately joined to the first row of blocks in the spillway chute, with the crest overlapping the first row of blocks to shield their leading edges from flow impact. Vehicular access may be required across the crest and, if so, this can be achieved by a bridge supported by piers or, where access is not essential during high flows, by a water-splash with shallow side slopes.

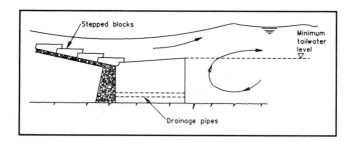

Fig. 4. Deflector block at the toe

25. The spillway toe and tailwater transition requires careful design so that the specific energy of the flow is dissipated without adversely affecting the stability of the spillway blocks or causing excessive scour downstream of the spillway. Three types of toe design have been studied. The first involves the construction of a deflector block at the downstream end of the sloping chute, as model tested and

shown in Fig. 4. The second option is to extend the spillway chute horizontally downstream of the toe of the dam and protect it with heavier stepped blocks than are used on the spillway, which should remain stable under a hydraulic jump. Thirdly, the blocks can be terminated upstream of a conventional in-situ concrete stilling basin.

CONCLUSIONS

26. In the authors' opinion, stepped blocks can provide a cost-effective form of spillway construction or protection to embankments subject to overtopping, particularly for upgrading spillway capacity at existing dams. Although stepped blocks themselves have been shown to be extremely stable under high discharges, the design of a stepped block spillway requires careful attention to detailing at the edges of the stepped blocks, in the underlayer and subsoil. The design of a stepped block spillway can only be partly standardised and the results of the CIRIA study will provide a framework within which a competent engineer can consider the use of stepped blocks at a particular site.

ACKNOWLEDGEMENTS

27. The authors gratefully acknowledge the support and advice of many organisations and individuals which have assisted in the study. Particular acknowledgement is given to Professor Y.P. Pravdivets of the Moscow Institute of Civil Engineering who pioneered the development of the blocks and advised the CIRIA study team, Mr. R.W.P. May of Hydraulics Research Ltd. who undertook the literature review and advised on the model testing, and Mr. M.E. Bramley of the National Rivers Authority who was the CIRIA research manager responsible for initiating the study in 1988. The work was funded by the Department of the Environment, the U.S. Army Research and Development Group UK, Severn Trent Water and the University of Salford.

REFERENCES

1. GRINCHUK A.S. et al. Test of earth slope revetments permitting flow of water at large specific discharges. Gidrotekhnicheskoe Stroitel'stvo, 1977, April, 22-26.
2. BAKER R. Precast concrete blocks for high velocity flow application. Journal of IWEM, 1990, vol. 4, December, 552-558.
3. BAKER R. Performance of wedge shaped blocks in high velocity flow. Laboratory report on CIRIA RP407, University of Salford/CIRIA, 1991, August.
4. GRINCHUK A. and PRAVDIVETS Y. Precast reinforced concrete revetment of earth slopes used to discharge water. Gidrotekhnicheskoe Stroitel'stvo, 1977, July, 19-23.
5. PRAVDIVETS Y. and BRAMLEY M. Stepped protection blocks for dam spillways. Water Power and Dam Construction, 1989, July, 49-56.

A review of spillway flood design standards in European countries, including freeboard margins and prior reservoir level

F. M. LAW, Institute of Hydrology, Wallingford

SYNOPSIS. As a cornerstone to its review of the ICE guide to Floods and Reservoir Safety, the Working Party has surveyed European practice. This paper summarises the outcome of a questionnaire that focussed on the key parameters of Table 1 in that guide. Standards continually evolve as they adjust to operational experience and incidents inside and outside any one country. Given the varied number, type and mean age of dams coming under any single legislation, it is not surprising to find that coverage differs. There is no reason to think that UK guidelines need change fundamentally unless and until joint probability research into the relationship between the key parameters warrants it. However overseas experience of ice sheets, log jams and similar rare events needs to be drawn upon more widely in the United Kingdom. The concept of a Probable Maximum Flood has only limited acceptance in much of Europe but so has hazard categorisation.

APPROACH
1. The other European country ICOLD committees that were represented at the UK Working Party meeting at the Vienna Congress site on 19 June 1991 were (in alphabetical order)

Austria		Norway		
France	*	Portugal	*	
Finland	*	Spain		
Germany	*	Sweden		
Ireland	*	Switzerland	*	
Italy		Yugoslavia	*	

Those asterisked have kindly provided written answers to a brief questionnaire. Otherwise notes were made of responses during that meeting or recourse has been made to previously published material (see Bibliography).
2. This is believed to have been the most representative European gathering on this specific subject. Naturally the interest and significance of responses is related to the number and type of dams in the countries concerned (Table 1). The UK 'reply' to each question is the author's interpretation.

RESERVOIR DESIGN AND CONSTRUCTION

Table 1. Dams in European nations

Nation	Number	Fill	Concrete/Masonry
Austria	133	30	103
Belgium	15	7	8
Czechoslovakia	146	99	47
Denmark	6	2	4
Finland	50	37	13
France	468	158	310
Germany (W/E)	261	174	87
Greece	13	9	4
Ireland	15	7	8
Italy	440	105	335
Luxembourg	3	1	2
Netherlands	10	10	0
Norway	245	132	113
Portugal	81	30	51
Spain	737	151	586
Sweden	141	96	45
Switzerland	144	28	116
United Kingdom	535	429	106
Yugoslavia	123	62	61
TOTALS	3566	1567	1999

Footnote: Source is ref 1.

3. The questionnaire avoided questions of methodology and concentrated on the target level of safety and associated assumptions.

LEGISLATION OR GUIDELINES ON RESERVOIR FLOOD SAFETY STANDARDS

4. Question 1 asked if a country had a legal or advisory document setting out these standards. Edited responses follow in alphabetical order.

Austria: Federal Water Law of 1959, as amended 1990 (ref. 2) for dams over 30 m high. 26 member Austrian Commission on Dams has formal role.
All lower dams controlled by local authority, not Federal Ministry of Agriculture and Forestry.

France: NO. Large dam design governed by two stage Expert Committee review. Small dams reviewed by regional authority. Existing legislation on dam failure evacuation procedures.

Finland: 1984 law. Dam safety code of practice 1985, revised 1991.

Germany: Standards document (in English) DIN 19 700 on Dam

LAW

	Plants (seven types)
	Part 10 (1986) General specifications (14 pp)
	Part 11 (1986) Dams (13 pp)
	Part 13 (1986) Weirs (4 pp)
	Part 14 (1986) Pumped-storage reservoirs (2 pp)
Ireland:	NO. ICE Guide influences practise. Environmental Impact assessment of reservoirs (under EC directives) would address reservoir safety standards.
Italy:	YES. Strict on freeboard.
Norway:	The Norwegian Regulations for Planning, Construction and Operation of Dams (of 14/11/80). English version 1986, Norwegian University Press, 132 pp. They apply to dams over 4 m high (defined as extending from the lowest point of the dam foundation on the upstream side of the core to normal TWL), or impounding over 500 ML.
Portugal:	Decree-Law 11/90 of 6 June 1990.
Switzerland:	NO. Swiss Federal Office for Water Economy has a Comissioner for Dam Safety.
United Kingdom:	The Institution of Civil Engineers "Floods and Reservoir Safety" 1989 edition is advisory. The framework legislation is the Reservoirs Act 1975.
Yugoslavia:	Standards and regulations for the design and safety of fill dams against floods. Regulatory document for warning people living downstream of a large dam.

SPILLWAY DESIGN FLOOD RARITY

5. There is always the possibility of confusing the hydraulic design capacity of spillweir/gates/spillway/stilling basin with the larger flood that the dam may be capable of containing before a failure onset. For simplicity the Question 2 read "In current dam design practise in your country, how is spillway design flood rarity chosen? Does it vary with the dam size or perceived hazard from the dam? Is it defined with reference to the flood inflow peak or the spillway outflow peak".

Austria:	5000 year inflow flood, whatever the downstream condition, tögether with dambreak emergency flood plan/map.
Finland:	Inflow flood rarity depends on dam hazard classifications (P, N and O).
France:	Inflow flood rarity rarer for fill dams than for concrete dams; potential destroying power of reservoir taken into account.
Germany:	Dams - 1000 year flood inflow peak. Flood control reservoirs - 100 to 1000 year flood, depending on storage size. PMF also taken into account 'if it is advisable and possible', but with zero safety freeboard
Ireland:	PMF used in checks on old earth dams.

RESERVOIR DESIGN AND CONSTRUCTION

Hydroelectric dams upgraded to handle at least the flood with 10^{-4} annual probability (all gates operational)

Norway: 1000 year design inflow flood, unless Directorate of Water Resources decides otherwise. Natural flood conditions downstream must not be impaired nor the flow exceed given limits. PMF must be computed and routed as a check.

Portugal:

Table 2. Portugese dam classification

DAM (height in metres)		SPILLWAY DESIGN FLOOD RETURN PERIOD (years)	
CONCRETE	FILL	HIGH POTENTIAL RISK	SIGNIFICANT POTENTIAL RISK
Over 100	Over 50	10000 to 5000	5000 to 1000
50 to 100	15 to 50	5000 to 1000	1000
15 to 50	Under 15	1000	1000
Under 15	-	1000	500

Spain: 500 year inflow flood (generally) with safety factor.

Switzerland: 1000 year flood, with check that 1.5 x 1000 year flood will not cause the dam to fail (eg no overtopping of fill dam crest).

United Kingdom: (see Table 3)

Yugoslavia: 0.01% annual probability flood inflow.

PROBABLE MAXIMUM PRECIPITATION

6. Question 3 "Has your country calculated Probable Maximum Precipitation for all or part of its area"?

Austria: PMP research well underway on regional grid basis (ref. 3).

Finland: NO.

France: NO.

Germany: Available for certain sites or states only (ref. 4).

Ireland: PMP only calculated for certain hydropower catchments but not published. However generalised FSR PMP maps available for all of Ireland in ref. 5.

Norway: YES (ref. 6).

Portugal: PMP available, but unpublished, for only very restricted areas (apparently at nuclear power sites).

LAW

Spain: As for Portugal.
Switzerland: PMP research underway at the Swiss Federal
            Institute of Technology, Lausanne.
United   YES (ref. 5).
Kingdom:
Yugoslavia: PMP only computed for a few major recent dams.

Table 3. UK reservoir flood guidelines

Category of Dam	General standard	Minimum standard if rare overtopping is tolerable
A. Reservoirs where a breach will endanger lives in a community	Probable Maximum Flood	0.5 PMF or 1000 year flood (take larger)
B. Reservoirs where a breach (i) may endanger lives not in a community (ii) will result in extensive damage	0.5 PMF or 10000 year flood (take larger)	0.3 PMF or 1000 year flood (take larger)
C. Reservoirs where a breach will pose negligible risk to life and cause limited damage	0.3 PMF or 1000 year flood (take larger)	0.2 PMF or 150 flood (take larger)
D. Special cases where no loss of life can be foreseen as a result of a breach and very limited additional flood damage will be caused	0.2 PMF or 150 year flood	Not applicable

In the case of Category B and C dams an alternative economic optimisation study is allowed for in which the chosen flood minimises (on a probability basis) the sum of spillway and damage costs. However the inflow flood taken may not drop below the minimum in the table above.

FREEBOARD
7. Question 4: "What provision is made for minimum wave run-up freeboard that should be available even at the design flood peak water level? Are those wave calculations made for a windspeed of stated duration and return period probability?"

RESERVOIR DESIGN AND CONSTRUCTION

Austria: Consideration given to different overtopping behaviour of fill and concrete dams and to log jam risk below heavily wooded catchment.

Finland: For both high hazard potential (P) and normal dams (N) freeboard to be:
(a) at least 'twice as high as the maximum wave height above the highest water surface elevation of the reservoir'; or
(b) at least the ice thickness with 10 year return period.
However for 'O' category dams freeboard is to be
(a) at least as high as max-wave height; or
(b) at least the 5 year ice thickness
Run-up is computed taking account of the dam's upstream slope and wind speed/direction/duration.

France: Calculations use the highest known wind velocity.

Germany: No specific wind duration or return period. DIN 19 700 gives generalised advice about run-up, surge, ice and variable safety margins but leaves choice to design engineer.

Ireland: Ref. 3 is used.
For sufficiently stable concrete dams flood water could reach crest level leaving the parapet wall (say 1 m high) to handle wave freeboard needs.

Norway: Fill dams - Sufficient freeboard 'to avoid water splashing over the dam through a combination of wave run-up, wind setup and standing waves'. Wind not less than 30 m/s; run-up calculated for twice significant wave height.
Concrete dams - Sufficient freeboard 'to prevent water flowing over the dam thereby causing damage to the dam or the area downstream'. Traffic across the dam taken into account.

Portugal: Article 48 requires a freeboard between maximum flood water level and dam crest. Calculations involve wave run-up, the hydrologic uncertainty and possible earthquake settlement in seismic areas.

Spain: Safety factor (maximum outflow ÷ spillway design flow) kept at about 1.3. This corresponds to freeboards typically of 1.0 to 3.0 metres.

Switzerland: 'Reasonable freeboard for waves and runup' (ref. 8); higher for fill dams than for concrete dams (ref. 9 for method related to flood volume).

United Kingdom: Category:
A and B Winter: maximum hourly wind once in 10 years.
Summer: average annual maximum hourly wind.
C and D Average annual maximum hourly wind
For storage ponds with no natural inflow winds 50% over the values given above.

Yugoslavia: A freeboard standard only applied to fill dams, using a wind velocity of 2% annual probability

LAW

(presumably of relevant duration). Alternatively the maximum observed wind may be used (typically 100-150 km/hour). Allowance is made for rip-rap roughness in the run-up calculation. Typically freeboard is 0.5 to 0.7 m.

8. Question 6 "Has your country any overriding minimum freeboard requirement relative to normal operating water level"?

Austria: NO (but note three class test of structural loading involves varying safety factors).
Finland: YES. High hazard (P)) Minimum 0.4 m from max
) flood level to core
Normal (N)) crest
Other (O)   Minimum 0.3 m
These figures apply after allowing for settlement.
France: NO.
Germany: Not in PMF check case.
Ireland: Not prescribed but ref 3 used informally, particularly for earth dams.
Norway: YES. Minimum 0.5 m where freeboard significant for safety, except where an overflow dam is designed for overtopping (when zero is applied). For set stonework run-up not less than 2.4 times significant wave height.
Portugal: Article 32 of Regulations for Safety of Dams creates a 'Specific Safety Standard' for each important dam.
Switzerland: NO.
United Kingdom: YES.

Category	Wave surcharge allowance
A & B	not less than 0.6 m
C	not less than 0.4 m
D	not less than 0.3 m

Yugoslavia: Recommended minimum crest height above normal TWL of 1.5 m (dams under 15 m height) or 2.0 m (all higher dams).

PRIOR RESERVOIR LEVEL

9. This often neglected criterion was tackled by Question 5 "What reservoir level is assumed at the onset of the design flood? Does this correspond with an assumed outflow condition?".

Austria: Normal retention level.
Finland: Normal operating level at onset of spring melt flood.
Higher levels to be assumed for summer rainfall floods.
"In special cases the inclination of the water surface of the reservoir due to flood resistance or wind has to be accounted for and ice cover effects ... can be of importance in other cases".
France: Maximum operating water levels with no initial spill.

RESERVOIR DESIGN AND CONSTRUCTION

Germany: Normal water level, with no other detail specified. (DIN 19 700 should read 'normal water level' not 'maximum water level' in the crucial clause 4.2.1).
Ireland: Reservoir full. Discharge follows owner-prescribed rules.
Norway: 'Unless otherwise specified, the water level at the start of the flood is assumed to be normal water level (NWL)'.
Portugal: Article 32 ensures a Specific Safety Standard for each important dam.
Switzerland: Normal water level.
United Kingdom: Category A & D dams - full and spilling mean daily flow
Category B & C dams - just full; no spill.
Yugoslavia: Maximum normal water level, whether spillway gates or not. In the case of flood control reservoirs designed to control the 2% (1 in 50 year) flood the flood control storage is presumed full before the spillway design flood arrives.

GATED SPILLWAYS

10. Question 7 asked whether it was assumed that a specified number of gates would stick closed during the design flood.

Austria: Preference for fixed crest weirs to avoid gate operation problems
Finland: NO, but gates have to be equipped with emergency opening systems that do not need electric power.
France: One gate closed is normally assumed. But the Expert Committee will countenance a spillway operation safety study giving the probability of having sufficient waterway through careful maintenance and an adequate gate control system.
Germany: NO, except where the water level always ranges within gate height and then the one with the largest capacity is assumed closed.
Ireland: NO, but with one gate failed the electricity authority's dams must be able to pass a lower check flood of 10^{-3} annual probability.
Norway: Gated spillways only used 'when possible malfunctioning of the mechnical components will not cause unacceptable consequences with respect to overall dam safety. Selection of mechanical equipment, design and operational procedures shall be based on an assessment of all possible extraordinary conditions which may occur' (ref. 10).
Portugal: At least two gates in any gated spillway. These must be capable of both local and remote operation, with manual and powered opening, the latter needing two separate electricity feeds.
If automatic gates are installed, they must have devices that check their automation and reliability

	at all reservoir levels.
Spain:	Controlled spillways to have maximum possible number of gates (always more than 2), no nappe interference, at least 2 power sources and manual operation option (ref. 11).
Switzerland:	Largest capacity gate taken as closed if 1000 year flood used.
United Kingdom:	(a) At least two gates in any gated spillway (b) Ability to pass 150 year flood if one gate closed (c) For Category A dams, automatic gate operating equipment with standbys and manual override option. (d) High maintenance standards and regular gate operation.
Yugoslavia:	NO, but on one key scheme one of the seven gates is a reserve.

BOTTOM OUTLET USE DURING FLOODS

11. The final question (8) asked ."Is allowance made for emergency outlet flows through the dam or associated structures during the design flood"?

Austria:	Neither bottom outlets nor turbines taken to be in use.
Finland:	Suitable dam sections may be opened in an emergency under a contingency plan.
France:	Emergency outlets are now avoided but exist in dams over 20 years old.
Germany:	Use of intake structure allowed if certain it will work; it is important that its discharge does not hydraulically interfere with that of the spillway.
Ireland:	Neither through turbines or scour gates; the former might trip and the latter be inoperable in severe conditions.
Norway:	'The discharge to the power plant and the capacity of bottom outlets is, as a general rule, not included in the capacity requirements for the spillway system' (ref. 10).
Portugal:	Spillway capacity must be adequate without needing bottom outlets, water intakes or other arrangements.
Switzerland:	YES, in particular cases (ref. 9).
United Kingdom:	Recommendation to ignore drawoff and releases; for hydroelectric reservoirs assume generators are unable to operate.
Yugoslavia:	NO, although at one dam the use of the 5 m diameter horizontal turbine during the design flood is being contemplated.

DISCUSSION

12. It is tempting to advance the thesis that each country's practice is the product of the best and the worst in its history, not least in dam engineering. On first sight

questionnaire responses show little real consensus except on starting full with closed bottom outlets; and yet they demonstrate the continuing need to address local issues.

13. Perhaps those countries which have relied traditionally on large freeboard but low return period spillway capacity are concentrated in seismically active zones. Thus the freeboard is large empirically to handle crest deformations but with the secondary benefit of greatly reducing the need for large spillway capacity by routing attenuation.

14. Again those countries showing most dependence on statistical techniques have amongst them a longer history of hydrometric data collection than the UK can muster, together with leading developers of statistical theory.

15. Nevertheless it is disappointing that no other country has brought together the necessary combination of criteria in the manner of Table 1 of the ICE guide. That Table may be a reason why current research into the joint probability of rain, wind, temperature and prior state has advanced further in Britain than elsewhere.

16. It remains to be demonstrated whether safety standards rise with time (as ever bigger floods hit a country), or with national wealth and the desire for security, or with the total number of dams (which inevitably raises the annual chance of a failure affecting a country).

CONCLUSIONS

17. As always it is wise to take the best practice from any source providing the setting remains relevant. So it is recommended that the following "under-studied" topics be reviewed in the UK:
(a) Ice sheet formation, thickness, loads; ice jams; frequency of occurrence; wave suppression by ice sheets - in each case drawing on Scandinavian practice.
(b) Any enhanced freeboard to handle settlements being used to lower spillway capacity needs.
(c) The Gradex method of flood run-off probability (ref. 12) - best compared with FSR estimates by a French researcher working in Britain using a full UK regional dataset in collaboration with local researchers.
(d) Floating debris, especially trees, and the prevention of spillway blockage.

18. Overall it seems conclusive that the UK is over-safe with its guidelines although slackening of them will require more certainty about flood estimation precision. Slackening appears more warranted for higher hazard category dams, paradoxical as this may seem. However community concern and panel engineer responsibility is unlikely to choose to run risks in a world with a non-stationary climate.

ACKNOWLEDGEMENTS

19. The co-operation of ICOLD committees is appreciated. The views expressed in this paper are solely those of the author. Reservoir flood safety research at the Institute of

Hydrology is funded by the Department of the Environment with support from both the Ministry of Agriculture, Fisheries and Food and the National Rivers Authority on flood assessment.

REFERENCES
1. ICOLD WORLD REGISTER OF DAMS. Update 1988, Table 1a.
2. KONIG, F. and SCHMIDT, E. "Construction of dams in Austria - authorization procedure" in Dams in Austria, p. 55-63. Austrian National Committee on Large Dams, 1991.
3. HAIDEN, T. et al. On the influence of mountains on large-scale precipitation: a deterministic approach towards orographic PMP. Hydrological Sciences Journal 1990, vol. 35, 5, p. 501 ff.
4. GERMAN ASSOCIATION FOR WATER MANAGEMENT (DVWK) Document 209/1989 (in German).
5. NATURAL ENVIRONMENT RESEARCH COUNCIL. Flood Studies Report, 5 vols, 1975.
6. NORWEGIAN METEOROLOGICAL INSTITUTE. Extreme precipitation values. 1984, Tech. Rpt. 3/84.
7. INSTITUTION OF CIVIL ENGINEERS. Floods and Reservoir Safety 1978, reprinted with amendments 1989.
8. MINOR, H. E. and SCHMIDIGER, R. Selection of spillway type giving attention to safety aspects. Proc. ICOLD San Francisco Congress 1988, Q63-R20, p. 310.
9. BIEDERMANN, R. et al. Safety of Swiss dams against floods; design criteria and design flood. Proc. ICOLD San Francisco Congress 1988, Q63-R20, p. 347.
10. ANON. The Norwegian Regulations for Planning Construction and Operation of Dams. Norwegian University Press, English Translation 1986, 132 pp.
11. GARCIA, J. D. Basic criteria for sizing large dam spillways. Proc. ICOLD San Francisco Congress 1988, Q63-R65.
12. GUILLOT, P. Structure de la relation stochastique non-linéaire averse-crue: consequences pour l'estimation des crues extremes. Rencontres Hydrologiques Franco-Roumaines, Paris 1991.

BIBLIOGRAPHY
CARLIER, M. A. The Design Flood: Guidelines. Report of ICOLD Committee on Design Flood, 1990. 166 pp.
REED, D. W. and FIELD, E. K. Reservoir Flood Estimation: Another Look. Report to the Department of the Environment 1991. 116 pp.
CZECHOSLOVAK STANDARDS CSN 736814 and CSN 736824 (for small dams).
SPANISH SPECIFICATION FOR LARGE DAMS.

The flood control works for the Cardiff Bay Barrage

DR P. J. MASON and S. A. BURGESS, Sir Alexander Gibb & Partners, and T. N. BURT, Hydraulics Research (Wallingford) Ltd

SYNOPSIS. The Cardiff Bay Barrage will impound the flow from the rivers Taff and Ely. During river floods the bay can become periodically tide locked leading to rises in bay level. The sluice structures designed by Sir Alexander Gibb & Partners Ltd will route such floods automatically such that flooding in Cardiff will not be adversely affected by construction of the barrage. Comprehensive back-water analyses were carried out by HR (Wallingford) Ltd.

INTRODUCTION.
1. The proposed Cardiff Bay Barrage will stretch across the mouth of Cardiff Bay from Queen Alexandra Dock to Penarth Head. It is designed to exclude the bay from the effects of the very high tidal range of approximately 12m, in the Bristol Channel and thus create a stable fresh water impoundment. The lake produced will improve the amenity value of the bay and the lower reaches of the Rivers Taff and Ely by stimulating land development opportunities and water recreation.
2. The design of such a structure requires the skills of both the marine and dam Engineer. The hydraulic structures required to discharge river floods from the reservoir are in particular, common to dam engineering. In the case of the Cardiff Bay Barrage, the structures provided and the gate/reservoir operating rules must accomplish this against fluctuating tidal levels in the Bristol Channel and in a manner which does not increase the risk of flooding through Cardiff. It is these aspects of the barrage design which are addressed in this paper.

HISTORY OF FLOODING IN CARDIFF.
3. Historically, low lying land in Cardiff has been flooded by water from the River Taff or by surge tides in the Bristol Channel. An exceptionally severe flood on the River Taff occurred in 1960 estimated at $850m^3/s$ with a return period of approximately 100 years. Subsequently the Welsh Water Authority commissioned a physical model study at HR Ltd. to examine methods of providing flood alleviation for the city. Soon after that study was completed in 1979 another high flow of $652m^3/s$ occurred on the Taff with a return period of approximately 40 years. A flood relief scheme was constructed as a matter of priority. During it's construction a surge tide of about +7.82m ODN occurred in the Bristol Channel and defence levels at the seaward end of the flood relief scheme were re-designed to provide protection for the city from a repetition of such tide levels. In February 1990 a high river flow occurred on the River Taff and a high surge

tide in the Bristol Channel as separate and unrelated events. The river flood had a peak flow of 468m^3/s with a return period of about eight years and the high tide attained about +7.8m ODN.

DESIGN CRITERIA FOR THE BARRAGE

4. The design criteria in respect of flood control for the barrage thus had to reflect the observed historical interaction of both the river flood waters and the tide. In the Feasibility Studies, the NRA had established the need for the barrage to safely withstand a nominal 1 in 100 year flood discharge (derived from both the Taff and Ely) in combination with a Highest Astronomical Tide and also the mean annual flood from the two rivers in combination with a 1 in 100 year surge tide. These two principal design events are termed the High Discharge Event and the Surge Tide Event.

HYDROLOGY STUDIES

5. In order to derive representative design flood hydrographs for the mathematical and physical modelling of the proposed scheme, a hydrological assessment of the catchment was carried out. Estimates of return periods of floods on the Taff and Ely were made using the unit hydrograph (UH) rainfall-runoff model technique and by statistical analysis of flood frequency. The methods used were broadly those of the Flood Studies Report and its Supplementaries.

6. Records of mean daily and monthly flows exist for gauging stations at Tongwynlais and Pontypridd on the Taff, and at St Fagans on the Ely, providing data variously between 1965 and 1989. A regression analysis of the combined data set allowed an 29 year series to be estimated for Tongwynlais. Two of the flood frequency curves derived and shown in Figure 1 are an extreme value type 1 and the FSR regional curve.

7. A third flood frequency curve was derived from the UH analysis. This analysis determined the design hydrographs. The UH parameters required were calculated using the FSR regression equations with referral to previous studies undertaken on the Taff for calibration of the model. The catchments of the Taff and Ely differ significantly, the former being large and principally mountainous whilst the latter is very much smaller and a lowland, with lower rainfall, drainage density and more permeable soils. The hydrographs derived for each of the required design return periods are shown in Figure 1.

8. For the Taff the UH estimates were initially somewhat lower than those predicted by the statistical method, while for the Ely they were considerably higher, being closer to the FSR curve. In each case, therefore, the UH estimates were adjusted to give approximate agreement with the FSR region curve.

9. A subsequent analysis of flood volumes confirmed the reliability of the two hydrographs.

TIDAL WATER LEVELS

10. Previous work and experience had already identified the important influence of the tide on flood levels for even minor floods. Studies were thus carried out to better define both the occurrence of still water levels and the tide shape.

11. The analysis of actual levels recorded at Cardiff Dock was supplemented by the 400 year old tide records from Avonmouth for which there is a reliable relationship for predicting extreme levels at Cardiff. From this enhanced data

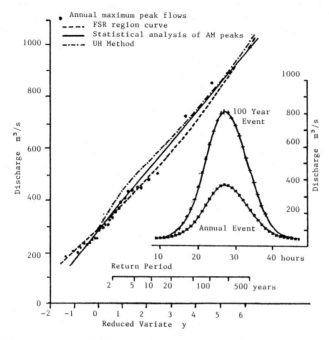

Figure 1.   Flood Frequency Curves and Design Hydrographs

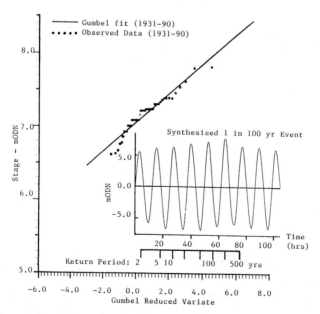

Figure 2.   Annual Maximum Sea Levels at Cardiff and Synthesised 1 in 100 year Event

RESERVOIR DESIGN AND CONSTRUCTION

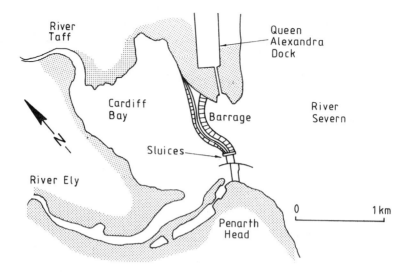

Figure 3. Key Plan

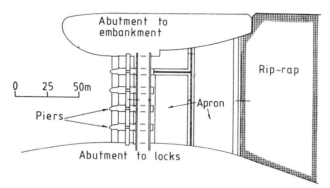

Figure 4. Plan on Sluices

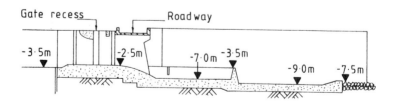

Figure 5. Longitudinal Section on Sluices

set it was possible to derive the frequency distribution of the maximum water levels, shown in Figure 2.

12. A "surge" comprises two components. The first is that attributed to the physical astronomical forces and which are predictable. The second is that caused by meteorological forces, eg. the wind and barometric air pressure, and the physical description of the site concerned. At Cardiff, as elsewhere in the Bristol Channel, surges demonstrate a distinct semi-diurnal oscillation which peak shortly before predicted high water. This is attributed to the interaction of the surge and tidal components as they propagate up the estuary.

13. To derive an estimate of the tidal conditions, the tide was simulated for the 100 year event, by combining the Highest Astronomical Tide with a synthesised surge residual scaled to give a peak water level equal to the return period at still water level, shown in Figure 2. The 100 year extreme event has a predicted maximum level of 7.96m OND.

CONCEPTUAL DESIGN OF SLUICES

14. Feasibility designs for the barrage were carried out in 1988/89. These featured sluice gates sited at either end of the barrage. Each location featured 3 No 18m long flap gates with high level cills at +2.0m OND. These were associated with 8 No culverts sited lower with cills at -2.5m OND for low level flushing. Only the high level gates were to be used for discharging floods.

15. Site investigations showed that the upper zones of foundation material were relatively compressible and that to achieve the strength of foundation needed for these important hydraulic structures it would be necessary to found them at approximately -10m OND.

16. As a result of a design review by GIBB, prior to the preparation of tender designs, it was considered appropriate to change the proportions of the gates to match the deeper structure which was going to be necessary. It was found that a superior discharge capacity would be provided by 5 No gates each 9m wide and with their cills at the lower levels of -2.5m OND. The lower cill level meant that the gates also served the function previously provided by the low level culverts.

17. The tender design features a single sluice structure located at the Penarth Head end of the barrage. The sluice gates will be housed in reinforced concrete headworks founded at approximately -10m OND, to ensure minimal settlement. Concrete joints have been arranged such that each gate is housed in a self contained monolith, to minimize structural defections. Immediately downstream of the headworks a reinforced concrete stilling basin is provided for energy dissipation. The design of this basin was optimised by hydraulic model testing. Rip-rap is located downstream of the basin. On either side of the gates and stilling basin, wide abutments will be formed using sand filled, steel sheet pile cofferdams. Long term corrosion protection will be provided to the sheet piling by concrete facings or epoxy paint.

HYDRAULIC MODEL TESTING

18. Hydraulic model tests were carried out at HR Ltd, both to confirm the discharge characteristics of the gates and sluice structure and to optimize the design of the stilling basin and headworks.

19. The derivation of appropriate discharge characteristics was complicated by the need to define both drowned and

undrowned conditions, due to the tidal tailwater. Suitable families of curves were however, developed for input into the computational model. The tests also showed that there were no hydraulic reasons to change the design of the headworks although some architectural changes were made.

20. The concrete stilling basin was originally designed to be at level -6.0m OND with a raised end sill at -3.5m OND approximately 80m downstream of the main gate. A suitable hydraulic jump formed on the apron but at low tides the raised end sill produced a zone of supercritical flow downstream. This was not considered to be acceptable.

21. The basin was redesigned as a stepped basin with an upstream, primary basin at -7.0m OND and a downstream, secondary basin at -9.0m OND. The two were separated by an intermediate weir at -3.5m OND and the end cill was lowered to -6.5m OND. Various other modifications were made at the same time to the gate structure adjacent to the fish pass as a result of ongoing dialogue with the NRA. The changes worked very well and it proved possible to shorten the secondary basin by 12m.

GATES

22. The ability of the gates to respond reliably and continually to fluctuating, fluvial flow and tidal requirements is a key feature of the sluices. The sluice gates envisaged for the Cardiff Bay Barrage are 5 No steel, double leaf, hook gates. Each will seal an opening 9m wide and 7.5m high. The gates are of a type which has had extensive world wide use for over 60 years and which provide constant flood control on many of Europe's major rivers such as the Danube. This type of gate was selected both because of it's proven record of reliability and because of the flexibility of operation afforded. The upper leaves can be lowered to give overflows, the lower leaves can be raised to allow underflows or both leaves can be lifted clear to give unimpeded flow with minimal headworks.

23. Operation of the gates will be by electrically driven hydraulic cylinders. Various back-up arrangements will be provided. Each gate will have a dedicated pumping unit with duplicate pumps, electrical supplies will be duplicated and back-up diesel generating sets will be provided. Further diesel sets will be able to pump the hydraulic cylinders directly and they could also be hand pumped. Normal operation of the gates will be by pre-set computer control with gate operation based on sensing water levels in the bay and estuary. These could be overridden if required.

FLOOD CONTROL AND OPERATING PHILOSOPHY

24. The logic required to operate the gates automatically was developed using mathematical models and computer simulation. Principal events studied were the High Discharge and High Tide events mentioned earlier. The gate operating logic developed was such that with the barrage in place, flood levels through Cardiff will be the same or lower than they are at present without the barrage. The tests were run assuming that one of the five sluice gates was out of action due to maintenance.

25. Additional runs were made to assess the effects of the 1 in 500 year river flood on the barrage structure, the effects of land reclamation in the bay and gate response to short return period floods.

26. Low flows entering the bay from the Taff and Ely pass

preferentially through the fish pass. If flows increase to amounts in excess of those which can be accommodated by the fish pass, the upper leaf of the gate adjacent to the fish pass operates preferentially to maintain a bay level of +4.5m OND. Floods which cause the bay to rise above 4.5m OND will trigger the raising of the lower leaves of the gates in a predetermined sequence depending on both the rate of bay level rise and the corresponding sea level.

27. At times of neap tide the bay level of 4.5m OND will always be above sea level. The gates will automatically open if bay levels are above 4.5m OND with continued rises triggering further opening stages. As the bay level drops the gates will close.

28. At times of spring tide the bay is tide locked for about 2½ hours of each tidal cycle. To avoid sea water entering the bay at these times the gates have to remain closed which means that all flood waters entering the bay are temporarily impounded causing bay levels to rise. At the point when the sea level drops below the bay level the gates begin to open by a predetermined set of operational rules. These cause the impounded water to be released and the bay level to fall. The variables in the programme are set to ensure that the rate of drop stays within a predetermined drawdown envelope.

29. It should be noted that the operational mode envisaged has considerable flexibility and adjustments can be made to procedures in the light of experience. It should also be noted that the system is designed to be fully automatic in that no prior flood warning is required before full operation takes place. Although the system can be manually overridden this would require positive intervention. If left, floods are routed continually based on the reservoir and sea levels pertaining at the time.

MATHEMATICAL MODELLING

30. Two principal mathematic models were used to simulate gate operation and resulting bay levels. The first, named GIBGATE, simulates the change in volume in the impoundment with stage caused by the inflow river hydrographs and outflow through the sluices. This outflow is dependent on the head discharge relationship and the operation of the gates. This model is a development of the HR BARPRO model and was also used to define the gate operational policy. Both BARPRO and GIBGATE are models specifically designed for the Cardiff Bay Project.

31. The second model, FLUCOMP, is a one-dimensional model, representing the change in discharge and water level along the length of the river, but assumes this is a constant at any one cross-section. The effects of bridges, weirs and sluices, etc are represented by various empirical formulae. The operational policy developed on GIBGATE was coded into FLUCOMP.

32. With these two models it was possible to analyse the response of the impounded bay level to the varying river and tide conditions and to assess the backwater effect on the river levels in the Taff relative to the existing flood defence levels.

FLOW ROUTING RESULTS

33. The results presented in Table 1 are from the FLUCOMP model and show the predicted levels in the bay and Taff for each principal design event considering the present, no-

barrage condition and the future, post-barrage condition. The existing flood defence levels are also given for comparison.

34. The river levels predicted are lower for the post-barrage condition than for the pre-barrage condition for both design events considered, enhancing the available freeboard to the existing flood defences.

Table 1. Flood levels along the River Taff – Existing and with Barrage in Operation

Chainage m	Location	Flood Defence	Levels m ODN			
			HIGH DISCH'GE EVENT		SURGE TIDE EVENT	
			Existing	Barrage	Existing	Barrage
1432	U/S of Barrage		7.17	7.08	7.95	5.79
3813	D/S of Clarence Bridge	8.30	7.14	7.06	7.95	5.80
4048	U/S of Clarence Bridge	8.30	7.10	7.01	7.95	5.79
5008	D/S of Penarth Bridge	8.20	7.30	7.23	7.97	5.89
5243	U/S of Penarth Bridge	8.30	7.51	7.45	8.00	5.99
5533	D/S of Woodst Rail Br.	8.30	7.64	7.60	8.03	6.06
5648	U/S of Woodst Rail Br.	8.60	7.77	7.73	8.05	6.13
6198	D/S of Cardiff Bridge	8.60	7.98	7.94	8.08	6.30
6298	U/S of Cardiff Bridge	9.00	8.36	8.32	8.14	6.50
7440	Pontcanna Fields	11.30	9.42	9.40	8.44	7.94
8075	Blackweir	13.50	11.41	11.41	9.79	9.79

BENEFITS FROM THE BARRAGE

35. Although the Cardiff Bay Barrage is being constructed primarily for development reasons, the mathematical studies carried out demonstrate that flood levels through Cardiff will be lower than at present and, more particularly the Barrage will provide an effective protection against high sea levels.

36. Predictions are always susceptible to the "unforeseen". However, the studies carried out by Gibb with HR have demonstrated the flexibility of the proposed scheme and its ability to accommodate a changing environment. The gate operational philosophy incorporates variables that allow the experience of operations to refine it, whilst the design philosophy means that should sea levels increase, as many predict they will, then the barrage will continue to provide protection to the City of Cardiff.

Numerical modelling of reservoir sedimentation

I. C. MEADOWCROFT, BSc(Eng), AMICE, **R. BETTESS**, BSc, PhD, and **C. E. REEVE**, BSc, PhD, HR Wallingford

SYNOPSIS. Sedimentation has a major impact on the useful life and economic viability of many reservoirs. Traditional methods of predicting reservoir sedimentation are empirical and do not take into account many features of a reservoir that are known to affect sedimentation. This paper shows how numerical modelling can be used to provide more detailed information than other methods. The recent application of the model to large Indian reservoir illustrates the use of the model and the results which can be obtained.

INTRODUCTION

1. Sediment deposits can have a dramatic effect on the storage capacity and useful life of a reservoir. The adverse effects caused by reservoir sedimentation include:

  a) Reduction in the storage available and hence a reduction in the yield provided by the reservoir
  b) Deposition at the head of the reservoir, leading to an increase in flood levels in the contributing rivers upstream
  c) Degradation downstream of the dam. This may threaten structures associated with the dam and lead to problems at structures further downstream such as bridges or intakes
  d) Increased evaporation losses for a given storage volume

2. It is important for the engineer to make an assessment of the likely quantity of sediment deposits throughout the design life of the reservoir. The location of sediment deposits may also be important in determining the serviceability of the reservoir. Studies into reservoir sedimentation are relevant to proposed and existing reservoirs: for proposed reservoirs, sedimentation studies may influence the design and siting of the dam, and the economic value of

RESERVOIR DESIGN AND CONSTRUCTION

the scheme; sedimentation problems at existing reservoirs may be partially alleviated by management of the catchment or reservoir, thus extending the useful life.

3. This paper describes a numerical model which can be used to simulate reservoir sedimentation. The model is based on physical principles of flow and sediment transport, and may be applied to a wide range of cases. The model can be used to investigate the effects of reservoir design and operation on the quantity and location of sediment deposits. The model predicts the quantity of deposition and therefore loss of storage capacity of the reservoir. The model also predicts changes in bed level at each location in the reservoir, and can therefore be used to estimate bed levels at the dam. This can give an indication of the time at which high sediment concentrations will begin to affect outlets or sluices at the dam. It can also be used to estimate the extent and quantity of upstream river sedimentation which can result in raised water levels, affecting bridges and other infrastructure.

4. The paper illustrates the use of the model in studies into sedimentation at reservoirs in India.

MODEL FORMULATION

5. This part of the paper will describe in outline the numerical model. The model is designed to predict siltation and water levels in reservoirs for many years into the future.

6. The model is one-dimensional. This means that only variations along the length of the reservoir are considered and all of the quantities calculated are averaged over the width of the reservoir. Consequently local effects, such as the concentration of flows adjacent to outlets, may not be accurately represented. On the other hand, the simplicity of the model means that it is feasible to carry out long term predictions.

7. The geometry of the reservoir is represented as number of cross sections. The model cross sections will generally be located at survey range lines.

8. The quantity and location of sediment within a reservoir is determined by the hydraulics of the flow into and within the reservoir. To model sedimentation, therefore, the details of the flow must first be determined. Having calculated flow conditions in the reservoir, calculations of sediment transport can be carried out. Bed level changes throughout the reservoir can then be found, based on the sediment calculations. The model described here is a time-stepping model: flow conditions computed at a particular timestep are used to carry out sediment transport and bed level calculations. The revised bed levels are then carried forward to the next timestep, when the procedure is repeated. Using this method,

calculations can be carried out to calculate flows and bed levels for any particular time in the future.

9. The following sections describe briefly the fundamental processes on which the model is based.

Flow calculations

10. An assumption is made that, since changes in flow conditions take place on a much shorter time scale than those of the bed level, the flow can be regarded as steady. The calculation of water flow therefore reduces to a steady flow or backwater calculation. The data required includes water levels at the dam, discharge, details of the reservoir geometry and the hydraulic resistance, or roughness, of the bed. If data on water levels is not available, which may be the case when modelling proposed reservoirs, a storage routing module may be used to generate time series predictions of water level based on known inflows, outflows, spillway characteristics and evaporation rates.

11. The water level, velocity and energy gradient are calculated at each cross section, starting at the section nearest the dam and proceeding upstream.

Sediment transport calculations

12. From the calculated velocities, depths and hydraulic gradients, sediment concentrations at each cross section may be determined. The method for calculating sediment concentrations depends on the size of the sediment. For sediments of sand sizes and larger, (conventionally assumed to be larger than 0.06mm diameter), the movement of the sediment is assumed to depend only on the local hydraulic conditions. As the sediment size decreases, the speed with which sediment concentration reacts to changes in flow conditions diminishes. For smaller sizes, in the silt range, the sediment concentration depends not only on the instantaneous local hydraulic conditions, but also on the history of the flow. The model takes account of this fundamental difference by carrying out separate calculations for sand and silt sizes, and combining the two to obtain a total concentration.

13. Sand concentrations are calculated using the Ackers and White sediment transport method (refs 1 and 2). This is one of several formulae for predicting sand concentrations, and has been shown to be at least as accurate as any other method. The method enables the sand concentration at each cross section to be calculated as a function of the velocity, depth and hydraulic gradient of the flow, and the sediment size. The inflow of sand into the reservoir is determined from the calculated concentration at the upstream cross section. This implies that the upstream boundary of the model should ideally be far enough upstream of the

head of the reservoir so that changes in bed and water levels resulting from reservoir sedimentation are insignificant at the upstream section.

14. The calculation of silt transport is based on functions which describe the advection, deposition and erosion of the silt. The rate of deposition depends on the effective fall velocity. This is a function of the size of the particles, the amount of turbulence and the silt concentration. If the shear stress is above a certain threshold value, then it is assumed that the turbulence generated is sufficient to maintain all of the silt in suspension.

15. Re-erosion at high flows of previously deposited silt is modelled based on an excess shear stress formulation.

Calculation of bed levels

16. The model solves a sediment continuity equation in order to calculate changes in bed level at each cross section. Essentially, if more sediment enters a particular control volume than leaves it during a timestep, then there will be deposition and a rise in bed level within that control volume. Similarly, erosion will result if more sediment is carried out of the control volume than is transported in.

17. The fact that the model is one-dimensional means that differential rates of deposition or erosion across the width of the reservoir cannot be calculated directly. The change in bed level may be assumed to be uniform across the wetted part of the cross section, or may be varied across the section as a function of water depth, according to a user-defined function. Alternatively, sediment may be deposited only in the deepest parts of the cross section to form a horizontal bed surface. The choice of how to distribute erosion and deposition should be based on the shape of the reservoir, the operating policy, and reservoir surveys which may indicate the pattern of deposition in existing reservoirs.

CASE STUDIES: RESERVOIRS IN INDIA

18. The model has recently been applied to several reservoirs as part of a programme of studies into reservoir sedimentation in India. The following sections summarise studies associated with Bhakra Dam. This illustrates the type of data required, the modelling procedure adopted and the types of results which are obtained.

Bhakra Dam

19. Bhakra dam on the River Sutlej in Northern India was constructed in 1959. The total catchment area is approximately 57,000 sq km. The impounded reservoir is approximately 80km in length, and the width is less

than 1.5km over much of this length. The dam is approximately 120m high.

20. The geometry of the reservoir was represented by 44 cross sections, coinciding with survey range lines. Fig. 1 shows a plan of the reservoir with the cross sections. Bed levels along these range lines have been regularly surveyed throughout the life of the reservoir. This meant that the model could be calibrated by comparing model predictions with measured sedimentation rates. Daily measurements of water level measured at the dam were input to the model, as were estimates of the total outflow at the dam due to power and irrigation abstractions. Suspended sediment measurements in the river upstream of the reservoir enabled a relationship to be developed relating silt concentrations to discharge. This relationship was used during model simulations to specify the quantity of silt entering the reservoir. Sediment sampling data was also available which enabled the physical properties of the sand and silt to be estimated.

21. The model was calibrated by simulating the period from 1966 to 1988. The aim of the calibration exercise was to confirm that the model was correctly simulating the observed sediment behaviour. The model should match the observed total sediment deposited, and also the distribution of deposits throughout the length of the reservoir. Fig. 2 shows, in longitudinal profile, the observed bed levels in 1966, and the observed and predicted bed levels in 1988. The figure illustrates that the model is successfully predicting the longitudinal distribution and overall rate of sedimentation. In particular, the model accurately predicts the observed growth of the top-set and fore-set slopes. The encroachment of the fore-set bed into the reservoir, between the chainages of approximately 12km and 22km, is a notable and important feature of the sedimentation.

22. The calibrated model was then used to predict sedimentation for up to 80 years into the future. Several operational scenarios were examined, including:

    i) No change in present operating conditions
    ii) Reservoir operated with reduced full reservoir level
and iii) Sediment input to the reservoir reduced.

For each scenario, the following results were obtained.

    a) Longitudinal profiles of reservoir bed level up to 80 years into the future. Fig. 3 shows predicted bed levels, assuming no change in reservoir operating conditions
    b) Loss of total reservoir storage capacity. Fig. 4 compares storage loss under existing

RESERVOIR DESIGN AND CONSTRUCTION

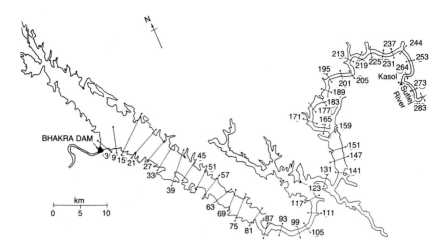

Fig. 1   Plan of reservoir at Bhakra, showing location of model cross sections

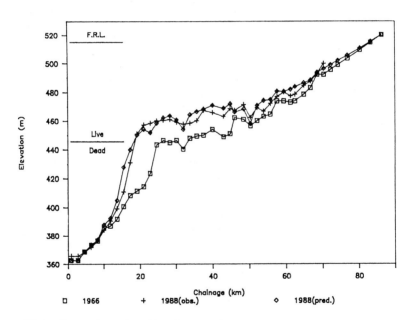

Fig. 2   Longitudinal section of reservoir at Bhakra, showing observed and predicted bed level changes during calibration period. (The dam is located at Chainage 0km)

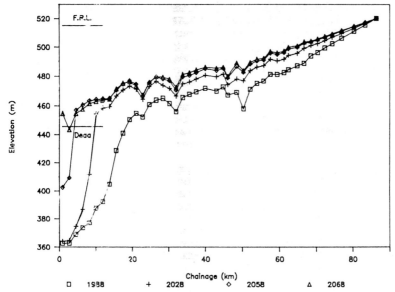

Fig. 3 Longitudinal section of reservoir at Bhakra, showing long term predictions of bed level assuming no change in reservoir operation.

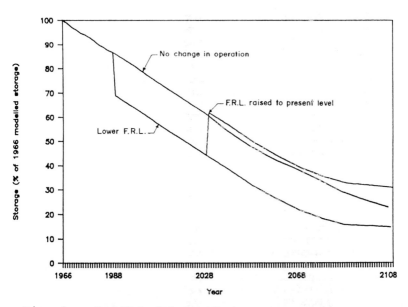

Fig. 4 Predicted loss of storage capacity of reservoir: effect of changing the full reservoir level (FRL) on long term storage loss.

c) Loss of live storage capacity.
d) The time at which the bed level at the dam will reach the levels of power and irrigation outlets.
e) Effect on water levels upstream of the reservoir. Water levels in the river resulting from flood flows were compared for each of the scenarios.

The range of results available illustrates the versatility of the modelling technique.

23. Previous studies at this reservoir had indicated that the time taken for the reservoir to loose 25% of its live storage capacity would be approximately 160 years (ref. 3). This estimate was based on assumptions about the location of future sediment deposits, combined with application of Brune's median curve (ref. 4). The corresponding time predicted by the numerical model was approximately 70 years. The difference is largely explained by the fact that the pattern of deposition will change in the future, and this cannot be accurately predicted from present trends. The numerical model, however, is physically based and provides more realistic representation of the future behaviour of the reservoir.

24. Sedimentation was found to be relatively insensitive to changes in the full reservoir level, and reduction in sediment input was found to be the most effective way of extending the life of this reservoir.

ACKNOWLEDGEMENTS

25. The work described in this paper was carried out as part of a project funded by the United Nations DTCD. The views expressed are those of the authors and not necessarily those of the UN. The work was carried out in collaboration with the Central Water Commission (CWC) in India, whose co-operation is gratefully acknowledged. Data for the modelling was kindly provided by Bhakra BEAS Board.

REFERENCES
1. ACKERS P. and WHITE W.R. Sediment transport: a new approach and analysis. ASCE, JHD, 99, HY 11, pp2041-2060.
2. Sediment transport: the Ackers and White theory revised. Report SR 237, HR Wallingford, April 1990.
3. Bhakra Reservoir, Sedimentation Studies, 1972 to 1989. Bhakra Dam Circle, Bhakra Beas Management Board
4. BRUNE G.M. Trapping efficiency of reservoirs. Trans AGU, vol. 34, No. 3.

Triggers to severe floods: extreme rainfall and antecedent wetness

DR D. W. REED, Institute of Hydrology, Wallingford

SYNOPSIS. A method is presented which explores the combination of rainfall and antecedent catchment wetness that is most effective in explaining the seasonal pattern of flooding observed on particular catchments.

INTRODUCTION
1. The estimation of floods for reservoir spillway design generally requires more than a statistical analysis of the data. Available record lengths are typically too short (in comparison with design return periods) to permit a direct analysis of flood peak data. Recourse is therefore made to the rainfall-runoff approach in which a catchment model is used to transform an extreme rainfall event into an extreme flood.
2. The rainfall-runoff approach to reservoir flood design is complicated. Even if the particular rainfall-runoff model used is itself quite simple, the approach relies on a correct choice of design "inputs" to achieve a target "output", usually in the form of a design water level at the dam. Reed and Anderson (ref. 1) outline a formal methodology for inspecting this choice. The methodology is both statistically complex and demanding in data; moreover, its outcome is still to some extent dependent on the particular model used to link the various inputs to the design output.
3. A central theme of the Joint Probability Studies for Reservoir Flood Safety (ref. 1) is to examine the partial dependence between environmental variables such as rainfall, wind, and antecedent wetness. Because of the complexity of the main approach, it is of interest to consider whether some limited understanding of the interaction of key variables can be achieved by simpler means.
4. This paper presents an empirical technique which explores the combination of rainfall and antecedent wetness that best explains the seasonal pattern of flooding on particular catchments. Because the method uses rainfall and date information alone, application to studies of rainfall-induced landslides is also possible (ref. 2).

ANTECEDENT WETNESS
Effect on flood formation
5. Antecedent catchment condition can have a profound effect

on flood response to rainfall. Typically, it is antecedent catchment wetness arising from prior rainfall that leads to a higher than normal flood response. However, antecedent catchment dryness can also be a strong influence. In most cases, the effect is to suppress or moderate the flood runoff; however, the cracking of clay soils may sometimes provide an exception (ref. 3).

6. Instances when extreme antecedent wetness had a marked effect on flooding include the floods of December 1972 in Northern England (eg. 5 December 1972 on the Aire) and those of January and October 1988 in Truro, South-West England (ref. 4). Also, had antecedent conditions not been exceptionally dry, it seems likely that the September 1968 floods in South-East England (eg. 14 September 1968 on the Bourne) would have been even more severe.

7. The principal mechanisms by which antecedent catchment wetness influences flood formation lie in the general ability of natural soils to accept significant rates of rainwater through infiltration, and in the predilection for major portions of a catchment to contribute to short-term flood runoff only when surface soils are saturated, local depressions filled, and throughflow routes developed to drainage channels.

8. It can be argued that the effect of antecedent wetness is likely to be proportionately less for a very extreme storm, when the depth and intensity of rainfall are likely to submerge any influence due to antecedent wetness. That very extreme floods arise only from very extreme storms is, however, a somewhat doubtful corollary.

Representation in flood estimation

9. The rainfall-runoff model commonly used in design flood estimation is a simple unit hydrograph method, packaged to generate estimates of the T-year event from a standard set of design inputs (ref. 5). While one of these inputs represents antecedent wetness, its design value is linked only to the general wetness of the area and cannot represent soil or drainage characteristics. The Flood Studies Report (FSR) rainfall-runoff model is necessarily simple in order that it can be applied at any site in the UK without reference to gauged data.

10. The FSR rainfall-runoff model uses a catchment wetness index built out of a 5-day antecedent precipitation index, API5, and an estimate of soil moisture deficit, SMD:

$$CWI = 125 + API5 - SMD . \qquad (1)$$

There are several limitations in this index. Firstly, the choice of a 5-day antecedent precipitation index is rather arbitrary. Secondly, it is unsatisfactory that, in some circumstances, the model permits a unit of rainfall to both neutralize a soil moisture deficit and contribute to the API, thus raising CWI by two units. Thirdly, the method does not readily represent inter-catchment differences in the significance of catchment wetness to runoff generation.

11. It is to be expected that the dynamic contribution of catchment wetness to flood runoff volumes will be influenced by

factors such as soil type, land cover and slope, the effect of antecedent conditions being greater on normally permeable catchments than on impermeable or heavily urbanized ones.

12. A more advanced approach to flood estimation might seek to simulate the catchment response to a continuous long-term record of rainfall, perhaps using a hydrological model with a stronger physical basis. Such models generally represent soil moisture variations explicitly, obviating the need for index methods. Attractive though this approach could be, its technical difficulties dictate that the "design event" approach will continue to be important for some years yet. It is therefore of interest to consider an empirical technique which explores the combination of rainfall and antecedent wetness that best explains the pattern of flooding experienced on a given catchment.

A NEW APPROACH
Background

13. Two particular factors have contributed to the approach presented here. The first is the "rainfall versus antecedent rainfall" plot presented by Crozier and Eyles (ref. 6) as a backdrop to understanding landslip occurrences. Figure 1,

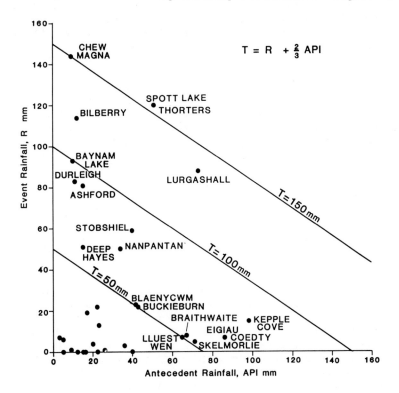

Fig. 1 Event rainfall and antecedent rainfall
for some UK dam safety incidents

taken from Reed and Field's recent review of reservoir flood estimation (ref. 7), uses the format to illustrate event rainfall and antecedent rainfall for some historical UK dam safety incidents. It is seen that a linear combination of event rainfall (R) and the antecedent rainfall index (ARI) provides a coarse guide to the severity of meteorological conditions. (However, Fig. 1 should not be taken to infer that heavy rainfall and/or antecedent wetness were necessarily implicated in the particular safety incidents cited.)

14. The antecedent rainfall index is defined by:
$$ARI_k = k R_1 + k^2 R_2 + \ldots + k^n R_n \qquad (2)$$
where k is the daily recession factor, n is the number of days over which the index is calculated, and R_i denotes the rainfall on the ith day preceding the event.

15. The equation:
$$T = R + w_1 ARI_k \qquad (3)$$
defines a simple "trigger" model; a higher value of T indicates that an incident or failure is more likely to occur. The relative importance of antecedent rainfall in comparison to event rainfall is indexed by the weighting parameter w_1. The value of 2/3 used in constructing the trigger lines in Fig. 1 was chosen subjectively.

16. A second stimulus to the approach came from completion of an extension to a major holding of flood peak data at the Institute of Hydrology: an archive of peaks-over-threshold flood data (ref. 8). Data are held for 859 UK gauging stations

Fig. 2

Catchments in peaks-over-threshold flood dataset

(Fig. 2). In contrast to annual maximum flood data, the peaks-over-threshold (POT) series present considerable information about the seasonal characteristics of flood response. The data comprise magnitude and date of independent peaks above a threshold, the threshold being chosen to provide an average of about three to five peaks/year. The criteria for independence are described in FSR Volume IV (ref. 5).

17. In the method that follows, the POT flood data are applied to adjust the parameters of a trigger model until a good match is achieved between the seasonal distribution of high trigger values and the seasonal distribution of floods.

The trigger model

18. While plots such as Fig. 1 provide a first appraisal, Crozier and Eyles (ref. 6) recognize that some representation of the longer-term water balance is desirable. A period of significant soil moisture deficit in summer/autumn is characteristic of temperate climates such as New Zealand and the UK. The approach taken by Crozier and Eyles is to employ estimates of potential evapotranspiration - derived from regional climate data - and a soil moisture extraction model applicable to standard land types, although it might be preferable to carry out a detailed catchment water balance using local climate station data. A broadly similar approach is followed here except that, to be still less demanding of data, only long-term average soil moisture deficit data are used.

19. The trigger model proposed has the general form:

$$T = R + w_1 \text{ ARI} - w_2 \overline{\text{SMD}} \tag{4}$$

where the overscore denotes an average value of SMD for the time of year.

20. It is reasonable that the parameters n and k in Equation 2 be chosen to be in harmony. For example, the FSR 5-day antecedent precipitation index has a daily recession factor of 0.5. By setting all the rainfall terms in Equation 2 to unity, it can be shown that a daily recession factor of 0.5 delivers almost 97% of its effect within five days. More generally, for an antecedent rainfall index with daily recession factor k it is reasonable to constrain n such that:

$$1 - k^n = 0.97 . \tag{5}$$

Application of Equation 5 eliminates one parameter from the trigger model.

21. A half sine wave (Fig. 3) was found to provide a reasonable fit to long-term average end-of-month SMD values calculated by MORECS (ref. 9), which provides estimates of soil moisture deficit on a 40 by 40 km grid across the UK. Whereas drawdown curves for soil water (like reservoir stocks) are typically skewed in particular years, the long-term average values show little asymmetry, with the maximum deficit typically occurring in late summer. MORECS values for drier regions such as Eastern England show a longer drawdown, with the period of soil moisture deficit typically beginning earlier and ending later than in higher rainfall regions.

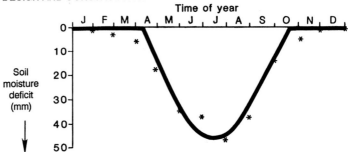

Fig. 3 Mean month-end soil moisture deficit (1961-1990), approximated by half sine wave

22. The parameters of the approximation to $\overline{SMD}$ (Fig. 3) define the typical onset of deficit, and the date and magnitude of the maximum deficit. Best-fit values were derived by optimization, using a least-squares criterion applied to the end-of-month mean SMD values.

Calibration

23. With the prior derivation of $\overline{SMD}$, the trigger model defined by Equations 2, 4 and 5 has three parameters: w_1, k and w_2. The parameter w_1 can be expected to take a value somewhat less than unity; a value of 1.0 would indicate that event rainfall was no more influential than antecedent rainfall, except in its timing. Because the $\overline{SMD}$ term is accounting for average seasonal effects only, there is little prior information about the likely magnitude of w_2.

24. Calibration of the trigger model for a particular catchment proceeds as follows. Firstly, m flood events are drawn from the POT archive and their dates converted to angles, to represent the seasonality of floods as a circular distribution (ref. 10). Trial parameter values are used to evaluate the trigger function for each day of the flood record, using daily rainfall data from a representative gauge. (Where necessary, rainfall data are compiled from more than one raingauge to provide a complete record. $\overline{SMD}$ data are taken from the MORECS grid square closest to the catchment, or from a weighted average of near-neighbours.) The highest m independent peaks are then abstracted from the trigger series and their seasonal distribution compared with that of the floods.

25. Initial work has used two objective functions to aid optimization of the trigger model. The first, obj_1, seeks to minimize the root mean square (RMS) of monthly differences in the frequencies of trigger and rainfall events. The second, obj_2, seeks to minimize the difference between the centroid of the seasonal distribution of trigger events to that of the flood peaks. Neither objective function is entirely satisfactory. The RMS criterion is sensitive to individual events joining and leaving the peaks-over-threshold series of trigger events; consequently, the "surface" representing obj_1 is insufficiently smooth to guarantee convergence to a global

minimum. The centroid-matching criterion, obj_2, suffers less from this effect but can be satisfied rather too easily, ie. any one of the three parameters can be fixed arbitrarily and a good match nevertheless obtained. The solution adopted was to seek parameter values that provide a reasonable fit under both criteria.

APPLICATION
The Rhondda at Trehafod

26. The Rhondda at Trehafod is a large, but quickly responding, catchment in South Wales. Flood flows at the gauged section are contained and the station provides data of generally high quality from June 1968. Extensive flooding occurred in the Rhondda in December 1960 and December 1979. Historically, the valley has been subject to some of the highest 1 and 2-day rainfall accumulations experienced in the UK (eg. 211.1 mm at Lluest Wen on 11 November 1929). The river has two main subcatchments, the smaller of which is impounded in its headwaters by Lluest Wen and Castell Nos reservoirs (Fig. 5). A reservoir safety incident occurred at Lluest Wen on 23 December 1969 (ref. 11), and it is one of the incidents for which rainfall and antecedent rainfall estimates are marked on Fig. 1. A catastrophic slide of a colliery spoil tip occurred at Aberfan, 6 km outside the catchment, on 21 October 1966.

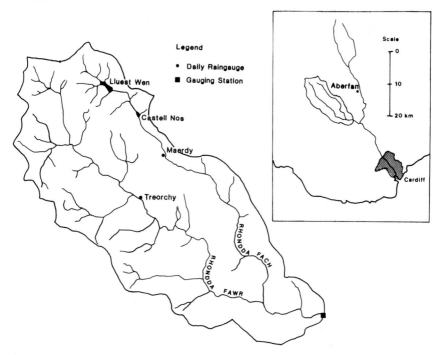

Fig. 4 Rhondda catchment to Trehafod

Results

27. The distribution of floods is strongly seasonal, with a mean date of 11 December and a standard deviation of 67 days (see. ref 10 for circular statistics formulae). The simplest trigger model is obtained by setting w_1 and w_2 to zero in Equation 4, so that the trigger variable is just the daily rainfall depth. For the Rhondda, the seasonal distribution of peak 1-day rainfall depths is fairly similar to that of the floods, with a mean date of 30 November but a rather less pronounced seasonality (standard deviation 85 days). The moderately good fit is confirmed by the quarterly event counts and objective function values in the second row of Table 1.

Table 1 Quarterly event counts and objective functions: Rhondda at Trehafod

	JFM	AMJ	JAS	OND	obj_1	obj_2	perf. index
Flood peaks	38	5	12	48			
R	35	10	20	36	2.8	0.18	73/103
R + 0.57$ARI_{0.75}$	38	5	18	40	2.2	0.05	71/103
R - 0.34$\overline{SMD}$	41	7	11	44	1.7	0.08	77/103

28. The third row of Table 1 illustrates the effect of introducing the ARI term into the trigger model. A weight (w_1) of 0.57 and a daily recession factor of 0.75 proved optimal. However, the improvement in obj_1 is modest given that two parameters have been optimized. In contrast, the $\overline{SMD}$ term performs well with only one parameter (w_2) optimized (see final row of table).

Verification

29. The empirical nature of the analysis makes it difficult to assess whether the trigger models have a sound physical basis. However, a test is available which provides some indication of the integrity of the three models. The above analysis uses only day numbers; the calibration pays no heed to the years in which the flood and trigger events occur. It is therefore possible to gauge performance by counting the number of simulated trigger events which occur within a day of an observed flood peak. This is the performance measure appearing in the right hand column of Table 1.

30. It is seen that the rainfall model (T = R) correctly identifies 73 of the 103 flood events. Thus, the marked correspondence between extreme 1-day rainfalls and extreme floods is confirmed. The rainfall plus antecedent rainfall model (T = R + 0.57$ARI_{0.75}$) performs less well, suggesting that antecedent rainfall may not be particularly influential on flood occurrence in this catchment. However, the rainfall minus mean SMD model (T = R - 0.34$\overline{SMD}$) correctly identifies 77 of the 103 flood events.

Kenwyn at Truro

31. A further catchment was selected for which it was anticipated that antecedent conditions might be rather more influential: the Kenwyn at Truro in South West England. The gauging station lies downstream of the recently constructed New Mills flood retention storage. The Kenwyn is known to be peculiarly sensitive to a combination of high groundwater levels, antecedent rainfall and event rainfall (ref. 4).

Table 2   Quarterly event counts and objective functions: Kenwyn at Truro

	JFM	AMJ	JAS	OND	obj_1	obj_2	perf. index
Flood peaks	46	3	3	29			
R	25	7	21	28	6.9	0.55	31/81
$R + ARI_{0.86}$	33	7	12	29	4.4	0.31	38/81
$R - 0.75\overline{SMD}$	42	2	2	35	3.1	0.02	41/81

32. From Table 2 it can be seen that the simplest model (T = R) performs much less well for the Kenwyn than for the Rhondda. A 1-parameter version of the $\overline{ARI}$ model is moderately effective, while the trigger model using $\overline{SMD}$ is again particularly helpful in matching the seasonality of floods. A composite model (T = R + $0.5ARI_{0.86}$ - $0.5\overline{SMD}$) performed better still, correctly identifying 46 of the 81 flood events.

33. Application of Equation 5 to the optimum daily recession factor (k=0.86) indicates that a 23-day antecedent rainfall index is appropriate, very much longer than that used in the FSR rainfall-runoff method.

SUMMARY AND RECOMMENDATIONS

34. The concept of a trigger model has been introduced in the context of floods. Preliminary results support the view that antecedent wetness is more important in relatively permeable and low rainfall catchments, where significant soil moisture deficits are commonplace. Other applications are envisaged.

35. Firstly, the trigger model may be useful in simplifying the "joint probability problem" faced in reservoir safety design (ref. 1). Secondly, by substituting an extreme value analysis of a trigger variable, it may be possible to reconcile rainfall frequency and flood frequency more fully. This could be valuable because of the much longer records available for daily rainfalls than for floods.

41. Thirdly, the method aids exploration of inter-catchment differences in flood response to heavy rainfall. It is planned to apply the technique to all catchments for which 100 or more events are held in the peaks-over-threshold flood dataset, with the aim of distinguishing groups of catchments for which the correspondence between rainfall and flood seasonality is

particularly good or particularly poor. One might then test the hypothesis that differential allowances for catchment wetness ought to be represented in generalized methods of flood estimation.

42. Current work is directed at establishing a more effective objective function, so that trigger model calibration can proceed efficiently and with greater objectivity. Because the method of calibration uses date information only, the approach seems particularly suited to the preliminary appraisal of rainfall-induced landslide hazard (ref. 2).

ACKNOWLEDGEMENTS
Pilot software for fitting trigger models was written by Mark Smith, under a NERC vacation employment scheme to encourage numerate students to consider a career in environmental science. The Joint Probability Studies for Reservoir Flood Safety are funded by DOE (contract no. PECD7/7/365), while collection and analysis of peaks-over-threshold flood data are funded by MAFF. Pamela Naden and Adrian Bayliss are thanked for suggestions directly relevant to the trigger analysis.

REFERENCES
1. REED D.W. and ANDERSON C.W. A statistical perspective on reservoir flood standards. British Dam Society Conference on Water Resources and Reservoir Engineering, Stirling, June 1992.
2. NADEN P.S., REED D.W. and CALVER A.F. Rainfall analysis and hydrological hillslope modelling for slope stability in mountainous areas. Proc. Frane Indotte da Piogge Nelle Aree Montuose e Collinari Mediterranee Europee, Potenze, March 1991.
3. ROBINSON M. and BEVEN K.J. The effect of mole drainage on the hydrological response of a swelling clay soil. J. Hydrology, vol. 64, 205-223.
4. ACREMAN M.C. Hydrological analysis of the Truro floods of January and October 1988. Hydrological Data UK 1988, Institute of Hydrology, Wallingford, 1989, 27-33.
5. NATURAL ENVIRONMENT RESEARCH COUNCIL Flood Studies Report (five volumes). NERC, London, 1975.
6. CROZIER M.J. and EYLES R.J. Assessing the probability of rapid mass movement. New Zealand Institution of Engineers, Proc. of technical groups, vol. 6, issue 1(G), 2.47-2.51.
7. REED D.W. and FIELD E.K. Reservoir flood estimation: another look. Institute of Hydrology, Wallingford, 1992, Report No. 114.
8. BAYLISS A.C. and JONES R.C. The peaks-over-threshold database at the Institute of Hydrology. Report to Ministry of Agriculture, Fisheries and Food, March 1992.
9. THOMPSON N., BARRIE I.A. and AYLES M. The Meteorological Office Rainfall and Evaporation Calculation Service: MORECS. Met. Office, Bracknell, 1981, Hydrological Memorandum No. 45.
10. MARDIA K.V. Statistics of directional data. Academic Press, London, 1972.
11. TWORT A.C. The repair of Lluest Wen Dam. J. Institution of Water Engineers & Scientists, 1977, vol. 31, 269-279.

A statistical perspective on reservoir flood standards

DR D. W. REED, Institute of Hydrology, Wallingford, and
DR C. W. ANDERSON, University of Sheffield

SYNOPSIS. Reservoir flood standards in the UK require estimation of the probable maximum flood in many cases. Clear though the procedure summarized in the Floods and Reservoir Safety Engineering Guide is, many engineers and hydrologists view PMF with suspicion. Some struggle to accept the concept of an upper limit flood. Others question whether the recipe in the Flood Studies Report is satisfactory, and this has been expressed particularly in respect of the allowance for snowmelt. The paper reviews the limited role that statistical analysis presently plays in reservoir flood standards in the UK. Nowhere is the absence of a formal statistical approach more keenly felt than in the combination of extreme rainfall, snowmelt, antecedent condition and wind speeds used to estimate maximum flood levels. The approach of the DOE-funded Joint Probability Studies for Reservoir Flood Safety is described.

INTRODUCTION
Reservoir flood standards
   1. Reservoir flood standards in the UK are set by Table 1 of the ICE guide to Floods and Reservoir Safety (ref. 1). This stipulates that earthen embankment dams sited above a community should be capable of passing the probable maximum flood (PMF). However, the table indicates that design floods less than PMF can be used for dams where a breach would pose less of a threat to life or where the structure is designed to withstand some degree of overtopping without breaching.
   2. For these cases the table provides two sets of design standards. One set is phrased as factors of the probable maximum flood (namely: 0.5PMF, 0.3PMF and 0.2PMF), while the other is set in terms of flood return period (namely: 10000-year flood, 1000-year flood and 150-year flood). The standards table indicates that the higher value should be used, although it is evident (ref. 2) that many Panel Engineers have a strong personal preference. Each set has its adherents and it seems that the choice is partly a matter of philosophy and partly of convenience.
   3. Reed and Field (ref. 3) report comparisons for 15 reservoirs selected to be broadly typical of UK impounding reservoirs, although reservoirs with a partly urbanized

RESERVOIR DESIGN AND CONSTRUCTION

catchment or a further reservoir within its catchment are not represented. They conclude that the two sets of design standards are well matched for design floods estimated by the procedures set down in the Flood Studies Report and Flood Studies Supplementary Reports (refs. 4 and 5).

4. In addition to specifying the design flood, Table 1 of the ICE guide specifies a concurrent design wind speed to be used in wave run-up calculations, and a design initial reservoir contents.

Approaches to reservoir flood estimation

5. At least four approaches to reservoir flood estimation can be distinguished:
- statistical analysis of flood peaks
- envelope of historical maximum floods
- rainfall-runoff: probable maximum flood
- rainfall-runoff: T-year flood.

6. The direct analysis of flood peak data, and extrapolation therefrom, is judged insufficiently reliable for use in reservoir flood estimation, given the relatively short periods of gauged flood data typically available. In an authoritative review of frequency analysis methods, Cunnane (ref. 6) advocates that estimates from flood data for a particular site should always be tempered by a wider analysis. Regional pooling of data can be expected to reduce inconsistencies in flood estimates between catchments, particularly where the design return period is rather longer than the period of gauged record. However, the difficulty remains that a catchment's extreme flood potential may not be apparent in the available flood record. Moreover it is to be expected that a proportion of flood growth curves will in fact be appreciably steeper than the regional average; disregard of this may be unacceptable given the public safety implications of reservoir flood design. These difficulties argue against a direct analysis of flood peak data when estimating very rare events. Nevertheless, the approach has been recommended for reservoir flood appraisal (ref. 7).

7. Envelope curve methods are still used for reservoir flood estimation in the Republic of South Africa (ref. 8). Their chief weakness is that the formulation attaches importance to catchment area but makes no explicit allowance for other physiographic characteristics known to be important. That envelope curves tend to drift upward, as new extreme events occur or come to light, is a weakness that applies in principle to any method embodying the argument "this much is credible, but no more". The criticism can be directed at UK methods also, in that just such arguments underlie the maps of 2 and 24-hour estimated maximum rainfall depths, albeit with a larger database.

8. The potential for an upward drift in maximum rainfall estimates can be illustrated by reference to the Calderdale storm of 19 May 1989. Much of the debate which followed (refs. 9, 10, 11 and 12) arose out of an inference from a daily raingauge reading that the storm had breached the mapped 2-hour

maximum rainfall for the district. While the validity of the inference remains in doubt, it is alarming that meteorologists made such an instinctive defence to the maximum rainfall map - even before detailed radar data had been examined! Had engineers and hydrologists not demonstrated that exceptional localized floods did occur on a number of north bank tributaries to the Calder, it seems likely that the observation of 193.2 mm would have been expunged from the record, rather than simply marked as suspect. The ability to detect spatial detail and timing is a reputation that radar monitoring of rainfall richly deserves. Let us hope that radar data continue to be used to augment, but not replace, conventional measurements of rainfall.

9. A supporting argument to use of a rainfall-runoff method for estimating a 1000 or 10000-year flood is that the statistics of extreme rainfall depths are better defined than those of extreme floods. This reflects both the richer statistical database (ie. more sites with long-term measurements) and the readier transfer of information about extremes from site to site (since ground influences are less acute). While few disguise the difficulty of constructing and calibrating a hydrological model that remains faithful in extreme events, the rainfall-runoff approach is promoted largely on the grounds that the physical processes known to be important in flood formation should be represented in concept at least.

10. The Flood Studies Report (FSR) approach uses a relatively simple rainfall-runoff model that lacks an explicit representation of snowmelt. One hope is that such errors as arise in its application may be reasonably systematic from site to site. This expectation reflects a feature that local catchment properties influence the parameters of the rainfall-runoff model but not specifically its design inputs.

The role of statistics

11. The above discussion highlights two areas in which statistical considerations appear in the present procedure for reservoir flood estimation. Firstly, there is a requirement to estimate design rainfall depths of high return period. Although the methods given in Volume II of the FSR remain in general use, advances have been made in methods of pooling data to arrive at estimates of very rare events in South-West England (ref. 13).

12. Secondly, statistical analysis is important in the choice of "inputs" to achieve a desired "output". In the FSR method, the inputs are a particular storm depth, duration and profile - and a particular antecedent catchment wetness - to achieve the output peak of stated rarity. However, the remainder of this paper is concerned with a related but more complicated "joint probability" problem, that of choosing a coherent set of inputs to yield a maximum reservoir water level of stated rarity.

RESERVOIR DESIGN AND CONSTRUCTION

THE JOINT PROBABILITY STUDIES PROJECT
Aims
13. The Joint Probability Studies for Reservoir Flood Safety have two aims. The first is to appraise statistically the present rules for <u>combining</u> design inputs in reservoir flood safety assessment. Having examined the dependence between hydrometeorological variables, the second aim is to advise on whether a more rational combination of design inputs is possible. For example, is it practical to recommend that exposure characteristics of a dam site be reflected in the particular combination of extreme rainfall and high wind used to define the design event?

Factors influencing reservoir flood level
14. The study is taking the water level, L, at the dam to be the key variable in reservoir flood safety assessment; this is the <u>output</u> variable. The level will vary through time in response to weather factors and reservoir usage. Naturally the interest centres on high values of L, brought about by episodes of wild weather that we will loosely term storm events.

15. At these storm events, L is thought to be a function of five main factors:

- R rain falling during the storm - conveniently expressed in the three components: depth, duration and temporal profile;
- C wetness of the catchment at the onset of the storm - a function of prior rainfall, and possibly also of wind, temperature and/or season;
- W wind speed and direction at the time that the reservoir water level is responding to the flood inflow;
- F reservoir fullness - a function of prior rainfall, reservoir usage and/or season;
- S snowmelt - principally a function of prior precipitation and temperature, and of concurrent temperature and wind speed.

The dependence of L on these factors is summarized in the equation:

$$L = f(R, C, W, F, S) \qquad (1)$$

where f defines the structure function (see Para. 28).

16. Snowmelt is placed last in this list not through unimportance but because the variable is complicated and difficult to treat in a comprehensive manner. In fact, snowmelt is less an input factor than an intermediate variable.

17. Precipitation is implicated in all but one of the Equation 1 factors. Thus the analysis of rainfall data is of particular importance to the study.

18. Certain components of the above factors are more generally influential than others in their effect on L; these are termed <u>primary</u> variables, in the sense that it is their <u>extremes</u> which are invariably implicated in achieving critically high water levels. In addition to the obvious choice of rainfall depth, it may be appropriate to assign catchment wetness and snowmelt as primary variables. In

contrast, it is plausible to treat rainfall duration and profile, reservoir fullness and wind direction as <u>secondary</u> variables, in that they are more incidental to production of an extreme value of L. The study is therefore concerned only with <u>concomitant</u> values of these secondary variables, ie. their values contingent on an event identified as extreme in terms of one or more of the primary variables. For the present, wind speed is being treated as a primary variable, although it may fulfil only a secondary role in reservoir configurations where the exposure to wave attack is very limited.

<u>Dependence</u>

19. Extreme storm depressions, squall lines, mesoscale convective systems (ie. large areas of interacting thunderstorms), and isolated thunderstorms are scenarios for an extreme storm event. In each case it is possible that very heavy rainfall will be accompanied by severe gales or downbursts (ref. 14). Thus it appears likely that there will be some dependence between the extremes of the rainfall and wind speed primary variables. Some dependence may also be anticipated between snowmelt and antecedent catchment wetness.

20. The study of dependence in multivariate extremes is demanding of both statistical skills and data, and the degree of dependence between variables may prove difficult to establish. Dales and Reed (ref. 15) examine the much simpler problem of dependence between values of a single variable observed at differing sites. Their model of spatial dependence in rainfall extremes can be used to estimate the risk of experiencing a given extreme rainfall at one of a network of N sites, be they raingauges or impounding reservoirs (ref. 16).

21. A similar model would be welcome for estimating dependence between river flows on neighbouring tributaries, so that flood frequency estimates for a main river site could exploit flood records from tributary gauging stations. However, the latter problem may well require a more formal "joint probability" approach, ie. construction of the structure function (para. 28) to relate flows on the main river to flows on the tributaries.

22. Many of the recent developments in multivariate extremes have made in coastal studies, particularly of the interaction between surge and wind (refs. 17, 18 and 19).

23. The joint probability problem faced in reservoir flood safety assessment is more complex in several respects. Firstly, there are few if any sites for which long-term records of the output variable (reservoir water level) and the primary input variables (rainfall, wind and temperature) are available. Secondly, the design return period of interest in reservoir safety assessment is longer than that generally met in river or coastal defence problems.

<u>Methodology</u>

24. Suppose $(X_1, X_2, \ldots, X_p)$ denote the primary variables and $(X_{p+1}, \ldots, X_{p+s})$ denote the secondary variables. A storm event is taken as an occurrence of a large value of any one of $X_1, X_2, \ldots, X_p$. Each of these variables will be subjected to

an extreme value analysis in the conventional (ie. univariate) sense. For analytical convenience, the statistical distribution fitted to the extreme values of a particular variable - termed the marginal distribution - will be standardized by transformation to a unit Frechet form.

25. The dependence between extremes of the primary variables will be modelled by point-process modelling (refs. 20 and 21). Consider the case of two primary variables X_1 and X_2, and restict attention to events in which either or both of X_1 or X_2 take unusually large values. A scatter plot of the resulting points (X_1, X_2) will then produce a cloud of points well away from the origin in the positive quadrant of the plane. Theory shows that the pattern of these points must be random, with the expected number of points in small areas with "polar" coordinates r and w (where $r = x_1 + x_2$ and $w = x_1/r$: not quite the standard polar forms) given by:

$$\lambda(r,w) = h(w)/r^2, \qquad (2)$$

where h is a function which fully summarizes the dependence between extremes of X_1 and X_2. (This is a so-called heterogeneous Poisson process model, with intensity density λ.)

26. For statistical modelling, a flexible parametric form for h is adopted, and it and the marginal univariate standardization are fitted by maximum likelihood. Since combinations of variables are treated in which one variable might be at an extreme level, whilst another might not be particularly large, the initial univariate marginal distributions in the model are described by the empirical distribution function in the main body of the data but by an extreme value form in the upper tail (refs. 20 and 21).

27. The secondary variables will be treated as concomitants attached to the points defined by the primary variables, and their conditional joint distributions modelled too, ie. conditional on extremes of the primary variables.

The structure function

28. The project must go beyond the description of multivariate extremes, however complex this first task may be. It is crucial to the resolution of any joint probability problem that a model is constructed to provide a fair representation of the way that the input variables influence the output variable (ref. 20). In some applications, such as the interaction of fluvial and tidal floods (refs. 22, 23 and 24), it is possible to construct a hydrodynamic model to represent the river between the upstream and seaward sites at which the primary inputs are observed. In particular cases it may be possible to summarize the outcome of many detailed modelling runs in a simple formula or graph which summarizes the way in which values of the primary variables interact to yield critical water levels at an intervening site of interest. Irrespective of whether a simple summary is possible, this is the <u>structure function</u> that is indispensable in tackling joint probability problems.

29. The structure function relates the value of a chosen output variable to two or more input variables on which <u>it</u>

depends. Note that the structure function does not inform whether there is any dependence <u>between</u> the primary input variables.

30. A simple example may assist. Consider that the output variable is the sum of throws of p dice (one throw per dice), and that the input variables (X_1, X_2, ... , X_p) are the individual throws. The structure function for this problem is $Y = X_1 + \ldots + X_p$. The equation applies irrespective of whether the dice are loaded or not; it simply represents that our concern is with the sum of the throws.

Sampling the output variable

31. Since input variables such as rain and wind are unpredictable, so are the corresponding values of the output variable, the reservoir water level L. The distribution of L will be determined by the distributions of, and patterns of dependence between, its input variables. (In the dice example above, if the dice are thrown independently, and there are more than about five or six of them, then the output variable will follow a normal distribution to a good approximation. But if we throw the dice in some way which guarantees that they all show the same number - perhaps by rolling the second and subsequent ones repeatedly until each shows the same number as the first - then the distribution of Y will give a probability of 1/6 to each of the particular totals p, 2p, ..., 6p, a quite different distribution.) As part of this inheritance by an output variable of the distributional properties of its inputs, the distribution of extreme values of the ouput variable will be governed by the joint probability distribution of extremes of the primary input variables (tpgether with the associated conditional extremes of secondary variables).

32. A method of sampling from the distribution of extreme water levels L is therefore to generate values from the extremal models governing the primary and secondary input variables, and then to evaluate the output variable (ie. structure function) at these values. The result will be an extreme observation on L with the correct probability distribution.

APPLICATION
The models

33. In the reservoir flood joint probability problem, a composite of models is required to define the structure function:
- a rainfall-runoff model to convert the rainfall (R) and catchment wetness (C) into flood runoff into the reservoir;
- a snowmelt subroutine, attached to the rainfall-runoff model, to simulate snowpack formation and melt;
- a reservoir flood routing model;
- a wave model to simulate wind-induced wave effects on the reservoir water level.

It is impractical within the present project to attempt to make

RESERVOIR DESIGN AND CONSTRUCTION

major advances in these modelling fields. Thus the approach being taken is to draw reasonable models from the relevant literature.

34. The choice of rainfall-runoff model and snowmelt subroutine must reflect the available data, while allowing the more important effects to be represented. One candidate is a nonlinear storage model plus a modified degree-day snowmelt formula (ref. 25) although a fuller description would be preferable. The choice of model to represent wind-induced wave effects is being sought in a relatively discordant literature; if there is an authoritative review, the authors have yet to find it! It is possible that the study will adopt the modified Saville method (ref. 26) which underlies the advice given in the ICE Guide. With regard to the representation of wind-induced surge effects, it is planned that the usual method of routing a flood through the reservoir will be modified to use a "slanted-pool" rather than "level-pool" formulation.

Data

35. The availability of data inevitably impinges on the design of the project. The interest in extremes dictates use of a substantial period of record. Because of the need to extract concomitant values of variables, continuous records of the key variables are required (as opposed to just peak values). While some progress can be made with daily rainfall data, it is desirable to have higher resolution information and this is essential when seeking to model the catchment/reservoir response to rainfall, wind and snowmelt. Because of the scale of investment in what is a complex study, it was appropriate to seek data of the highest quality.

36. The above requirements led to the purchase of 20 years of hourly data (1970-1989) for a premier Met. Office observing station at Eskdalemuir, in the Southern Uplands of Scotland. The data comprise rainfall accumulations, mean air temperatures, mean wind speed and average wind direction.

37. Initial work has appraised the quality of the data and investigated particular anomalies. Through reference to daily rainfall, and hourly "weather" observations, it proved possible to fill in short periods for which hourly rainfall readings were missing or discrepant. Following the Met. Office change to use of an unheated tipping-bucket raingauge linked to a solid-state event recorder, routine reconciliation of hourly and daily rainfall observations ceased in 1987. Consequently it was possible to derive a complete hourly rainfall record only for the 17-year period 1970-1986.

38. Wind is one of the most difficult hydrometeorological variables for which to achieve systematic records. The site at Eskdalemuir was initially well exposed to winds from most directions but afforestation in the early 1970s led to pronounced sheltering effects on the record, there being a suggestion in the data that the plantations matured as effective windbreaks in the early 1980s. Interpretation of the wind speed data is further complicated by an instrument change in 1980 and a major change in processing techniques in March

1981. Combined with shifts in regional wind patterns associated with climatic change or variability, it is proving difficult to achieve a consistent wind record for the study period.

39. The hourly air temperature data were found to be of generally high quality and required no specific treatment prior to analysis.

Catchment/reservoir systems to be simulated

40. The approach being taken is to use the Eskdalemuir data as the primary inputs to an array of hypothetical catchment/reservoir systems. The models used will be spatially lumped, so that very little catchment information will need to be specified; parameter values will be chosen to be typical of small upland catchments.

41. Representation of snow accumulation will necessarily be simple. A straightforward approach is to assume that precipitation at air temperatures below $0^{\circ}C$ occurs as snow, that at air temperatures above $3^{\circ}C$ occurs as rainfall, while a mixture occurs in between. A simple representation of altitudinal effects will be attempted using air temperature lapse rates and (possibly) wind speed growth rates.

42. Several hypothetical reservoirs will be considered, of contrasting sizes and shapes. For each, the lake will be re-oriented in a number of directions so that the influence of full or partial exposure to the dominant regional wind can be assessed. Reservoir fullness will be characterized through a hypothetical reservoir inflow calculation, based either on simulated runoff from the catchment or (more simply) on daily mean flow data scaled from a gauging station in the Esk basin.

43. The above approach will lead to a large number of hypothetical catchment/reservoir systems to consider. It is anticipated that attention will focus on particular examples which are different enough to test the hypothesis that a single set of design inputs may not suffice, but not so different as to be atypical of UK catchment/reservoir systems.

Generalization

44. Some generalization of conclusions from the above investigation will be assisted by other studies which place the rainfall, wind and temperature characteristics of Eskdalemuir in a wider context. However, it is recognized that the degree of dependence between input variables seen at Eskdalemuir could be idiosyncratic. If marked dependence is found, it may be necessary to extend the study to investigate data from other sites. Verification may also be required if the study finds that extremes of the primary input variables are statistically independent, simply because of the widely held view to the contrary!

CONCLUDING REMARKS

45. The DOE-funded Joint Probability Studies for Reservoir Flood Safety project is exploring the application of multivariate extreme value analysis techniques to the problem of choosing design inputs for reservoir flood safety appraisal.

RESERVOIR DESIGN AND CONSTRUCTION

46. With general regard to joint probability problems met in environmental engineering, there is an aspect that makes the authors pessimistic that general solutions will be readily achieved. It appears that the choice of statistical technique required is dependent on the relative importance of the primary input variables in their influence or joint influence on the output variable. Thus it is necessary to know the structure function, at least approximately or implicitly.

47. Further research may eventually lead to simpler techniques suitable for use by the non-specialist. For the present, it is perhaps enough for the practicing engineer to be wary of simple formulae that purport to offer the right combination of inputs to achieve a desired output rarity.

ACKNOWLEDGEMENTS
The Joint Probability Studies for Reservoir Flood Safety project is funded by the UK Department of the Environment (contract no. PECD7/7/365). The authors acknowledge the guidance and assistance provided by Dr Jonathan Tawn and Mr Saraleesan Nadarajah. University of Sheffield research on joint probability problems is also funded by the Ministry of Agriculture, Fisheries and Food. Meteorological data for Eskdalemuir were supplied by the Met. Office's Edinburgh Climate Office.

REFERENCES
1. INSTITUTION of CIVIL ENGINEERS Floods and Reservoir Safety: an engineering guide. ICE, London, 1978 and 1989 (2nd edition).
2. CLARKE C.L. and PHILLIPS J W Discussion of Floods and Reservoir Safety Guide and the Reservoirs Act 1975. Proc. Institution of Civil Engineers, 1984, pt. 1, vol. 76, 834-838.
3. REED D.W. and FIELD E.K. Reservoir flood estimation: another look. Institute of Hydrology, Wallingford, 1992, Report No. 114.
4. NATURAL ENVIRONMENT RESEARCH COUNCIL Flood Studies Report (five volumes). NERC, London, 1975.
5. INSTITUTE of HYDROLOGY Flood Studies Supplementary Reports. IH, Wallingford, various dates.
6. CUNNANE C. Statistical distributions for flood frequency analysis. World Meteorological Organization, Geneva, 1989, Operational Hydrology Report No. 33, WMO Publ. No. 718.
7. KUUSISTO E. Some thoughts on the design flood. UNESCO seminar on dam safety, Rovaniemi, Finland, August 1988.
8. KOVACS Z. Regional maximum flood peaks in southern Africa. Dept. of Water Affairs, Pretoria, 1988, Tech. Report No. 137.
9. ACREMAN M.C. Extreme rainfall in Calderdale, 19 May 1989. Weather, 1989, vol. 44, 438-446. [Discussion: vol. 45, 156-158.]
10. COLLINGE V.K., ARCHIBALD E.J., BROWN K.R. and LORD M.E. Radar observations of the Halifax storm, 19 May 1989. Weather, 1990, vol. 45, 354-365.

11. COLLIER C.G. Problems of estimating extreme rainfall from radar and raingauge data illustrated by the Halifax storm, 19 May 1989. Weather, 1991, vol. 46, 200-209.
12. ACREMAN M.C. and COLLINGE V.K. The Calderdale storm revisited: an assessment of the evidence. Proc. British Hydrological Society Symp., Southampton, September 1991, 4.11-4.16.
13. REED D.W. and STEWART E.J. Focus on rainfall growth estimation. Proc. British Hydrological Society Symp., Sheffield, September 1989, 3.57-3.65.
14. GOLDEN J.H. and SNOW J.T. Mitigation against extreme windstorms. Reviews of Geophysics, 1991, vol. 29, 477-504.
15. DALES M.Y. and REED D.W. Regional flood and storm hazard assessment. Institute of Hydrology, Wallingford, 1989, Report No. 102.
16. INSTITUTE of HYDROLOGY Collective risk assessment for sites sensitive to heavy rainfall. IH, Wallingford, 1988, Flood Studies Supplementary Report No. 18.
17. TAWN J.A. and VASSIE J.M. Extreme sea levels: the joint probabilities method revisited and revised. Proc. Institution of Civil Engineers, 1989, pt. 2, vol. 87, 429-442.
18. TAWN J.A. Estimating probabilities of extreme sea levels. Applied Statistics, 1992, vol. 41, 77-93.
19. COLES S.G. and TAWN J.A. Modelling multivariate extreme events. J. Royal Statistical Society, 1991, ser. B, vol. 53, 377-392.
20. COLES S.G. and TAWN J.A. Statistical methods for multivariate extremes: an application to structural design. To appear in J. Royal Statistical Society, ser. B.
21. JOE H., SMITH R.L. and WEISSMAN I. Bivariate threshold methods for extremes. J Royal Statistical Society, 1992, ser. B, vol. 54, 171-183.
22. THOMPSON G. and LAW F.M. An assessment of the fluvial tidal flooding problem of the river Ancholme, UK. Proc. International Union of Geodesy and Geophysics Symp. on Assessmnet of Natural Hazards, Hamburg, August 1983.
23. BERAN M.A., JONES D.A., HARPIN, R. and SMITH A.P.L. Stage frequency estimation for the tidal Thames. International Association of Hydraulic Research, Paper to Symp. on Stochastic Hydraulics, Birmingham, August 1988.
24. WORLD METEOROLOGICAL ORGANIZATION Hydrological aspects of combined effects of storm surges and heavy rainfall on river flow. WMO, Geneva, 1988, Operational Hydrology Report No. 30, WMO Publ. No. 704.
25. REED D.W. A review of British flood forecasting practice. Institute of Hydrology, Wallingford, 1984, Report No. 90.
26. INSTITUTION of CIVIL ENGINEERS Reservoir flood standards. ICE, 1975, Floods Working Party discussion paper.

Trapping efficiency of reservoirs

C. E. REEVE, BSc, PhD, HR Wallingford

SYNOPSIS. A numerical reservoir sedimentation model is used to investigate the factors which determine the trapping efficiency of reservoirs. Procedures for the calculation of reservoir trapping efficiency are developed.

INTRODUCTION
1. It is the nature of reservoirs to cause a reduction in both the velocity of flow in a river and the water surface slope. This reduces the capacity of the river to transport sediment and encourages the deposition of sediment in the reservoir. The accumulation of sediment reduces the amount of water storage available and hence the utility of the reservoir. In extreme cases effectively all the useful storage may be lost due to sedimentation. The rate at which sediment accumulates has a major impact on the useful life of a reservoir and so is significant in assessing the economics of a proposed reservoir. There is, therefore, the need to be able to assess sedimentation when a reservoir is being planned.
2. In considering the impact of sedimentation on a storage scheme it is important to know the loss of available storage after a given time period as this directly affects the yield of the reservoir. The distribution of the sediment deposits affects the stage / storage curve and so may have an impact of the operating rules of the reservoir. For the designer, therefore, there is a need to be able to predict both the amount and distribution of sedimentation.

IMPACT OF RESERVOIR SEDIMENTATION
3. The adverse effects caused by reservoir sedimentation may include,

- a) a reduction in the storage available and hence a reduction in the yield provided by the reservoir,
- b) degradation downstream of the dam. This may threaten structures associated with the dam and lead to problems at structures further downstream such as bridges and intakes,

c) deposition at the head of the reservoir leading to an increase in flood levels in the contributing streams upstream,
d) increased evaporation losses for a given storage volume.

TOOLS FOR INVESTIGATING RESERVOIR SEDIMENTATION

4. Until recently reservoir sedimentation could only be assessed using simple, empirical methods. To estimate the volume of deposited material the notion of trapping efficiency was introduced. The trapping efficiency of a reservoir is defined as a ratio of the quantity of deposited sediment to the total sediment inflow.

5. References 1 to 3 provided simple graphical means to determine trapping efficiency and these have been used extensively. Since, however, the trapping efficiency must depend upon the sediment size, the flow through the reservoir, the distribution of flows into the reservoir and the way that the reservoir is operated, it follows that such estimates of trapping efficiency can only provide approximate values which may, on occasions, be seriously in error.

6. More recently, however, numerical models of sedimentation in reservoirs have been developed and these have provided a means of studying reservoir sedimentation in much greater detail. Unfortunately such models tend to be complicated to use and expensive to apply.

INSTANTANEOUS TRAPPING EFFICIENCY

7. Trapping efficiency may be measured by measuring any two of the following three quantities; the incoming sediment load, the sediment deposited in the reservoir and the outgoing sediment load (Ref. 1). More frequently the volume of sediment deposited in the reservoir is determined by comparing surveys of the reservoir separated by a number of years. Thus the trapping efficiency quoted in work such as that by Brune represents the trapping efficiency determined over a number of years.

8. It is possible, however, to define an instantaneous trapping efficiency, that is, the value of the ratio of deposited to incoming sediment for a particular combination of inflow, outflow and reservoir level. When reservoir levels are low and the outflow is zero this may be large but when reservoir levels are high and the outflow is large the instantaneous trapping efficiency may be small. Thus the trapping efficiency can be regarded as varying throughout the year as the inflow and reservoir level varies. The trapping efficiency determined over a number of years is thus the instantaneous trapping efficiencies averaged in some appropriate way over a period of a number of years.

9. The notion of an instantaneous trapping efficiency is useful for the application of numerical models to

the problem of reservoir sedimentation. While it is
difficult to deal in some aggregate way with inflows
and outflows over a number of years it is much easier
to determine instantaneous conditions, perhaps on a
daily basis, and then perform an averaging process on
the results to determine the trapping efficiency over a
year or number of years.

DESCRIPTION OF STUDY
10. A numerical reservoir sedimentation model was
used to investigate the trapping efficiencies of a
number of reservoir geometries and characteristics in
order to improve the understanding of the factors
determining trapping efficiency. This involved
changing a number of variables such as reservoir
length, width, depth, discharge and determining the
corresponding trapping efficiencies. From this
information a procedure was developed to determine the
instantaneous trapping efficiency of a reservoir.
11. The numerical reservoir sedimentation model used
for this study is described in Reference 4.

USEFULNESS OF DEVELOPED PROCEDURE
12. The developed procedure can be used to consider
the trapping efficiency of a reservoir at a number of
different instants during a year so that the average
yearly trapping efficiency could be determined.
13. By utilising this procedure the engineer will be
able to quickly estimate the trapping efficiency, and
thus active life, of existing reservoirs.
14. The procedure will also be of use when assessing
sites for new reservoirs. Using the procedure the
trapping efficiencies of reservoirs at different sites
can be assessed. This will enable certain sites to be
eliminated at an early stage of the appraisal process
thus reducing the number of sites which must be
analyzed in detail.
15. Once a particular site and reservoir has been
selected, a numerical reservoir sedimentation model can
be used to identify the location of deposition together
with subsequent changes in bed level and loss of
storage.

NUMERICAL SIMULATIONS
16. A series of numerical experiments were carried
out with different reservoir geometries, flow
conditions and sediment fall velocities.
17. The simulated reservoirs had idealised
geometries. In plan they consisted of an isosceles
triangle with the river inflow at the vertex and the
dam at the opposite side. In longitudinal profile the
reservoir took the form of a right angled triangle with
the right angle at the dam crest. The length, mean
width and mean depth were used to characterise the
geometry of these idealised reservoirs.
18. The fall velocities of sediments were used to
represent variations in sediment size. The single

particle still water fall velocity of different size sediments, at different temperatures, are given in Reference 5.

PRESENTATION OF RESULTS
19. The trapping efficiencies of the simulated reservoirs were calculated from the results of the numerical experiments. The trapping efficiency was taken to be the percentage of the sediment input at the upstream limit of the reservoir that was deposited in the reservoir. These trapping efficiencies are presented on non-dimensional diagrams. In this way the results are generally applicable and independent of the actual geometry, flow condition or sediment fall velocity.

DIMENSIONAL ANALYSIS
20. There are five dimensional variables which should be considered when investigating reservoir sedimentation: omega, Q (or v), b, d and L. Representative values for b and d must be chosen to characterise a reservoir's geometry; the mean values of b and d were chosen as the representative values. For the idealised reservoirs studied in the numerical simulations the mean values of b and d are the values of these variables at the point mid way along the length of the reservoir.

21. The five dimensional variables were combined to give three non-dimensional variables,

$$Z1 = \frac{bd\omega}{Q}$$

$$Z2 = \frac{b}{L}$$

$$Z3 = \frac{d}{L}$$

where : b is the average reservoir width (m)
d is the average reservoir depth (m)
L is the reservoir length (m)
ω is the sediment fall velocity (m/s)
and Q is the discharge (m^3/s)

22. These variables were used on the non-dimensional plots of trapping efficiency.

EFFECT OF SEDIMENT FALL VELOCITY ON INSTANTANEOUS TRAPPING EFFICIENCY
23. Increases in sediment fall velocity, and thus increases in Z1, are expected to result in an increase in trapping efficiency. The change from zero to 100% trapping efficiency occurs over a two to three order of

magnitude change in Z1 and thus a two to three order of magnitude change in sediment fall velocity. A one order of magnitude change in sediment size results in a two order of magnitude change in sediment fall velocity.

24. At trapping efficiencies below 20%, large changes in sediment fall velocity result in only small changes in trapping efficiency. At trapping efficiencies from 20% to 90%, the trapping efficiency is more sensitive to sediment fall velocity. The increase in trapping efficiency from 20% to 90% occurs over a one order of magnitude change in fall velocity. Thus, a five times increase in sediment size can result in an increase in trapping efficiency from 20% to 90%. For trapping efficiencies in excess of 90%, the trapping efficiency again becomes less sensitive to sediment fall velocity.

25. The dependence of trapping efficiency on sediment fall velocity and thus sediment size makes it imperative that great care be taken in the selection of a representative sediment size when estimates of the reservoir trapping efficiency are to be made.

EFFECT OF MEAN DEPTH TO LENGTH RATIO (Z3) ON INSTANTANEOUS TRAPPING EFFICIENCY

26. For a given value of Z1, the reservoir trapping efficiency increases as the mean depth to length ratio decreases. Since changes in Z3 values are due to changes in reservoir length, this implies that for a constant mean depth the trapping efficiency of a reservoir increases as the reservoir length increases.

27. Numerical experiments were carried out with reservoirs that had the same Z3 values but different absolute mean depths and lengths. These experiments showed that the mean depth to length ratio, Z3, is important rather than the absolute length or mean depth.

EFFECT OF MEAN WIDTH TO LENGTH RATIO (Z2) ON INSTANTANEOUS TRAPPING EFFICIENCY

28. For a given value of Z1, the trapping efficiency of a reservoir is independent of the mean width to length ratio. The reason for this lack of variation is that the flow in a reservoir is one dimensional along the axis from the river outlet to the dam. Hence it is to be expected that variation in mean width relative to length will have minimal effect on trapping efficiency.

29. Numerical experiments were carried out for reservoirs with a large number of different mean width to depth ratios. Provided the mean depth to length ratio was not altered, all the trapping efficiency curves plotted at the same position on the Z1-trapping efficiency plot. This confirms the result outlined above that the trapping efficiency is independent of the mean width to length ratio.

RESERVOIR DESIGN AND CONSTRUCTION

EFFECT OF DISCHARGE ON INSTANTANEOUS TRAPPING EFFICIENCY

30. Analysis similar to that described above was carried out to investigate the effect of discharge on instantaneous trapping efficiency. The results of this analysis implied that the trapping efficiency of the simulated reservoir is dependant on the discharge. Since the method of presentation of the results was chosen in order to eliminate dimensional dependence this result is unreasonable; it implies that the non-dimensional variable Z1, which should have eliminated the dimensional dependence of discharge, has been incorrectly chosen.

Shear velocity correction applied to Z1

31. The apparent dimensional dependence on discharge is due to the value of sediment fall velocity that has been used to calculate Z1. In calculating the variable Z1 the single-particle still-water fall velocity is used. In the numerical model this fall velocity is modified since if the shear velocity is above a given threshold, the turbulence generated is sufficient to maintain all the material in suspension. This threshold value is known as the critical shear velocity, v_{*CRIT}. As the shear velocity reduces the fall velocity tends to the still-water value. It is this modified fall velocity that is used to calculate the trapping efficiency in the numerical model. This inconsistency between the fall velocity used in the model and that used to calculate the variable Z1 is the reason for the apparent dependence of trapping efficiency on discharge.

32. In order to overcome this problem a modified form of the non-dimensional variable Z1 was used as follows,

$$Z1 = \frac{bd\omega\left(1 - \frac{v_*}{v_{*CRIT}}\right)}{Q}$$

where v_* is the representative shear velocity and v_{*CRIT} is the critical shear velocity above which no deposition occurs.

33. To be consistent with the choice of representative values for the width and depth the representative value of shear velocity has been chosen to be the value that occurs at the mid-point of the reservoir. In HR's experience the critical value to shear velocity above which no deposition occurs is 0.01 m/s.

34. At small discharges, the shear velocity is small relative to the critical shear velocity and the correction applied to Z1 is negligible. Thus at low discharges there is no change to Z1.

35. At higher discharges the shear velocity becomes

larger relative to the critical value, and the correction applied to Z1 becomes significant. This means that at high discharges the modified value to Z1 is smaller than the original value.

36. When the modified form of the non-dimensional variable Z1 is used, the apparent dimensional dependence on discharge is eliminated.

37. The procedure outlined above works very well providing the shear velocity is less than the critical shear velocity throughout the reservoir. As soon as this condition is violated the procedure breaks down.

Effective length correction

38. As the discharge increases so does the shear velocity. As this happens shear velocity at the upstream end of the reservoir increases. Eventually a value in excess of the critical shear velocity occurs. As the discharge increases further, the position at which the critical shear velocity is exceeded encroaches further down the reservoir. No deposition occurs when the shear velocity exceeds the critical value. Hence if the critical shear velocity is exceeded at any point in the reservoir, the active length of the reservoir will be less than the actual length of the reservoir. Under these circumstances the active length of the reservoir needs to be used when assessing the trapping efficiency. In this paper the active length of a reservoir is referred to as the reservoir effective length. When the effective length is less than the absolute reservoir length the shear velocity correction to Z1 still needs to be applied.

39. The mean width, depth and shear velocity used to calculate Z1, Z2 and Z3 must relate to the active length of reservoir. When the active length is less than the total length, the mean width and depth of the active reservoir will, in general, be greater than the mean width and depth of the total reservoir.

PROCEDURE FOR ESTIMATING RESERVOIR TRAPPING EFFICIENCY WITH KNOWN VARIATION OF SHEAR VELOCITY

40. Based on the results of the analysis described above a procedure has been developed which accurately determines reservoir trapping efficiency without using the numerical model. In order to apply this procedure a backwater calculation must be carried out to determine the variation of shear velocity throughout the length of the reservoir.

41. The procedure comprises two parts :

(1) Determine Z1 and Z3.
(2) Using the calculated values for Z1 and Z3 estimate the trapping efficiency using the curves shown in Fig. 1. Linear interpolation can be used to find the trapping efficiency of reservoirs for which a Z3 is not shown.

RESERVOIR DESIGN AND CONSTRUCTION

PROCEDURE FOR ESTIMATING TRAPPING EFFICIENCY WITHOUT CARRYING OUT A BACKWATER CALCULATION

42. The procedure described above for estimating the instantaneous trapping efficiency of a reservoir requires the distribution of shear velocity along the length of the reservoir to be known. This information is required in order to determine the effective length of the reservoir. For all practical problems such a calculation requires software on a computer; this may not always be available. A modified procedure which involves hand calculation of the effective length and shear velocity has been developed.

Estimating effective length

43. To be consistent with the approach taken to estimate instantaneous trapping efficiency the procedure for estimating effective length must be non-dimensional.

44. One obvious non-dimensional variable to use is the ratio of effective length to absolute length; this is referred to as the effective length ratio and is defined as,

$$Z4 = \frac{EL}{L}$$

where: EL is the effective length.

45. The effective length of a reservoir will decrease as the discharge increases; hence discharge must be included in a second non-dimensional variable. The effective length is defined as the distance from the dam to the point where the critical shear velocity is exceeded. Clearly this threshold value must also be included in this second non-dimensional variable. In order to obtain a non-dimensional variable incorporating discharge and critical shear velocity, parameters with dimensions L^2 are required; the square of the mean width was chosen. Hence the second non-dimensional variable was:

$$Z5 = \frac{Q}{bbv_{*CRIT}}$$

46. In order to establish the variation of the effective length ratio, Z4, with Z5 a certain effective length ratio was assumed; the discharge that gave this effective length ratio was estimated and the corresponding Z5 calculated.

47. The variation of the effective length ratio with Z5 was established for a number of reservoirs. This analysis revealed that the form of the relationship

between these two non-dimensional variables depended on a further non-dimensional variable defined as,

$$Z6 = \frac{b}{d}$$

48. The effective length ratio, obtained from values of Z5 and Z6, depends on the value of roughness length.

Estimating the shear velocity v_*.
49. Provided the mean width, mean depth, discharge, critical shear velocity and roughness length are known the procedure outlined above can be used to estimate the effective length ratio of a reservoir. Since the absolute reservoir length will also be known the active of a reservoir can then be calculated.

50. In order to estimate the shear velocity, the Colebrook-White equation and is applied to a reservoir which has the same values for Z3 and Z6 as the real reservoir and an effective length calculated by the procedure outlined above. The mean width and mean depth of this reservoir, referred to here as the effective mean width and effective mean depth can be defined as,

$$EMW = ELR \times b$$
$$EMD = ELR \times d$$

where : EMW is the effective mean width
and   EMD is the effective mean depth

51. The mean slope of this reservoir can be calculated using the Colebrook-White equation in the following form,

$$S = \frac{Q^2}{(EMW \times EMD)^2 \left[32g \times EMD \left(\log_{10} \frac{k_s}{14.8\, EMD} \right)^2 \right]}$$

52. The shear velocity v_* can then be calculated using,

$$v_* = \sqrt{g \times EMD \times S}$$

Modified procedure for estimating reservoir trapping efficiency
53. The techniques outlined above allow hand

RESERVOIR DESIGN AND CONSTRUCTION

calculation of the effective length and shear velocity for a reservoir. Using these a modified, desk, procedure for estimating reservoir trapping efficiency has been devised. This procedure comprises four parts,

(1) Determine the effective length of the reservoir
(2) Determine the shear velocity
(3) Determine Z1 and Z3
(4) Using the calculated values of Z1 and Z3 estimate the trapping efficiency using the curves shown in Fig. 1. Linear interpolation can be used to find the trapping efficiency of reservoirs for which a Z3 value is not given.

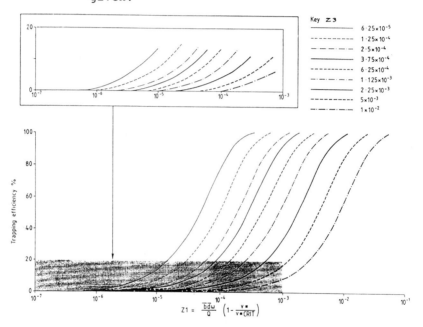

Fig. 1 Trapping efficiency as a function of Z1 and Z3

APPLICATION OF DESK PROCEDURE TO REAL RESERVOIRS
54. The analysis described in this paper has been carried out on idealised reservoir geometries. In reality reservoirs do not have this idealised shape. In order to assess the accuracy of the derived procedure when applied to real reservoirs the procedure described above was applied to four real reservoirs. The estimated trapping efficiencies were compared with the results of the numerical model simulations of the same reservoirs with the same flow and sediment conditions and also the trapping efficiencies estimated by using Brune curves. The four real reservoirs chosen

for study were reservoirs in Zimbabwe which had previously been analyzed at HR. These reservoirs had geometries that were increasingly divergent from the idealised geometry.

55. Table 1 compares the trapping efficiency estimates by the three methods for each of the four reservoirs. The table shows that for each reservoir the developed procedure (TRAPEF) and model estimates of trapping efficiency are close to each other and that both estimates are within the confidence limits associated with the Brune curves.

Table 1 Comparison of Brune curves, numerical model and new procedure (TRAPEF) trapping efficiencies for real reservoirs with variable discharge

Reservoir	Brune curve trapping efficiency	Numerical model trapping efficiency	TRAPEF trapping efficiency
Gozho	95%+4%	94%	99%
Siya	98%+2%	97%	100%
Manjirenji	98%+2%	97%	96%
Maynard	60%+12%	69%	66%

CONCLUSIONS

56. Procedures have been developed for estimating instantaneous trapping efficiency of a reservoir. These procedures can be used when a backwater calculation has been carried to estimate the variation of shear velocity along the reservoir and also in the absence of such calculations.

57. When the procedures are applied to real reservoirs estimates of trapping efficiency within the confidence limits of Brune curves are given.

REFERENCES
1. BRUNE G.M. Trapping efficiency of reservoirs. Trans AGU. Vol. 34. No. 3, 1953.
2. CHURCHILL M.A. Discussion of Gottschalk's paper on Analysis and Use of reservoir sedimentation data. Proc. FIASC, USDA (Washington), 1948.
3. GOTTSCHALK L.C. Analysis and Use of reservoir sedimentation data. Proc. FIASC, USDA (Washington), 1948.
4. MEADOWCROFT I.C., BETTESS R.and REEVE C.E. Numerical Modelling of Reservoir Sedimentation. Proceedings of Conference on Water Resources and Reservoir Engineering, University of Stirling, 1992.
5. GIBBS R.J., MATTHEWS M.D. and LINK D.A. The relationship between sphere size and settling velocity. Journal of Sedimentary Petrology. Vol. 41 (1). 1971.

Small embankment-type reservoirs for water supply and amenity use

B. H. ROFE, C. G. HOSKINS and M. F. FLETCHER, Rofe, Kennard and Lapworth

SYNOPSIS. An increasing number of small embankment-type reservoirs will be constructed in the future in response to the growing trends for increased landscaping/amenity lakes and changing agricultural practice. Whilst some will be sufficiently large to come within the scope of the Reservoirs Act 1975, most are likely to fall in the capacity range of 2500 - 25000m^3 stored above ground level. Many design aspects will be similar to those of reservoirs within the Act, but the reduction in size may change the emphasis; environmental constraint may also be significantly different. Present guidance for these small reservoirs in the existing MAFF guide has become outdated and too narrow in its approach. Thus CIRIA have recently commissioned preparation of a new guide for issue in the near future.

BACKGROUND

1. It is estimated that about 2500 reservoirs come within the scope of British reservoir safety legislation, with perhaps a similar or greater number of smaller above ground water storage reservoirs not subject to any control. Few large reservoirs for public water supply, river regulation, flood alleviation or hydro-electric power are likely to be constructed in this country in the forseeable future. It is estimated, however, that a substantial number of small reservoirs storing less than 25,000mm^3 will be constructed annually over the next few years.

2. In the past small reservoirs were constructed primarily for irrigation and other farm purposes. Nowadays a wider use is evident, with functions as diverse as fish farming, fishing, fire fighting, conservation and amenity use. Increasingly, small reservoirs have been included as part of commercial, industrial and leisure developments to provide a water feature. A significant number of small reservoirs fulfil a dual or multiple role.

3. Reservoirs designed, built and maintained for the Water Service Companies (formerly the Water Authorities), British Waterways Board and many other large concerns have generally

been substantial and have been handled by experienced in-house staff or specialist consulting engineers. Many of the smaller reservoirs have been developed by persons who are unfamiliar with accepted techniques of dam and water engineering and the standards of design and construction sometimes have been far from adequate. Often planning and design has been undertaken by contractors, landscape architects, or other practitioners with limited knowledge of dam and water engineering and an inadequate understanding of the hydrological aspects. Construction has sometimes been carried out by contractors with little experience in reservoir construction. Earthworks and fill placing have been treated just as an uncontrolled muckshifting exercise without effort being made to achieve adequate fill placing and compaction standards. Standards of supervision have often been limited or absent whilst contractual arrangements have frequently been undefined and left much to be desired. Thus well intentioned efforts by the client to obtain a low cost reservoir may result in high construction costs, operation and maintenance problems, environmental insensitivity or legal problems.

NEED FOR GUIDANCE
4. A need therefore exists for a guidance document on all aspects of small water storage reservoirs from conception, through the design and construction stages, to maintenance. A previous guide (Ref 1) was produced by the Ministry of Agriculture, Food and Fisheries (MAFF) in 1967 and concentrated essentially on farm irrigation reservoirs. This was last revised in 1977 and is now in need of updating to take account of current practice and to allow for the increased usage and wider functions of these small water storage reservoirs.

5. An updated guidance document is to be produced by CIRIA to give information on the current technical and other aspects associated with the creation of a small embankment type reservoir. Rofe, Kennard and Lapworth are acting as a research contractor for the preparation of the guide, whilst funding for the project has been provided by the Department of the Environment, the Water Service Companies and the Forestry Commission.

FACTORS TO BE INCLUDED
6. The design and construction of small reservoirs can be affected by most of the problems influencing larger reservoir construction, albeit on a reduced scale. Certain features may also necessitate particular consideration for a small reservoir development which would often be less critical for a larger reservoir. In the past, such features which might have a significant effect on the design and construction have not always been appreciated at a sufficiently early stage, leading to difficulties and increased costs at a later date.

Thus clear indication of the major criteria and site specific
constraints to be developed form an essential part of the
guide. Good practice in many areas necessarily involves
judgement and experience and a guide of this nature can only
draw attention to the various aspects which need to be taken
into consideration or where specialist advice should be
sought.

7. The construction of a reservoir is likely to require
planning permission and it is necessary to discuss this and
other statutory aspects with the relevant body at an early
stage. Certain licences and consents are also normally
required from the National Rivers Authority in England and
Wales for river abstraction or impounding and obstructing a
watercourse and these must be obtained prior to construction.
Requirements may differ in Scotland and Northern Ireland and
specific advice should be sought in these areas.

8. The increasing importance of the environmental aspects
must be fully studied during the early stages of planning and
design to allow the full effects of the reservoir to be
assessed. Although some changes to the existing environment
must be accepted there is considerable opportunity for
potential landscape enhancement and the creation of
supplementary habitats for flora and fauna. It is important
that these should be taken into account in the design and
construction.

9. It is anticipated that the final document will be well
illustrated, both with figures and photographs and will
contain practical design and construction details.
Sufficient detail will be provided to allow the relatively
inexperienced user of the guide to gain some basic knowledge
and understanding of the principles involved in all stages of
small embankment-type reservoir creation. More detailed
information will be included in a series of appendices,
together with sources of information and specialist advice.
Some technical terms will differ from those in everyday use
and a glossary containing definitions of these and other key
technical terms used will be a necessary part of the guide.
In general, technical terms and notation will conform to
civil engineering convention.

SCOPE
10. The guide will be aimed at the client or engineer who
has limited experience in the field on small reservoir
construction. It will attempt to explain the procedures and
principles which form good practice through all the phases of
conception, planning, design, construction, maintenance and
remedial works. In particular, the guide will seek:-

- to identify the engineering, economic and environmental
  factors to be taken into account in the planning and

design stage, and to describe good design practice. Careful consideration must be given to the safety of the structure, protection of the public, reliability of supply and environmental acceptability
- to promote good standards of construction
- to draw attention to problems which have in the past detracted from the effective functioning of low-cost water storage reservoirs and, where appropriate, to indicate remedial measures
- to set out a practical and cost-effective approach to maintenance.

11. The guide generally will apply to both impounding and non-impounding small water reservoirs, although certain sections are not relevant to the latter. It will not deal, however, with the following:-

- reservoirs with a capacity in excess of $25,000m^3$
- masonry and concrete dams or service reservoirs
- hazards to downstream areas as a result of controlled or uncontrolled discharge from the reservoir
- tailings and slurry lagoons.

12. River training embankments and sea defences will also be beyond the scope of the guide. Many of the aspects discussed, however, are applicable, although certain features require a different approach (e.g. wave protection on river training embankments, overflow works on tailings/slurry lagoons).

MAJOR CHANGES OR EXTENSIONS TO PREVIOUS GUIDE

13. The proposed guide is presently at a final draft stage with a view to production later in 1992. Detailed comment on the various sections is premature at the present, but the major changes and extensions to the previous guide will include:-

(a) the need for an effective and adequate, yet simple, site investigation is stressed, including practical comments on the required degree and methods of ground investigation

(b) adequate and early consideration of the environmental aspects, including the opportunities for landscape and conservation enhancement. It should be noted that one planning authority has stated that a limited environmental assessment might be necessary as part of the planning procedure

(c) increased discussion of the hydrology which has been divided into two sections on yield and reservoir operation and a second section on floods, waves and reservoir safety. It is envisaged that the design flood assessment would interface with the rapid method given by the Institution of Civil Engineers (Ref 3)

with allowance for soil type, catchment slope and urban development on the catchment. The maximum inflow would be routed by a simple method and a multiplier applied to assess the design flow over the overflow arrangements
(d) a change of phraseology for the overflow arrangements, with a main overflow designed for a specified percentage of the maximum flood inflow with the remainder passed over an auxiliary spillway. This may be a grass spillway designed in accordance with Hewlett et al (Ref 2)
(e) comments on the use of geotextiles
(f) revision of materials and methods for lining reservoirs, including the use of geomembranes
(g) drainage and seepage control measures
(h) recommended slopes for various fill material and foundation conditions assessed on a conservative basis
(i) recommended planting details for trees and other vegetation, including constraints adjacent to the crest, drains, auxillary spillway, etc
(j) brief references to the recommended contractual arrangements for reservoir design and construction and for any specialist advisers
(k) the need for regular visual monitoring and any necessary maintenance work
(l) comment and advice on remedial work on recently constructed embankments, with discussion of their applicability to existing reservoirs.

OTHER ASPECTS
14. Despite changes in legislation in recent years and the formation of the National Rivers Authority as successor to the Water Authorities, it is apparent that some small reservoirs are continuing to be built without the necessary consents and licences.

15. Typically these are the very reservoirs where the technical standards and degree of competence are least. Whilst, in theory, their construction is a misdemeanor and powers exist to have them removed, it is usually necessary to prove a definite hazard is present. This is not straightforward and most are permitted by default. More stringent powers to control their creation appear necessary in conjunction with the updating of the guide.

REFERENCES
1. MAFF Water for irrigation. Bulletin 202, HMSO, London, 1977.
2. Hewlett H.W.M., Boorman L.A. and Bromley M.E. Design of reinforced grass waterways. CIRIA Report 116, CIRIA, London, 1987.
3. Floods and reservoir safety: an engineering guide. ICE, London, 1989.

Rock for dam face protection and the CIRIA/CUR manual on rock in coastal engineering

J. E. SMITH, BSc, FICE, Babtie Shaw and Morton, and J. D. SIMM, MEng, BSc, MICE, Robert West and Partners

SYNOPSIS. The Dutch group CUR and the British organisation CIRIA have recently published a Manual on the Use of Rock in Coastal and Shoreline Engineering. Whilst primarily intended for engineers concerned with marine works, there is much of value to the dams engineer. This paper reviews the current state of the art in the design of rock or "rip-rap" protection presented comprehensively by the manual as it affects dams. It also highlights some of the unexpected and interesting features that came to light during preparation of the manual.

INTRODUCTION
1. Over the last few years there has been a large increase in knowledge and understanding of wave processes and also their interaction with rock faced boundaries. There became a need for an up-to-date publication to draw all this knowledge together in the form of a design manual (Ref. 1). Existing contact and dialogue between Dutch and British engineers led naturally to a combined project in which their skills were brought together to create the new manual on using rock in coastal and shoreline protection.
2. This paper sets out to present to the reader interested in dams some of those aspects of this manual that concern him, and provide commentary on areas of particular concern, including highlighting new features brought out during the manual's preparation.

THE MANUAL
3. The manual has been written to take the reader logically through all the stages from initial concept planning, to data collection on to design and then finally maintenance.
4. It is unlikely that the engineer concerned with protection to the face of dams will need to study the whole manual. Many features dealt with in detail are not applicable, whilst others will have been included already within the normal design procedures for the embankment and reservoir themselves.

RESERVOIR DESIGN AND CONSTRUCTION

5. Other areas will be of significance and of direct interest. This paper highlights some of the major items.

6. The section of the manual on dams forms a small part of the chapter devoted to Structures and is provided to form a point of reference for dams engineers. The full list of chapter headings in the manual is listed below:-

  1. Introduction
  2. Planning and Designing
  3. Materials
  4. Physical Site Conditions and Data Collection
  5. Physical Processes and Design Tools
  6. Structures
  7. Maintenance

There are also detailed appendices including model specifications, materials test procedures, measurement for payment, and monitoring.

PURPOSE OF PROTECTION

7. Face protection to a dam is principally to prevent the face being eroded by wave action. The risk posed by wave damage from any one incident to the integrity of a dam itself is probably low. However, continual attack and the normally high hazard rating of a dam means that any damage must be contained. Face protection provides this containment by providing a barrier to damage that ideally lasts as long as the dam. Hence such facing must be virtually maintenance free and - in its function - indestructable. It is unlike many sea defence systems, where some damage, or profile adjustment, is acceptable.

WINDS AND WAVES
Physical Features and Techniques

8. There is a large difference here between conditions at dams and in the sea. There are probably four major features:-

  (a) For new dams, there is no pre-existing site wave data, unless the dam has only a minor modifying effect on an existing lake. Hence waves must be based on (predicted) winds.
  (b) Local topographic features can influence winds speeds and directions, sometimes significantly, again affecting wave patterns. (Ref.2)
  (c) The presence of the dam and lake may modify the local wind behaviour, and hence waves.
  (d) Water level changes usually occur slowly. They are almost invariably out of step with wind storms. Hence storm attack is primarily at one level. (Ref.2)

9. The manual gives some guidance on where to seek meteorological data to overcome these problems. Seldom can satisfactory sources be found, however, and considerable ingenuity is needed to fill the gaps. There is a severe lack of data on winds, or waves, on reservoirs or lakes in the U.K. Previous work by Hydraulics Research, searches for this manual, and current searches for the work on blockwork protection have all shown we know little about strong or extreme winds on lakes. This is unfortunate, as now it is relatively easy and cheap to install self recording weather stations at a dam, and the records obtained would prove valuable in increasing our understanding. At present it seems that adhoc judgment on local wind speed and direction needs to be applied, and further research in this area is required.

10. There seems to be enough background knowledge to assume that steep sided valleys will align the wind locally to blow directly along the lake. From the 1987 gales, the converse appears true for gentler topography, in that valley direction has little influence on wind direction.

11. Having made the judgement on the probable design wind conditions, this has to be translated into prediction of wave height. The best system for reservoirs seems to be the JONSWAP spectrum. This is based on North Sea data, and so fetch and exposure and wave spectrum are different from lakes. Nevertheless, it appears to give reasonable answers when combined with Donelans' work on fetch lengths. He proposed that fetch length should be measured along the wave direction and not wind direction. (Ref. 3)

12. This presupposes that the wave direction is known or can be calculated. For long, narrow water bodies the wave direction will probably be along the water body axis for a wide range of wind directions. For fetches of general shape, Donelan assumed that the predominant wave direction was that which produces the maximum value of wave period (for a given wind speed). From the JONSWAP formula, the maximum wave period is obtained when the product:

$$\cos(\phi_* - \theta)^{0.4} F_{*0.3}$$

reaches its peak within the range $/\phi_* - \theta/ = 90°$. Here θ is the predominant wave direction and F_θ is the fetch along that direction. For any irregular shoreline, and a given wind direction, the value of ϕ satisfying this condition can only be determined by trial and error.

13. Where wind data is estimated, and direction is not accurately known, it is suggested that the refinement of reducing windspeed to its component acting in the direction of the maximum period waves be omitted.

14. The Saville method of fetch calculation (Ref. 4) is also discussed. This is probably the method most familiar to dams engineers. Whilst in itself reliable, it is important to note that wave prediction from Saville's fetch should only be carried out using the S.M.B. method. It has been found that combining Saville with JONSWAP approach results in a serious underestimate of wave height. It is consequently unsafe to use this combination.

15. The prediction of significant wave height, H_s, by the JONSWAP approach is plotted in Figs. 1 and 2 taken from the manual. It can be seen that a fully developed sea is never likely to be reached on a reservoir. Likewise the significant wave height is never likely to exceed 2 m in the U.K.

16. It is also instructive to see that, apart from very large reservoirs, it does not take long for the peak wave period, T_p, to develop. Maximum wave height can build up very quickly and storm winds of short duration only are needed to produce a damaging sea. It follows therefore that it is advisable to check the effect of shorter duration strong winds and possible associated higher waves as well as longer period attack of 1 hour or greater. Although the number of waves, N, is reduced, the combined effect may be more damaging than the longer storm and larger number of waves (see also paragraph 23 ff).

Special Effects on Wave/Shore Interaction

17. <u>Refraction</u>. For most reservoirs, wave refraction, and the effects of shallow water will not be relevant. Occasional special cases can occur, however, such as when the dam is not at right angles to the predominant fetch or wind direction. Wave build-up can then be influenced when wind/waves are sub-parallel with a shore line. The guidance in the manual should be followed in these special cases.

18. <u>Reflection</u>. Reflection can occur, especially from side flow spillways located close to the ends of a dam. Wave heights can be increased significantly locally.

19. <u>Diffraction</u>. Whilst reflection and refraction of waves is usually not of great significance, it was found during the preparation of the manual that diffraction is potentially significant. Results from a small survey by questionnaire seemed to indicate that damage at several dams to both rip-rap and to block work was concentrated or initiated in areas close to free standing valve towers or overflows. Study of limited data on individual cases indicated that there was reasonable correlation between the areas damaged and the likely area of wave height increase caused by interaction of wave diffracted around free standing valve towers. Fig. 3 from the manual illustrates the phenomenon.

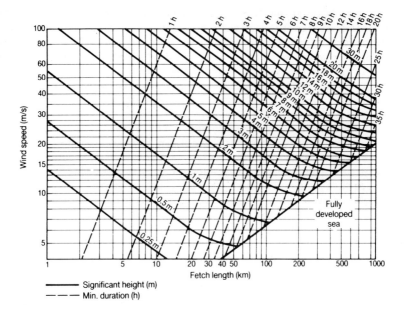

Fig. 1: Prediction of significant wave height from wind speed for JONSWAP spectrum.

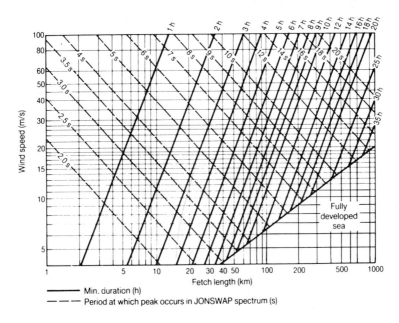

Fig. 2: Prediction of peak wave period from wind speed for JONSWAP spectrum

RESERVOIR DESIGN AND CONSTRUCTION

20. As identified in the manual, existing references on predicting diffraction should be used with caution. Although the theoretical data is limited, it is apparent that wave height can be increased substantially in the "wake" of a valve tower (15-20% is not unlikely). For new works, model tests could determine the severity of this effect, and also the optimum tower position to limit wave increase. It seems that keeping the tower well offshore, or conversely very close, avoids the areas where wave heights increase. For existing structures it indicates that heavier rock protection may well be necessary in the critical areas. The size increase is likely to be substantial. It would appear essential that diffraction effects be carefully considered on all new works and in any upgrade/maintenance programme for dams with towers or spillways upwave of the dam face.

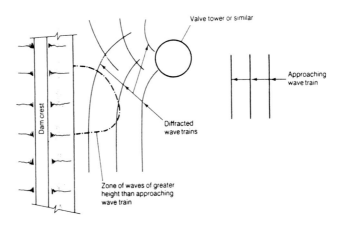

Fig. 3: Diffraction pattern from upwave obstruction

WAVES AND ROCK SLOPES

21. The behaviour of a wave on a slope, that is, the way it breaks, has a major effect on the ability of the slope to resist damage. This behaviour is described by the surf similarity parameter, or Iribarren number, which is the ratio of slope steepness to wave steepness:-

$$\zeta = \tan\alpha/S^{1/2} = \tan\alpha/(2\pi H_s/gT^2)^{1/2}$$

22. There are principally four types of breaking wave on a beach or slope. Each can be identified by the range of its surf similarity number. For most inland lakes, certainly in the U.K., the surf similarity parameter will probably range between 0.5 and 2.0. This means that the majority of storm waves breaking on a rip-rap face will be of the plunging type. Only on steeper faces will the action become collapsing. Surging waves are not likely to occur.

23. Thompson and Shuttler (Ref. 5) established a valuable relationship between wave action and protection rock size. This work was extended by van der Meer (Ref. 6) with further model tests. He found that the form of the breaking waves affected the resistance of the armour to damage, the relationship requiring two separate equations (Fig. 4) as follows:-

For plunging waves: $H_s/\Delta D_{n50} = 6.2 P^{0.18} (S_d/\sqrt{N})^{0.2} \zeta_m^{-0.5}$

For surging waves: $H_s/\Delta D_{n50} = 1.0 P^{-0.13} (S_d/\sqrt{N})^{0.2} \sqrt{(\cot\alpha)} \zeta_m^P$

The notation for which is given at the end of this paper.

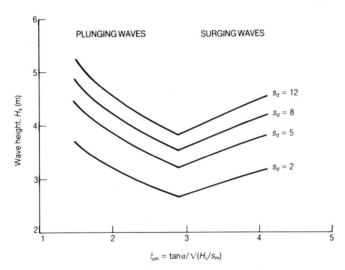

Fig. 4: van der Meer's equations shown with varying damage levels

24. As noted above, most inland reservoirs will show plunging waves, and the left hand half of the curves apply.

25. The method requires the permeability of the structure to be assessed. The structural response between a breakwater of predominantly large rocks compared to that of a face protection to an impermeable embankment is very different. All dams, including rockfill placed under modern criteria, can be taken as effectively impermeable to wave action. A notional permeability factor, P, of 0.1 must therefore be used.

26. The manual compares the van der Meer formulae with the traditional Hudson formula, used by many dams engineers. It points out some significant changes in recommended practice and supersedes CIRIA Report 61 (Ref: 6). It is probably well recognised now that the Hudson formula as originally set down gives too light a rip-rap as it relates to more permeable structures. This is demonstrated in the manual in chapter 5.

27. The manual gives good guidance on the effects of varying protection and hence range of damage suffered. The need to keep reservoirs in service and the difficulties of access to the upstream face to execute repairs will normally inhibit any design approach that allows statistically frequent damage. It follows therefore that the dimensionless damage number, S_d must be low - a value of 2 is recommended, which corresponds to Hudson's "no damage" condition.

FACE MATERIALS
Materials and Sources

28. There are two major differences between most coastal schemes and inland reservoirs over material sources. They are:-

(a) Transport by water is not usually feasible
(b) Frequently, especially in overseas projects, site-specific quarries are required, or local quarries have to be used.

The main consequence of these two constraints is that material choice is frequently restricted.

29. Where well established quarries are available the manual gives valuable advice on adopting standard gradings. The quarries will be used to providing these standards and can probably, therefore, supply at competitive rates as the rock is produced as part of their normal production cycle. Where site, or dedicated, quarries are being used, guidance is given on deriving non-standard gradings tailor-made to the quarry output.

Grading of Materials

30. Chapter 3 covers in detail all aspects of choice of rock materials, and the section on grading should help engineers in this difficult problem.

31. The concept adopts conventional grading approaches, but relies essentially on specifying upper and lower weight or

"sieve" sizes. Between these, the grading is controlled by specifying the mean weight of the rock, W_{em} between set limits. To avoid bias by small material or large, extreme limits to sizes are set.

Figure 5 illustrates the concept, where:-

ELCL	-	Extreme Lower Class Limit
LCL	-	Lower Class Limit
UCL	-	Upper Class Limit
EUCL	-	Extreme Upper Class Limit
W_{em}	-	Effective mean weight (This is lighter than W_{50}, the "percentage passing" mean weight and the manual gives guidance on conversion between the two weights)

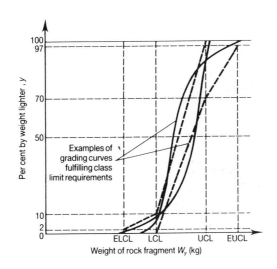

Fig 5: Grading class limits for a standard grading

32. An advantage of the system is that weighbridge weighing of part samples can be used, easing the problem of handling large, weighty materials.

33. Grading widths should not be unnecessarily restricted. Tests by van der Meer and others show no difference in stability between grading ranges for two layer armour (typical for dams) up to D_{n85}/D_{n15} of 2.5 (W_{85}/W_{15} = 16.0) and all the standard gradings proposed in the manual fall well within this limit. Further tests carried out as part of the preparation of the manual suggest that very wide "quarry run" gradings with D_{n85}/D_{n15} of between 2.5 and 5 should not be used for two layer armour systems, as progressive damage can occur.

34. The manual also draws attention to the aspect ratio of the rock, and to ensure predictable performance it will be advisable to ensure conformity with these limits. Limits which have been imposed in the past may be unduly severe. For example, research carried out in parallel with preparation of the manual has shown that tabular rock placed in a random two layer facing performs at least as well as more equant rock.

Rock Deterioration

35. Rock degradation is covered comprehensively. The rip-rap on dams will be in a fresh water environment. High pH or low salts content of the stored water need to be carefully examined. Degradation of limestone can be relatively rapid while soft waters can leach out cementing agents in other sedimentary rocks such as sandstones. Case hardening from the presence of salts is unlikely to occur on dams. Freeze thaw cycles need full consideration. Evaporation and drying in hot climates may need careful assessment, both from salt deposit aspects and temperature variations. These effects are of greater significance in large rocks as blocks can be substantially reduced in size if there are any defects in which frozen water or salts can collect.

QUALITY CONTROL AND TESTING

36. There is extensive guidance on quality control and testing. With regard to control of grading, there is no easy approach. The testing procedures given in Appendix 2.2 of the manual give valuable guidance, using the grading system noted earlier. They need to be carefully and systematically applied to ensure reliable results from the small number of tests usually feasible.

37. To assess materials for face protection, BS 812 has commonly been used (Ref. 7). The tests are not always relevant to the size of rock being used. The manual describes the relevance, where appropriate, of existing tests for aggregates, and gives suitable limits. To this it adds new methods and tests when existing tests are unsuitable due to the increased size of rock used.

38. For construction control, the use of stockpiles of known, measured rock size as reference material, and of a carefully prepared area of the work known to be within acceptable limits as a verification panel, are suggested, and have been used successfully on many dams.

CONCLUSIONS

39. Whilst the manual has been produced primarily for coastal and sea engineers, there is much of value for the dams engineer and project manager within its covers. This paper has concentrated on describing the following subjects provided by the manual.

   (a) Derivation of wave heights.
   (b) Wave/rock slope interaction.
   (c) Rock selection, specification, testing and control.
   (d) Design tools for selection of rock protection.

40. Production of the manual has also revealed or highlighted several aspects of great importance to dams engineers:-

   (a) New and improved methods for predicting wave conditions and for design of hydraulically stable rock facings.
   (b) Recommendations to improve the reliability of earlier design methods.
   (c) The limited amount of wind and wave data for inland lakes and reservoirs.
   (d) The significant influence on wave climate at the facing from diffraction around valve towers and other up-wave structures.

41. Further research is required on:-

   (a) The correlation between regional and local wind speeds at dam sites. Regular wind data collection of dam sites is an essential first step.
   (b) The influence of valve towers and other upwave structures on waves at the dam face.

ACKNOWLEDGEMENT

The authors acknowledge the encouragement and permission of both CIRIA and CUR to the publication of this paper.

NOTATION

D_{n50}	=	Nominal diameter of median on the mass distribution curve.
F	=	Fetch length in wave direction.
L_o	=	(deep water) wave length = $gT_p^2/2\pi$
H_s	=	Significant wave height (average of highest one third of wave heights).
N	=	Number of waves in a storm.
P	=	Notional permeability factor.
S_d	=	Dimensionless damage number.
s	=	Wave steepness = H_s/L_o = $2\pi H/gT2$
s_m	=	$2\pi H_s/gT_m^2$

T_m = mean (zero crossing) wave period ≈ 0.87Tp for a JONSWAP spectrum.
T_p = Peak wave period.
α = Slope of upstream face at water level.
Δ = Relative buoyant density.
θ = Mean wave direction.
ξ = Surf similarity parameter (Iribarren number).

REFERENCES

1.   CUR/CIRIA. Manual on the Use of Rock in Coastal and Shoreline Engineering. CIRIA, London 1991. CIRIA Special Publication 83/CUR Report 154.
2.   CARLYLE W.J. Wave Damage to Upstream Slope Protection of Reservoirs in the U.K., BNCOLD Conference 1987.
3.   OWEN M.W. Wave Prediction in Reservoirs: Comparison of Available Methods. Hydraulics Research, October 1988, Report EX 1809.
4.   SAVILLE et al. Freeboard Allowance for Waves in Inland Reservoirs. Proc. Am. Soc. Civ. Eng., 1962.
5.   THOMPSON D.M. and SHUTTLER R.M. Design of Rip-Rap Slope Protection against Wind Waves. CIRIA 1976, Report 61.
6.   van der MEER J.W. Rock slopes and Gravel Beaches Under Wave Attack. Delft Hydraulics, November 1988.
7.   BRITISH STANDARDS INSTITUTION. BS812 - Testing Aggregates - Parts 1 to 124 B.S.I, London (dates vary)

Design, construction and performance of Mengkuang Dam

E. H. TAYLOR, BSc, DIC, FICE, FGS, C. M. WAGNER, BSc, MSc, MICE, and J. H. MELDRUM, BEng, MICE, Mott MacDonald Ltd

The design of Mengkuang dam involved unusual solutions for both the embankment, which was founded on weak alluvial clays, and the combined spillway and draw-off works. This paper describes the design approach adopted, particular features of the dam construction and reviews its performance over the 7 years since commissioning.

INTRODUCTION
1. Mengkuang dam was constructed in 1982-84 by the Penang Water Authority (Pihak Berkuasa Air Pulau Pinang - PBA) as part of the Mengkuang Pumped Storage Scheme to provide an additional 370 000 m^3/day of raw water for supply to Penang State in Malaysia.

2. The Mengkuang Pumped Storage Scheme augments an existing scheme where a run of river supply is obtained from the Muda river just upstream of a tidal barrage. The new scheme pumps excess wet season flow from the Kulim river and stores it for use in the critical dry season months, February-May. In a future phase water will also be pumped from the Muda up to the reservoir.

3. The reservoir is located near Bukit Mertajam on Peninsula Malaysia on the Mengkuang river, a tributary of the Kulim. Water released from the reservoir is fed to the Sg Dua Treatment Works, utilising the same pipeline over a large part of the distance. The scheme is a typical example of pumped storage. Because no natural dam site was available on the main river the economic solution was to build a dam in a small catchment and pump the water up to it. The cost of pumping is offset by the low capital cost of the reservoir which requires only a small spillway and minimal diversion works. Water is also abstracted from the rivers at or near the tidal limit so use is made of water which would otherwise go out to sea and the ecological impact on the rivers is minimised.

4. There are also positive environmental benefits. The reservoir itself is a broad bowl set in rubber plantations and secondary jungle within a range of high hills, but also

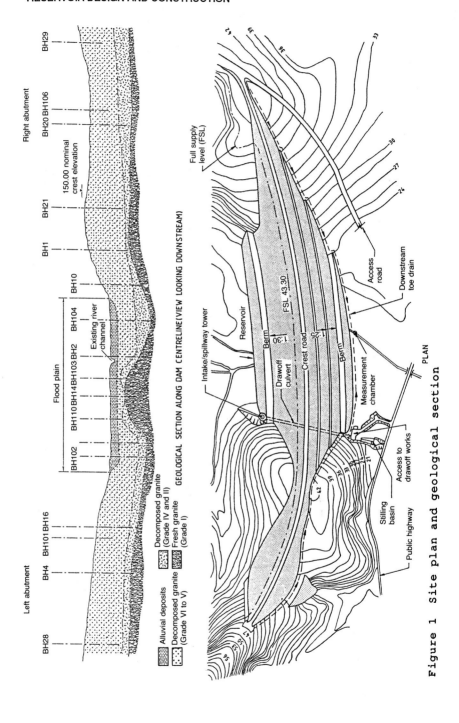

Figure 1 Site plan and geological section

close to major centres of population. The amenity potential was identified by PBA at an early stage and a particular emphasis was placed on the architectural appearance of structures and landscaping. The reservoir area has recently been declared a National Park by the Government of Malaysia.

GENERAL DESCRIPTION

5. The dam is a semi-homogeneous embankment, 1 050 m long and 31 m high, constructed of the granitic residual soils from the extensive overburden in the reservoir area. The dam is curved downstream to correspond with the topography, in particular a low saddle on the right abutment, but also to enhance the bowl shaped reservoir. The spillway and draw-off works are combined in a single intake/spillway tower and draw-off culvert. A plan of the dam is shown in Figure 1.

6. Around the reservoir rim there are two other saddles which drop below top water level. These necessitated two subsidiary dams one 11 m and the other 12 m high. The combined length of these two subsidiary dams is 580 m.

7. The dam was originally required to be constructed in two phases and the design had to allow for future raising by 6 m. Early in the Phase 1 construction, PBA decided to construct the dam directly to the Phase 2 height, which required design changes, including new slope stability analyses for a more rapid fill placing programme.

TOPOGRAPHY AND GEOLOGY

8. The valley at the dam site is a broad U shape with a 350 m wide flood plain and flanks rising at between 2° and 5°. On the right abutment there is a broad saddle at a level of 37.5 m. A geological section along the dam centreline is shown in Figure 1. The dam crest level is at 45.7 m.

9. Bedrock is a coarse grained granite typical of the formations forming the spine of Peninsular Malaysia. Also typically, depths of overburden due to tropical weathering are high, generally in excess of 15 m and often considerably higher. At the dam site Grade I to IV granite is located at depths below ground level of 10 to 20 m across the flood plain and between 20 and 35 m on the abutments. Across the flood plain the rock is overlain by decomposed granite soils generally between 9 and 15 m thick. This layer is further overlain by alluvial and colluvial soils up to 8 m thick. Generally the transition zone of partially weathered rock, Grade III to IV, is between 3 and 6 m thick and is highly fissured and fragmented.

10. Permeabilities of the granite bedrock were found to vary between zero and 100 lugeons with the higher results

Table 1
Permissible factors of safety used in design

Loading condition	Seismic coefficient		
	0	0.05g	0.10g
Long term phreatic phreatic condition (reservoir full)	1.50	1.25	1.00
Short term phreatic phreatic condition (end of construction, rapid draw-\down)	1.25	1.00	n/a

occurring, as to be expected, in the more fissured rock. Sound granite was generally found to be tight, but with occasional permeabilities up to 10 lugeons.

FOUNDATION SOILS

11. The properties of the various foundation soils are summarised in Table 1. The sandy clays/clayey sands on the abutments showed results typical for residual decomposed granite soils in Malaysia. The same decomposed granite soils underling the alluvium across the flood plain were found to be relatively unconsolidated, due to leaching out of feldspars over the course of time, but with similar strengths to the abutment soils once consolidated. There was also a sufficient fines (clay/silt) fraction remaining to make these soils relatively impermeable once consolidated, which had an influence on the design of the cutoff works.

12. The alluvium comprised highly variable and ill defined layers of loose sands or white/grey soft clays and silts. The layers varied throughout the flood plain with no obvious stratification. Once consolidated the CU and CD triaxial tests showed these soils to have a reasonable strength.

13. Mention should be made of organic material, ranging from undecayed wood/tree trunks to black organic soils, which was encountered in several boreholes at depths between 2.5 and 3.5 m within the alluvium. These deposits were considered to be in isolated pockets, rather than layering which could form potential failure planes. This was later confirmed during excavation of the cutoff trench. The undecayed wood was carbon dated at Cambridge University and found to be approximately 38 000 years old. A sample is shown with the poster presentation, indicating the resilience of this forest hardwood under anaerobic conditions.

CONSTRUCTION MATERIALS

14. The deep decomposed granite soils in the reservoir area were the obvious source of earthfill, with the more clayey material from the upper horizons used in the nominal core zone and the underlying coarser material in the shoulder zones. Properties of the fill material are given in Table 1.

15. The only natural source of filter material was a fine sand from nearby disused tin mine "palongs" and this was usable for fine filters at locations such as the drainage blanket filter where a clay/silt fraction was acceptable.

16. There were commercial quarries within a few kilometres and it was assumed that these would be used for rockfill, riprap and filter materials. Use of the quarry won material enabled the grading envelopes for the respective filter materials to be narrowly specified, which permitted the number of different zones to be reduced.

HYDROLOGY

17. The dam was classified as "major" under Snyder's classification (ref. 1), primarily in view of the failure potential, which could have serious consequences for the economy of Penang State. The spillway works were thus sized to discharge the Probable Maximum Flood (PMF).

18. The PMF hydrograph was determined by empirical methods based on a 24 hour Probable Maximum Precipitation (PMP) of 560 mm. The small catchment area of 4.2 km^2 combined with a short time of concentration resulted in a very peaky flood hydrograph of 420 m^3/s with the time base of the flood lasting only 4 hours. The total volume of the flood was 2.2 million m^3 and, with a reservoir surface area (Phase 2) of 179 ha there was every benefit, in attenuating the flood in the reservoir up to the limit dictated by a severe follow-on flood.

EMBANKMENT DESIGN

19. With deep tropical weathering of the abutments and soft clay within the alluvial and colluvial deposits on the valley floor the only practical dam was an embankment type. The decomposed granite was a source of readily exploited and easily worked fill material. The site investigations showed only a small variation in characteristics between different potential borrow areas so the embankment was designed essentially as a homogeneous type but modified by drainage and the lowest permeability fill was placed in a nominal core zone.

20. The proposed future heightening was intended to be carried out whilst the dam was in service with all the additional fill being added to the downstream slope. A sloping core zone with its inclined chimney drain was thus

RESERVOIR DESIGN AND CONSTRUCTION

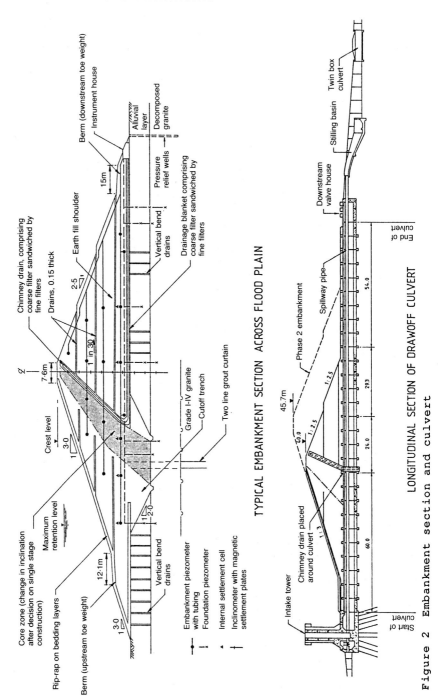

Figure 2 Embankment section and culvert

required. When the decision was taken to build the dam in a single phase it was possible to dispense with the arrangements for extending the core zone and chimney drain and to simplify the layout of the cross section. This decision was taken when the Phase 1 embankment was some 3 m above flood plain level and the inclination of the core zone modified accordingly.

21. A typical section of the final "as constructed" embankment is shown in Figure 2. With the high cost of quarry won filter material, the chimney drain and drainage blanket were designed to have relatively thin coarse high permeability filters sandwiched by fine filters. The width of each chimney drain layer was 0.6 m. The thicknesses of the drainage blanket coarse filter were 0.3 and 0.6 m on the abutments and across the flood plain respectively and the fine filter sandwich layers each 0.3 m thick. The original design for the heightened embankment also included a crack sealing filter upstream and a interceptor drain downstream at the contact between the Phase 1 and 2 core zones. This was deleted following the decision to heighten directly.

22. Earlier studies had recommended removal of the weak alluvial clays and sands involving excavation and subsequent filling of approximately 250 000 m^3 of material. Compared with a total volume of fill in the main embankment of 800 000 m^3 (excluding backfilling the flood plain) there were clear cost advantages in avoiding this excavation, albeit possibly using shallower embankment slopes. An extensive programme of slope stability analyses, both circular and non-circular, confirmed the practicality of this requiring the incorporation of a vertical band drain system through the alluvium and upstream and downstream toe weights reflecting the lower strength of the foundation soils in relation to the fill. This is described below.

23. The cut-off works across the flood plain consisted of a trench up to 7.3 m deep through the alluvium to enable the core zone to be keyed into the underlying decomposed granite and a grout curtain beneath this layer over a 180 m length where the thickness of decomposed granite soils was relatively thin. Elsewhere and on the abutments the decomposed granite soils were considered to be sufficiently impermeable and thick to provide an adequate blanket against seepage with no need for a grout curtain in the bedrock.

SLOPE STABILITY ANALYSES
24. Circular and non circular slope stability analyses were carried out for the usual long term and transient conditions. The area is of low seismicity and nominal ground accelerations of 0.05g and 0.10g were used in pseudo-static analyses for earthquake loadings. The design

Table 2
Material strength and consolidation parameters used in original design for heightening in two stages

Material	c' (kN/m²)	o'	c_v (cm²/min)
Embankment fill	21	32	0.50
Alluvial clay	14	23	0.10
Decomposed granite (flood plain)	21	33	0.70

permissible factors of safety are given in Table 1 and the shear strength and consolidation parameters used in the design are given in Table 2.

25. Preliminary slope stability analyses indicated that upstream and downstream embankment slopes of 1:3.0 and 1:2.5 respectively were feasible, combined with toe weights, if the build-up of pore pressures within the foundation and shoulder fill during construction was limited to a pore pressure ratio, r_u, value of 0.45, approximating to a phreatic surface at the embankment slope and equivalent to the rapid draw-down condition for impermeable fill shoulders, and in the core zone the r_u was limited to 0.60. These values were considered to be conservative.

26. The sensitivity of the design to possible variations in both the pore pressure ratio and the effective cohesion was examined. This work showed that the dominant influence was the pore pressures and that it was necessary to control this during construction in order to produce a well balanced and economical design. Horizontal drainage blankets were therefore introduced in both the upstream and downstream shoulders in order to shorten the drainage paths and so speed construction.

26. Particularly of value in the study of the process of pore pressure dissipation in the foundation and embankment was the study by Bishop and Vaughan on Selset reservoir in 1962 (ref. 2), which indicated a logical approach to the study of pore pressure build-up in relation to placement moisture contents and also on the decay of pore pressures. For the analyses a computer program was developed to examine the pore pressure history of the embankment and foundations with time at specific locations in relation to the fill placing programme envisaged. In the program it was assumed that each increment of load was instantaneously carried wholly by excess pore pressure which then decayed based on Terzaghi's consolidation equation.

27. The dissipation of pore pressures in the alluvium took the conservative approach of assuming the whole layer was a soft clay and ignoring the sand lenses which could not be regarded with confidence as being free draining. The need for radial drainage with vertical drains was identified and the adopted spacing of 3 m assumed 90% consolidation by the end of construction. In the case of the unconsolidated decomposed granite soils underlying the alluvium, two way drainage upwards to the alluvium and downwards to permeable Grade III-IV rock was assumed and it was found that the required pore pressure dissipation could be achieved without artificial drainage. The computer analysis also indicated the need for thin horizontal drainage layers in the embankment at approximately 6 m height intervals.

28. Whereas at Selset in the early 1960s large diameter sand drains were the most practicable solution for drainage of the embankment foundation, the development of vertical band drains comprising a thin preformed polypropylene core wrapped in a geotextile, which could be punched into the foundation, provided a considerably more rapid and economic solution. In addition the band drain had the advantage of being flexible under foundation settlement, thereby avoiding hard points.

29. Because of the presence of low permeability clays in the foundations the design was checked for progressive failure. With such a large length of the critical non-circular slip surface lying within the clays a significant variation in the strains along its length could be expected. However, the triaxial test results showed that strains of 12% to 18% were achieved without significant decrease in peak strength so the clays could be classified as non-brittle. Nevertheless the possibility of progressive failure was considered using post peak strength parameters of $c' = 0$, $\phi = 20°$ as an average along the failure surface within the clay. The strength parameters for the fill were not, however, reduced. Some relaxation was made to the design criteria whereby unusual loading conditions in combination with post peak strength parameters were considered extreme events for which a factor of safety of unity was acceptable.

EMBANKMENT HEIGHTENING

30. The early construction of the main embankment to full height necessitated a revised embankment section, as shown in Figure 2, and a reappraisal of the embankment and foundation loading under more rapid construction. The material strength parameters were also reappraised in the light of further investigations during the initial embanking. The revised strength parameters are given in Table 3.

Table 3
Material strength parameters
used in revised design for heightening in one stage

Material	c' (kN/m^2)	ϕ'
Embankment fill	19	30
Alluvial clay (weakest layer)	14	22
Decomposed granite (flood plain)	19	29

31. The build up in pore pressures during the initial embanking had been less than used in the initial design, particularly in the foundation where it was found that the sand layers were free draining and hence assisting in the dissipation of the pore pressures. In the redesign it was possible to take advantage of this knowledge and it was considered safe to reduce the r_u values from the initial conservative design assumptions to enable the more rapid construction.

SPILLWAY/DRAW-OFF WORKS

32. Several alternative arrangements for the spillway and draw-off works were studied leading to the adoption of a drop inlet spillway incorporated into a 29 m high intake tower and a single large articulated draw-off culvert carrying both the spillway and draw-off flows in separate 1 350 mm diameter pipes.

33. A longitudinal section of the draw-off works is shown in Figure 2 and the arrangement of the intake tower in Figure 3. The arrangement of these works was dictated primarily by (a) the proposed two stage construction, with the need to avoid operating works at the top of the Phase 1 tower, (b) the considerable depths of overburden dictating a piled foundation and broad base to the tower for stability purposes, (c) the low spillway discharge which enabled the flow to be piped thus avoiding free discharge in a culvert through the dam foundation.

34. The required broad base to the tower provided a suitable location for an operating chamber, containing the control and guard valves for the four different draw-off levels and for emergency draw-down. The latter discharges into the spillway pipe. Access to the chamber is through the culvert.

35. There are three draw-off shafts around the spine of the tower with the fourth used as a ventilation duct. Regulation of the draw-offs is by horizontally mounted

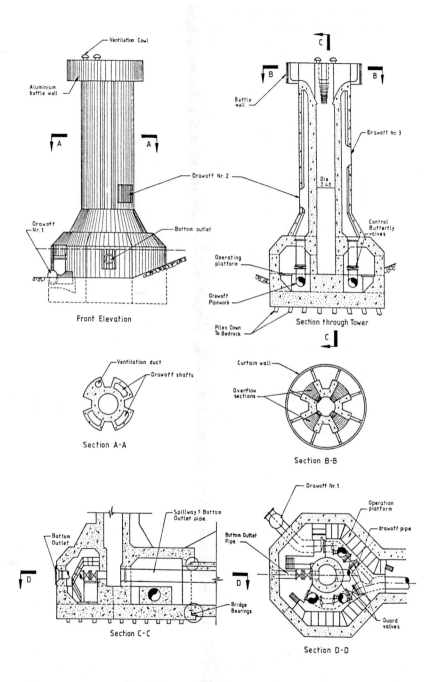

Figure 3 Intake/spillway tower

butterfly valves with metal to metal seating to minimise maintenance.

36. The spillway design discharge is 4 m³/s with a crest length of 5.2 m. A baffle wall in structural aluminium prevents large debris entering the spillway and an aeration nib is included at the nappe to reduce the diameter of the bellmouth. Energy dissipation at the end of the culvert is by a conventional stilling basin.

37. The 5.8 m wide x 6.6 m high reinforced concrete culvert is 170 m long and articulated in 27 Nr 6 m long segments to allow for settlement under the embankment loading. Each segment is separated by a movement joint, 50 mm wide, with structural collars to prevent differential movement. Each joint can take 2° deflection and the culvert was provided with a longitudinal camber, up to a maximum of 0.7 m, to achieve the required fall in the downstream direction "post construction". The pipework is all 1 370 mm ductile iron in 6 m lengths with the spigot and socket joints located to correspond with the culvert joints. Details of a typical culvert segment are shown in Figure 4.

38. There was concern about possible excessive opening of some of the pipe joints due to culvert settlement. Each joint was marked on installation to enable monitoring during and post construction.

CONSTRUCTION

39. The construction contract was awarded to Loh and Loh (Sdn) Bhd, a local Malaysian contractor with previous experience of dam construction in Malaysia and Singapore. The original contract period was 24 months, but this was extended to 30 months when the decision was made to construct directly to the Phase 2 crest. The value of the contract was £8.1 million. The dam was constructed to price

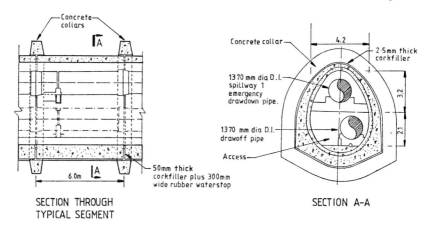

Figure 4 Culvert Details

and on time, attributable to close day to day liaison between contractor and consultant, rapid decision making by the Client, and a contractor who was keener to construct the dam quickly and efficiently than to chase every possible claim.

40. With the very small catchment diversion works were straightforward, initially by a channel on one side of the flood plain and then, as the embankment heightened, through the draw-off culvert.

41. Grouting works were carried out from original ground level, after completion of the cutoff trench, by drilling and grouting in a 50 mm diameter PVC pipe through the trench backfill and underlying decomposed granite soils. On completion of each hole the pipe was backfilled with a weak cement/ silt slurry. Much of the granite bedrock showed no takes but occasional locations gave high takes, up to 80 kg/m.

42. No particular problems were encountered with the fill placing except for continual ruts/surface cracking caused by the dump trucks driving over already compacted fill. As a result frequent re-scarifying and re-compaction was necessary. The degree of handling necessary for placing the coarse filter in the narrow chimney drain resulted in some initial segregation and it was decided to adopt a more single size, 20- 5 mm, grading than originally specified.

43. Placing of the 4 073 Nr vertical band drains with a total length of 24 100 m was carried out by Cementation Ltd, UK. Installation took six weeks after initial mobilisation. It was found possible to punch the drains through the alluvium and a bill item for pre-boring in harder ground was not required.

INSTRUMENTATION READINGS

44. Monitoring of the performance of the main embankment has been carried out with piezometers, settlement cells, inclinometers, extensometers and surface movement indicators.

45. The main instruments are located on four principal monitoring sections. Three on the flood plain and the other on the line of the culvert. One of these sections is shown in simplified form in Figure 2.

46. The total settlements recorded to the end of 1991 by the settlement cells located on the main instrument sections show that at least 90% of this settlement occurred before the end of construction and 98% had occurred within a further 12 months. The maximum settlement recorded of

RESERVOIR DESIGN AND CONSTRUCTION

1.2 m is close to the calculated maximum of 1.3 m. The readings for the cells on Section B-B (located centrally on the flood plain) are shown in Figure 5.

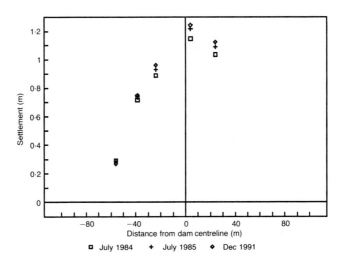

Figure 5   Settlement at Section B-B

47. Settlement in the culvert has been measured by levelling the floor adjacent to the joints between the culvert segments. The settlements to the end of construction and those to the last year are shown in Figure 6. The settlement has again occurred mainly during the construction period. Monitoring of the pipeline joints, as described in Paragraph 39, has shown a maximum opening of 49 mm 7 years after construction. The maximum permissible opening being 50 mm.

48. Pore pressures recorded in the foundation generally showed very little excess pressure resulting from the embankment construction loading; piezometers upstream of the core closely mirrored the reservoir water level; and piezometers downstream showed either similar piezometric levels before and after construction and impounding or a nominal increase after. Figure 7 shows samples of the piezometric levels recorded during construction, impounding and for the 12 months afterwards. Instrument locations are shown on Figure 2. Since commissioning the reservoir has been held consistently close to full and as a result there has been no opportunity to measure the piezometer response under draw-down conditions.

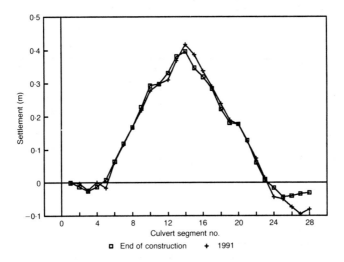

Figure 6   Culvert settlement

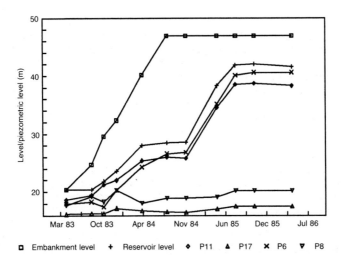

Figure 7   Piezometric levels recorded on Section B-B (P11 - in upstream alluvium foundation, P17 - in downstream alluvium foundation, P6 - in lower core zone, P8 - in downstream lower shoulder zone)

RESERVOIR DESIGN AND CONSTRUCTION

49. The results from both the settlement cells and piezometers indicate that the band drains and drainage layers operated satisfactorily with relatively consolidation. The more rapid dissipation of pore pressures than predicted during design is attributed partly to more extensive free draining sand layers in the foundation alluvium and coarser fill material with a lower clay fraction used in the embankment shoulders.

50. During the period of March and April 1984, when the embankment was at a level of 42 m, readings from the settlement cells gave rise for concern as heave was indicated in the core and upstream areas over 3 of the instrumented sections. The readings varied during the two month period and the residual heave was up to 0.15 m. No further heave was recorded as the embankment was brought to full height and piezometer cells in the area of the settlement cells generally continued the trend of their readings. No associated surface movement was observed.

PERFORMANCE OF STRUCTURES

51. A recent visit was made to the dam to review its performance. The dam has been well maintained by PBA staff with operating and maintenance instructions closely followed including instrumentation readings.

52. There were no cracks on the embankment crest and similarly no evidence of surface movement on the downstream slope. The draw-off culvert and operating chamber were completely free of water with a few minor leakage points self healed by leachate.

53. Seepage through the embankment and foundation has been consistently between 8 and 10 l/s and the flow is mainly coming from 3 pressure relief wells on the right abutment near the draw-off culvert. Although this flow is within the compensation release requirement, this seepage is higher than estimated during the design and is suspected to be due to the difficulties of completely sealing the Grade III-IV rock by drilling through the overlying overburden. It may also be due to an insufficient extension of the grout curtain into the right abutment and groundwater flow from the abutment area.

REFERENCES

1. SNYDER F.F Hydrology of Spillway Design Floods- Large Structures, Jour. Hyd. Div., Proc. Am. Soc. Civ. Eng, May 1964.
2. VAUGHAN P.R. and BISHOP A.W. Selset Reservoir, Design and Performance of the Embankment, Proc. Inst. Civil Engrs., Vol.21, London, February 1962.

Implementation of the Reservoirs Act 1975 and monitoring of dams

D. C. BEAK, BSc, MICE, Scottish Hydro-Electric Plc

SYNOPSIS

1 The paper describes the alterations that Hydro-Electric have made to their supervision of reservoirs, production of the Record and the monitoring and instrumentation of dams following the implementation of the Reservoirs Act 1975.

2 This has involved modifying existing inspection procedures, establishing a computer database for the reservoirs and the automation and computer analysis of monitoring and instrumentation of the dams.

INTRODUCTION

3 The North of Scotland Hydro Electric Board (the Board) was formed in 1943. With the principal aim of developing the hydro electric potential of the Highlands of Scotland.

4 In 1948 the UK electricity supply industry UK was nationalised and the Board took over the assets of all the public electricity supplies within its area of operations. These included a number of existing hydro developments such as those of the Grampian Electricity Co in the Tummel Valley but most of the hydro assets were built by the Board during the period from the late 1940s to the mid 1960s.

5 With the introduction of the Reservoirs Act 1975 the Board altered their policy with respect to maintaining the records, the instrumentation and the monitoring of the 76 large raised reservoirs for which they are the Undertakers by introducing computer systems. These systems have been further developed by Hydro-Electric as successors to the Board and are broadly described in this paper.

IMPLEMENTATION OF THE ACT

6   Since its formation the Board has had a policy of implementing all aspects of Reservoir Safety legislation in spirit as well as to the letter.

7   Routine inspections using a number of their own civil engineers who were experienced in the development, operation and maintenance of hydro works was always normal practice for the Board. In the early 1970s the reporting of these inspections became more formalised with the introduction of a standard format and checklists and there have been further gradual developments since then.

8   When the form of the 1975 Act became apparent, prior to implementation, the Board decided to "jump the gun" by implementing a number of measures which would be required under the new Act while the 1930 Act was still in force:

   A   Panel A Engineers appointed to inspect the reservoirs were "independent" having no connection with the Construction Engineer or the Undertaker.

   B   A number of younger engineers were trained with a view to their being suitably qualified to act as Supervising Engineers. By the time the 1975 Act was implemented, in 1986, the Board had a team of engineers with the appropriate experience and who successfully applied for membership of the Supervising Engineers Panel. Since then some of these engineers have retired and other engineers have been appointed to the Panel. By the time they are appointed all engineers have attended a number of inspections in the company of Inspecting and Supervising Engineers and have been involved in the decision making process.

   C   A change was introduced in the policy of having consecutive routine inspections carried out by a different engineer in order to provide different viewpoints on aspects of the dam. The 1975 Act requires that each Large Raised Reservoir has a Supervising Engineer appointed to watch over it. Accordingly procedures were altered to bring them in line with the Act and changes in Supervising Engineer appointments are now only made when circumstances make it necessary. In this way administration for all parties is kept to a minimum but, more importantly, the requirement of the Act for each structure to have someone with detailed familiarity is recognised.

D After studying the time and effort spent in maintaining the existing record and the additional work which was going to be required to comply with the 1975 Act it was decided that all statutory and many other items of information should be stored in a computer database.

9 When the format of the Record to be used under the Act was published HE took advice and were informed that while the general format and information to be recorded were specified in the Statutory Instrument it was not a requirement that the book published by Thomas Telford be used. This was a change from the 1930 Act which required the use of Form F published by the Stationery Office.

10 All the statutory information required to be included in the Record together with additional information which is considered valuable to record is stored on the database. The pages of the formal Record are produced in loose leaf format with amended pages being produced whenever required.

11 There are currently 7 Engineers within the Company who are appointed to the Supervising Engineers Panel, with another member of staff who has applied for membership. For some of these Engineers their Reservoir Safety duties are a major proportion of their responsibilities, while others are principally employed to carry out other duties. For this reason the numbers of reservoirs for which they have been appointed as Supervising Engineers varies between 3 and over 20.

PRESENT POSITION

12 Since the construction of the hydro schemes the method of recording the reservoir levels has been for local staff to read the water level from a gauge board and the details entered on a card which was sent to the Engineering Section at a central location. Under the 1930 Act this information was entered by hand into Section C of Form F. With the new database the user interface allows that water levels can be entered in either metric or imperial units, relative to either Ordnance Datum or the spill level of the dam and automatically converted to metric levels relative to Ordnance Datum in which the Record is maintained. The recording

of maximum and minimum levels for each week were a requirement of the 1930 Act but is no longer required under the 1975 Act. However Hydro-Electric has found it a useful item of information as have Inspecting Engineers and the practice of recording them has been continued.

13   A control room has been established in Dingwall controlling most of the reservoirs North of the Great Glen and another is being established near Pitlochry which will serve those in the South of the Company's area. In these control rooms the water level of all the reservoirs is monitored at half hourly intervals using pressure transducers in floatwells at each dam. When the Control Room monitoring systems have been fully developed this information will be downloaded automatically to the PC on which the database is recorded. Automatic interrogation of the data will extract the appropriate data for the official record. The remainder of the level information will be retained in a main frame computer archive, along with many other variables recorded simultaneously, to enable detailed studies to be carried out if required. The Record not only includes the tabular form of the levels but for each year the information is transferred onto a graph to enable the information to be scanned quickly (Fig 1).

14   The level is still checked manually at the gauge board regularly to ensure that accurate levels are being recorded at the control room. The database for each reservoir contains a maximum and minimum level which would not expect to be exceeded. If a level outwith this range is entered an alarm is raised and its accuracy verified.

15   A number of standard scans have been written into the database to enable monitoring of items such as dates for Statutory Inspections and actions "In the Interest of Safety". Any of the fields in the database can also be scanned to identify reservoirs with common features. This allows a quick check to be made on all dams which have the same feature or combinations of features and is particularly useful when a dam is identified as behaving in an unexpected manner. Local staff can be instructed to inspect similar features on other dams and report back to the Reservoir Safety Section.

16   The database is used to monitor the dates of all inspections. A search procedure has been established which enables a routine check to be carried out on a number of aspects of the administration of the reservoirs. This list details for each reservoir the name of the Supervising Engineer, the number of Supervising Engineers inspections required each year, the date of the last Supervising

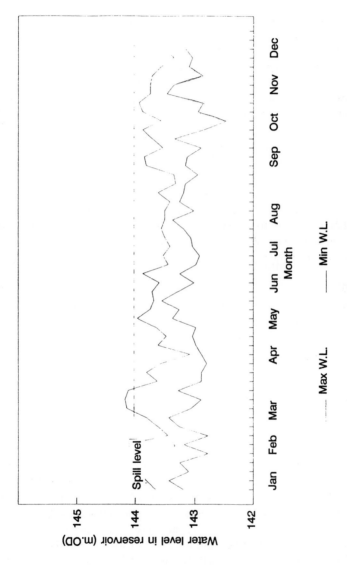

Fig. 1. Loch Tummel: water levels for 1990

Engineers inspection (and highlighting if overdue), the date of the last Supervising Engineers Annual Report, the date of the last and the next Inspecting Engineers Inspection, whether there are any actions "In the Interest of Safety" and whether there is any statutory instrumentation at the dam and if so, whether this has been carried out within the required period. This report is a simple yet powerful way of ensuring that the Company's obligations under the 1975 Act are being met.

INSTRUMENTATION AND MONITORING

17 Instrumentation of the dams has been carried out since they were constructed. Originally this took the form of continuing the instrumentation established by the Consultants who had designed the dams. As experience and confidence in the behaviour of the dams were established the frequency of monitoring was reassessed. The current policy is that in addition to any monitoring instructed by Inspecting Engineers or where a Supervising Engineer identifies an aspect requiring investigation, dams exceeding the following heights should have routine instrumentation, based on their categorisation from table 1 of "Floods and Reservoir Safety"

DAM CATEGORY	MINIMUM HEIGHT OF CONCRETE DAM	MINIMUM HEIGHT OF EMBANKMENT DAM
A	25 m	10 m
B	30 m	15 m
C	40 m	25 m

18 Where a number of years experience of instrumentation has shown that a dam is behaving in a predictable manner the frequency of monitoring is often reduced to carrying out instrumentation in the Spring and Autumn once every five years.

19 Most dams with pressure relief drainage systems have vee notch boxes established at the outlets to the systems to enable checks to be kept on any leakage. Wet areas on the banks downstream of the dam which could be leakage are also collected and channelled into measuring boxes. The flows in these boxes are normally read weekly by the local staff and the results forwarded to the Reservoir Safety Section. At any dam where there is cause for concern the vee notch is read daily and the figures telephoned in. These vee notch readings are frequently compared with the reservoir level and rainfall to determine whether the flow is leakage or surface runoff.

20 Originally most instrumentation took the form of crest levelling, alignment using a collimator, pendulums, crack and joint measurements read with callipers, Demec gauges or micrometers, temperature sensors cast into the concrete and piezometers. All these instruments were manually read and transferred to charts.

21 In many locations these methods of monitoring are still in operation. However new techniques have been gradually developed using more automation and computer reduction of the results.

22 Crest levelling is still used, but collimation has been largely superseded by EDM monitoring. This involves establishing base stations on bedrock downstream of the dam and measuring electronically the distance between the base station and targets established on the dam. For some years this has been carried out using a Geodimeter 112 with an accuracy of about +/- 3mm. The Company have recently purchased a Wild Distomat DI2002 and T1600 electronic total station which together give an accuracy of better than +/- 1mm + 1ppm. This should considerably enhance the accuracy of the monitoring, when it is introduced in the 1992 season and will dispense with the need to bring in outside assistance for the few locations where the Geodimeter was not considered sufficiently accurate.

23 One of the more valuable developments has been an automatic pendulum. This consists of a pendulum suspended down a shaft in the dam with a weight suspended in a reservoir of oil. Above the weight there are a pair of displacement transducers mounted at right angles. These measure movement in both directions. Crack and joint measurement is frequently measured using displacement transducers attached over the joint. Concrete temperature measurement are made by drilling into the core of dams and grouting in thermocouples. Longitudinal movement in dam galleries over lengths of up to 60 metres have been carried out using an invar wire rigidly fixed at one end with a weight over a pulley wheel at the other end. A displacement transducer is attached to the wire near the pulley to measure any movement. Vibrating wire strain gauges have been attached to concrete surfaces in arrays to enable strains to be measured in a plane.

24 It would be possible to record these readings and transmit the information via a land line to a remote computer. However since many of the locations are remote from power supplies or data links, it is more normal for these instruments to be monitored simultaneously and linked to a battery powered data logger. This

RESERVOIR MONITORING AND MAINTENANCE

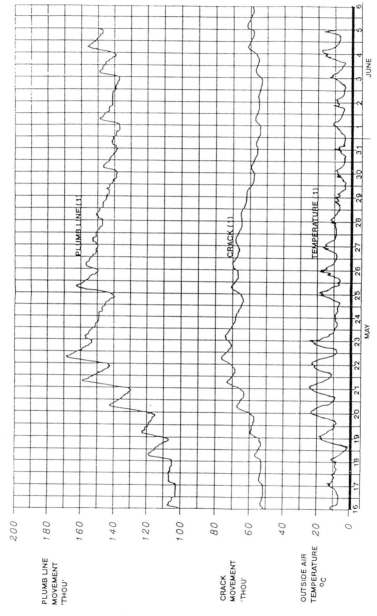

Fig. 2. Mullardoch Dam: Mulres 1 & 2, 16 May to 6 June 89

information is stored on the logger for prolonged periods of time, downloaded onto a hand held computer and transferred to a PC for analyzing and graphical display. The readings are taken at regular intervals often of 1/2 an hour. They can be arranged so that one of the functions triggers a set of very much more frequent readings. This has been used for example to closely monitor a number of functions when a flood gate comes into operation.

CASE STUDIES

Mullardoch

25 Monitoring of horizontal lift joints, vertical joints between blocks, a pendulum, crest levelling and EDM measurements had been carried out at this dam in Glen Cannich for many years.

26 Following the sudden increase in leakage through lift joints in the dam during 1986 an enhanced programme of monitoring was introduced to assist in establishing the cause of the problem and to monitor the behaviour during and after remedial works had been carried out. Details of the remedial works, which took the form of installing 28 post tensioned anchors of 1000 tonne capacity, has been well reported elsewhere and is not repeated here.

27 The instrumentation took the form of an automatic pendulum, crack displacement transducers, and thermocouples measuring concrete and air temperature. These were all read automatically at half hourly intervals and recorded locally on a data logger. In addition an invar wire was installed longitudinally in the gallery to investigate for expansion over a length of five blocks of the dam. Vibrating wire strain gauges were also installed in arrays in a number of locations to monitor strain changes due to seasonal variations and during the remedial work. These are read manually at weekly intervals and were also checked immediately before and after post tensioning operations.

28 When the half hourly readings were established it was discovered that there were significant diurnal variations in several of these functions in addition to seasonal variations. The trigger appears to be air temperature, when the temperature increases there is a lag of about four to five hours following which significant pendulum, and crack movements are noted. (Fig 2)

29 The value of automating instrumentation to enable diurnal cycles to be observed should be noted. Prior to installing this capability

RESERVOIR MONITORING AND MAINTENANCE

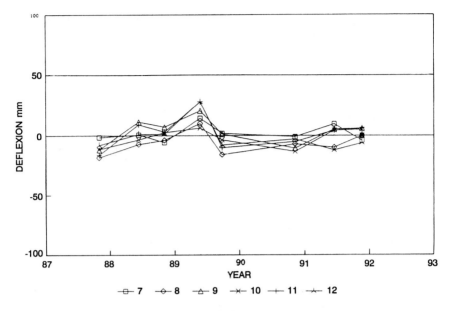

Fig. 3. Quoich Dam: laser levels of DS face - 2

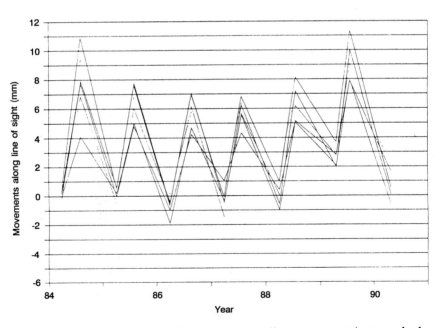

Fig. 4. Geodimeter results for Monar Dam: on line movts:temp/pres method: sheet 3

apparent random changes in observation were unexplained but subsequently it was evident that they were due to variations in the time of observations with respect to the diurnal cycle.

Quoich

30   This is a rolled rockfill dam in Glen Garry of some 40 m height retaining a reservoir of 380,000,000 Cu M. Levelling had been carried out since construction on to pins set into the rockfill in lines established along the downstream face of the dam. This had involved setting up the level at stations on the face of the dam which is at a slope of 1:1.4.

31   The Inspecting Engineer who was appointed in 1985 instructed that the number of monitoring points should be greatly increased to include hollows and humps, which had been noted over many years.

32   It was considered that handling instruments on the steep face of the dam was no longer satisfactory and alternatives were sought. The dam crest is straight with the downstream face more or less in one plane. It was decided to obtain a Spectra Physics EL1 rotating laser level and modify it to isolate the self levelling mechanism so that it would operate with the instrument at an angle. A stainless steel base with a machined face was attached to the base of the dam and the level mounted on it. The rotating level then establishes a plane parallel to and about 0.5 to 1.5 m above the plane of the downstream face. Pins have been inserted into rocks on the face at the points to be monitored.

33   In order to carry out the monitoring the only item of equipment required to be carried onto the face of the dam is a lightweight staff with a detector for the laser beam to measure the displacement at the selected locations. This technique is not as accurate as the precise levelling previously used but is considered adequate for the purpose and is considerably faster and safer. The 34 points monitored on the face of the dam can be checked in the course of about two hours. Previously levelling on 12 points took several hours to carry out. The technique of carrying out this measurement has had to be developed by experience and the repeatability of the results is being improved as experience is gained. (Fig 3)

RESERVOIR MONITORING AND MAINTENANCE

Monar

34 Levelling, EDM and inverted pendulums have been read over many years at this dam in Strathfarrar. The results have indicated that the crest moves upstream in the summer months and downstream in the winter as the concrete expands and contracts (Fig 4).

35 The inverted pendulums consist of a vertical hole cored into the bedrock from the base of the dam. A tank is mounted over and sealed to the top of the hole. The tank and the hole are filled with oil and a stainless steel wire anchored at the foot of the hole is connected to a semi submerged float in the tank. The centre of the top of the float is marked and there is a viewing panel with crosshairs in the top of the tank. By measuring the position of the mark on the float relative to the crosshairs, any movement between the anchorage in the bedrock and dam foundation can be monitored.

36 Although the instrumentation results have been consistent over the years it has been decided that monitoring of this dam should continue to be carried out twice every year since it the company's only double curvature concrete arch dam and one of very few in the UK.

Pitlochry

37 At Pitlochry there are a number of cracks in reinforced concrete towers at the ends of large floating drum gates. These cracks have existed for many years and do not appear to be changing significantly. However the mechanism by which the cracks have propagated is not clearly understood. For a number of years the cracks have been monitored by measuring across studs on either side using callipers. It was decided following a Statutory Inspection that the monitoring should be intensified to try and establish the cause of the cracking.

38 Displacement transducers were attached over 12 of the cracks, another attached to measure the position of the drum gate and a sound channel used to add a commentary as the monitoring took place. The transducers were linked to a 14 channel tape recorder and continuously recorded for a working day. During the test the drum gate and generators were operated to determine whether either of these had an effect on crack movement. A number of the channels were also monitored on chart recorders so that any movement could be observed as it occurred. Although movements

were observed these were so minor that they were not considered to be significant.

39   Following this intensive monitoring an automatic pendulum is being established and monitored together with 5 of the cracks and external air temperature. These will be measured at half hourly intervals over the course of a year to determine whether there is a diurnal or seasonal variation .

40   Similar methods of monitoring will be established on other dams whenever there is a need perceived to carry out monitoring of various aspects of their behaviour.

CONCLUSION

41   In this paper I have outlined the way in which Hydro-Electric has changed its reservoir surveillance and instrumentation policy from a system of manual recording and monitoring to one where much more use is made of automated equipment and computer programmes.

42   The adoption of the database has enabled us to produce a Reservoir Record which provides all the data required under the 1975 Act together with additional information that we wish to record about each reservoir. It is also used as a tool to reassure management that the Company's statutory obligations are being met.

43   The automation of instrumentation and monitoring has enabled us to carry out much more intensive surveillance than has previously been possible without the need to increase manpower resources.

Response of a clay embankment to rapid drawdown

C. J. A. BINNIE, MA, DIC, MConsE, FICE, FIWEM, FEANI,
D. J. SWEENEY, BSc, MSc, DIC, MICE, and M. W. REED, BSc, MICE,
WS Atkins Consultants Ltd

SYNOPSIS. As part of a regular Reservoirs Act inspection piezometers were installed in the upstream slope of Bartley Reservoir embankment dam. The piezometers were monitored closely during two periods of fairly rapid drawdown. Whilst piezometer levels in the stone weight zones behaved as expected, $\bar{B}$ values in the clay fill were below unity. This resulted in a significantly lower than expected factor of safety for the rapid drawdown situation. This paper describes the field installations, shows typical examples of field data obtained, and discusses the reasons for the unexpected results.

INTRODUCTION
    1. Bartley Reservoir is situated 11km south-west of the centre of Birmingham. It is required as standby storage for the water supply to Birmingham in the event of maintenance or operational problems at the Elan Reservoir or the 120km Elan aquaduct from Wales to Birmingham. The reservoir has a negligible natural catchment. The reservoir was formed by constructing an embankment across a small valley and has a storage capacity of 2.1Mm^3. The embankment is approximately 500m long, with a maximum height of 20m. Construction started in 1925 and finished in 1930.
    2. The area of the reservoir and dam is underlain by red marls with subordinate sandstones and minor limestones of the Keele Beds formation. These are generally covered by a mainly thin layer of glacial and post-glacial deposits substantially derived from the underlying or similar marls. A sandstone unit, locally termed the 20 foot Sandstone for convenience, underlies much of the area and outcrops beneath and downstream of the embankment.
    3. The embankment was constructed with a thin, vertical, lightly reinforced concrete core wall located centrally within it. The clay fill material forming the embankment was taken from the higher portion of the reservoir basin and transported to the dam in rail mounted skips. The specification required the fill to be

deposited in layers and well rammed and consolidated using heavy wooden pounders and rollers. However from construction photographs no evidence of compaction could be seen. In addition, at intermediate construction stages the surface of the fill sloped towards the concrete core and resulted in water ponding on the surface of the fill. The more clayey and adhesive material was specified to be retained within 1 to 1 batter lines from the top of the embankment. However, a ground investigation was unable to detect any significant variation in the fill material throughout the embankment. The upstream slope of the embankment was to be formed at an inclination of 1 vertical to 3 horizontal and the downstream slope at 1 vertical to 2½ horizontal. Figure 1 shows a cross-section through the designed upstream embankment slope.

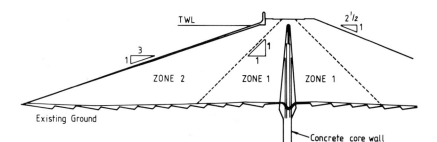

Figure 1 Cross-section through designed upstream embankment slope

4. During the construction of the embankment a number of stability problems occurred, with the upstream slope suffering three slips. The first slip occurred when the embankment was approximately 10m high, the second at approximately 15m height and the third at 18m height. Remedial works were carried out after each slip occurred and in total comprised:
° two rows of ferro-concrete piles driven along the toe of the embankment slope with a concrete waling located at the top of the piles,
° a concrete wall built to a height of 1.2m above ground level either side of the zone of piling, extending up the valley sides for some distance, parallel with, but

just clear of the original toe of the embankment,
- trimming of the toe bulges,
- a substantial amount of stone fill deposited on the slope to form the present benched profile, with an overall slope of 12° (1:4.7). Figure 2 shows the constructed upstream slope profile

ZONE	SOIL TYPE
1	Stone fill
2	Clay fill
3	Quartzite and sandy clay
4	Selected clay fill
5	Original ground [Sandstone and marl]

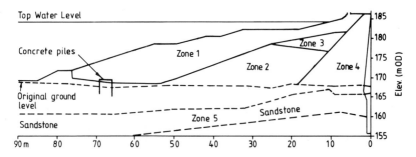

Figure 2   Cross-section through the constructed upstream embankment slope

5. The reservoir was filled in 1930 and has remained full, except for occasional drawdowns of less than 2m. The drawdowns were a consequence of routine maintenance work being carried out either at the Elan Reservoir or Elan Aquaduct. Bartley Reservoir would then be used to supplement the water supply for Birmingham, thus reducing the water level in the reservoir.

6. In October 1986, Severn-Trent Water appointed Mr C J A Binnie of WS Atkins Consultants Limited as Inspecting Engineer for the reservoir to carry out the statutory ten year inspection required by the 1975 Reservoirs Act.

7. An investigation consisting of a desk study and site inspections was carried out. This included consideration of the stability of the embankment slopes. The interim report recommended that a ground investigation be undertaken, to determine if adequate factors of safety exist in the embankment slopes.

8. In March 1989, Severn-Trent Water asked their Dam Review Panel, consisting of Mr Roy E Coxon, Mr John M McKenna and Dr John Newberry, to advise them on the investigations and studies now being done on Bartley Reservoir.

GROUND INVESTIGATION

9. The ground investigation took place in 1988, from March to September. During the investigation a total of 29 piezometers were installed in the upstream slope. These comprised:
- 5 Casagrande standpipe piezometers
- 10 Acoustic piezometers
- 6 De-airable acoustic piezometers
- 7 Pneumatic piezometers
- 1 Standpipe piezometer

The location of the piezometers are indicated on the embankment cross-section in Figure 3.

KEY
c - Cosagrande standpipe piezometer
a - Acoustic piezometer
p - Pneumatic piezometer
de - De-airable acoustic piezometer
s - Standpipe piezometer

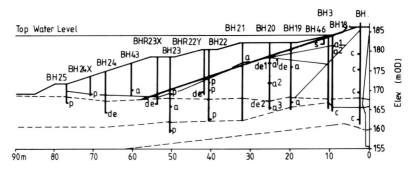

Figure 3   Location of piezometers in the upstream embankment slope

10. The piezometers were installed to provide information on pore water pressures in the upstream slope, especially during drawdown conditions. A particular requirement of the piezometers during the drawdown was to give a fast response to expected changes in pore water pressure. For this reason pneumatic and acoustic piezometers were used. Acoustic piezometers were placed where particularly rapid and accurate responses were required to small changes in pore water pressure.

11. An inclinometer was installed in borehole 21 to detect any movements in the upstream embankment slope during the drawdown.

METHANE

12. As the investigation progressed, and readings taken from both upstream and downstream piezometers were reviewed, a number of erratic changes in piezometric

levels were noted. The water levels in the standpipe piezometers fluctuated widely in some instances. The changes in level usually occurred over a period of less than a week, with the water level returning to original level quickly or up to several months later.

13. Methane migrating through the embankment was suspected as a possible cause and was tested for using a methanometer. Methane was detected in more than half of the piezometers usually with concentrations of less than 1%, although some piezometers were found to have concentrations between 20% to 90%. Nearly all the piezometer standpipes in the upstream slope contained some methane.

14. Samples of the methane collected from standpipes were analysed to determine whether the gas was of 'geological' or recent 'biogenic' origin. The results indicated that 82% of the methane sampled was of recent origin.

15. It is therefore probable that most of the methane was derived from the decomposition of organic matter within and beneath the dam. This material was probably plant remains which were left in the fill when the dam was constructed. A smaller proportion of gas was of geological origin and probably reached the site by being carried in groundwater or migrating up fault zones from a deep source.

16. In view of the erratic behaviour of the piezometers, which was believed to be due to the presence of methane, six de-airable acoustic piezometers were installed in the upstream embankment slope. The installation of these acoustic piezometer tips was carried out by a slightly unconventional method. The piezometer tips were pushed into prebored holes of marginally smaller diameter than the tips, thus eliminating the usual sand filter. It was considered that methane gas could collect in the sand filter and influence the pore water pressure recorded by the piezometer.

17. The effects of methane on the piezometer tips could be eliminated by de-airing the piezometer at intervals throughout the inspection and immediately prior to commencing an instrumented drawdown. These piezometers would then supply a base set of readings and act as a control for the remainder of the piezometers.

RAPID DRAWDOWN TEST

18. During 1988 and 1989 the reservoir was drawndown on two occasions by approximately 2m under rapid, but controlled, conditions.

19. In September 1988 the reservoir water level was lowered by 2.13m over a period of 5 days, for operational and test purposes. Piezometers were read manually at

regular intervals throughout the day for the duration of the drawdown period. During the refill period, which took 29 days, the frequency of monitoring was reduced to weekly.

20. The reservoir drawdown took place soon after the de-airable acoustic piezometers were installed (early August 1988). Part of the installation procedure was to de-air the piezometers after they had been placed in position. Therefore, it was considered unnecessary to de-air the piezometers a second time with the drawdown occuring soon after installation.

21. In November 1989, as part of Severn-Trent Water routine maintenance work, the supply of water to the reservoir was to be stopped and consequently the reservoir would be drawn down. Advantage of this drawdown was taken to make comparison with the data collected during the 1988 drawdown.

22. The reservoir water level was lowered by 1.85m over a period of 4 days, a similar rate to that of the 1988 drawdown. Data logging equipment was used to monitor both the acoustic and pneumatic piezometers. The acoustic piezometers were monitored at hourly intervals until the drawdown reached its lowest level, at which time the frequency was reduced to 2 hourly intervals. The pneumatic piezometers were monitored at 2 hourly intervals throughout. All monitoring of piezometers ceased 7 days after refill of the reservoir commenced. Manual readings of acoustic and pneumatic piezometers were taken at intervals throughout the monitoring period for comparison with the data logging results.

23. Prior to the commencement of the drawdown the de-airable acoustic piezometers were checked for the presence of gas in the piezometer tips. The piezometers were de-aired and the volume of gas evacuated noted. See Table 1 for the volumes of gas removed from the piezometer tips. On the morning of the start of the drawdown the de-airable piezometers were de-aired again but insignificant gas content was detected.

Table 1 Results of De-airing Test

Piezo-meter No.	Pressure Before mH_2O	Applied Pressure mH_2O	Applied Vacuum mH_2O	Vol of H_2O	Vol Air/Gas	Return to stability
24de	18.4	2.0	-6.0	600cc	"Few bubbles"	1:05hr
R23xde	16.0	2.0	-6.0	600cc	"Few bubbles"	2:30hr
R22yde	13.3	2.0	-6.0	600cc	1.5cc	0:30hr
$20de_1$	11.6	3.0	-6.0	600cc	2.0cc	0:20hr
$20de_2$	6.4	3.0	-6.0	600cc	2.0cc	0:30hr
19de	8.6	5.0	-6.0	600cc	4.0cc	0:40hr

RESULTS OF DRAWDOWN TEST

24. Figures 4 and 5 show typical piezometer levels recorded throughout the drawdown period

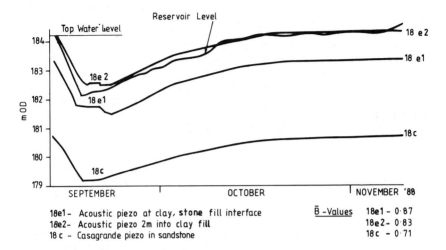

18e1 - Acoustic piezo at clay, stone fill interface
18e2 - Acoustic piezo 2m into clay fill
18c - Casagrande piezo in sandstone

B̄ -Values
18e1 - 0.87
18e2 - 0.83
18c - 0.71

Figure 4 Piezometer response to reservoir drawdown

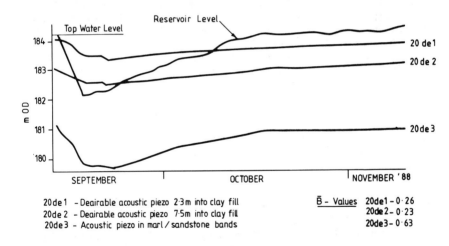

20de1 - Desirable acoustic piezo 2.3m into clay fill
20de2 - Desirable acoustic piezo 7.5m into clay fill
20de3 - Acoustic piezo in marl/sandstone bands

B̄ - Values
20de1 - 0.26
20de2 - 0.23
20de3 - 0.63

Figure 5 Piezometer response to reservoir drawdown

25. As expected, the piezometric pressure in the stone fill was highly responsive to variations in reservoir level. The response in the clay fill was for most piezometers far less.

RESERVOIR MONITORING AND MAINTENANCE

26. For use in embankment stability, analyses $\bar{B}$ values were calculated for all piezometers in the upstream slope based on the responses measured during both of the drawdowns. $\bar{B}$ is the ratio of the change in pore water pressure to the change in the total vertical stress. As the drawdown was not instantaneous some dissipation of pore water pressure will have occurred during drawdown and it would therefore be more correct to speak in terms of apparent $\bar{B}$ values.

27. The $\bar{B}$ values were determined from the best fit sloping line taken from the plots of time against piezometric level for the drawdown phase of the test. By this means, the effect of dissipation on the $\bar{B}$ values would be reduced, along with any other anomalies which may have occurred. Figure 6 shows the $\bar{B}$ values calculated for the piezometers in the upstream embankment.

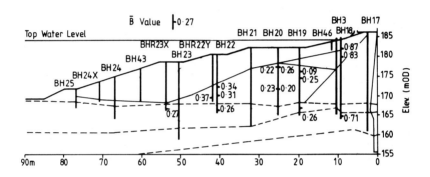

Figure 6 Upstream cross-section showing $\bar{B}$ values

28. $\bar{B}$ for the overlying stone fill was found to be about 0.9 and for the underlying sandstone band on average about 0.75. The clay fill responded with values in the range of 0.2 to 0.4, much lower than would be expected for a fully saturated upstream embankment, as would be expected beneath a reservoir maintained near top water level for nearly 60 years.

29. During the drawdown the inclinometer was monitored for movement in an upstream direction. Movements detected were negligible and scarcely within the range detectable by the instrument.

DISCUSSION OF UNUSUAL RESULTS IN THE MARL FILL

30. Gas accumulating in the tip of a piezometer or in the vicinity of the tip can cause misleading pore water pressure measurements to be obtained. In order to address

this problem selected upstream piezometers were installed without sand filters which could have acted as accumulation points for gas. As a further precaution de-airing facilities were included in these installations.

31. It is believed that use of the de-airable piezometer tips eliminated the potential problem of gas affecting the readings and that the changes in pore water pressures recorded during the reservoir drawdown truly represented the porewater pressure changes in the surrounding soil.

32. The presence of gas within the soil pores will cause a reduction in soil stiffness and an associated 'soft' pore water pressure response to loading or unloading. For a fully saturated soil, during drained loading or unloading, virtually all the changes in load are carried by the pore water as this is much stiffer than the soil skeleton. This is the most likely explanation of the less than expected reductions in pore water pressure during drawdown.

33. This raises the question of why the upstream embankment soil is not fully saturated after sixty years of reservoir impoundment. In part, this may be because the upstream slope was not sufficiently compacted during construction and pore air has not yet dissolved out of the fill.

34. The methane generated within the marl fill could be contributing significantly to maintaining this partial saturation. As pore air dissolves out of the fill it may be being replaced by methane gas, preventing full saturation. It is worth noting that it is possible for pore gas to be completely surrounded by pore fluid. In this state it will cause a reduction in stiffness of the soil without creating soil suction forces which require menisci acting on soil particles.

35. The degree of saturation of samples of clay fill was calculated using measured water content, bulk density and specific gravity. Where measured values of specific gravity were not available an assumed figure of 2.8 was used. The degree of saturation is plotted against depth below ground level in Figure 7. About 48% of the samples tested were partially saturated. Most of the partially saturated samples were below the phreatic surface and show a random distribution of degree of saturation within the embankment fill. Clearly it is not possible to determine degree of saturation very accurately in this way, but the results are indicative of a generally less than saturated state in the fill.

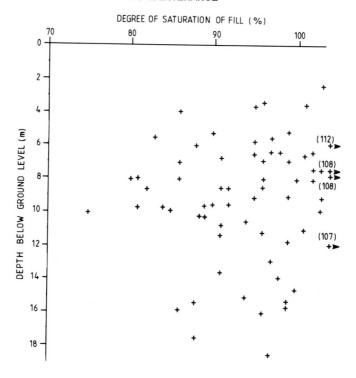

Figure 7. Degree of saturation plotted against depth, upstream embankment slope.

IMPLICATIONS FOR SLOPE STABILITY

35. In carrying out slope stability analyses for rapid drawdown conditions over a clay embankment, it is normal practice to assume that the soil is saturated and that the pore water pressure parameter $\bar{B}$ is close to unity. However, field data from Bartley Reservoir has demonstrated that $\bar{B}$ values as low as 0.2 can occur in clay soil as a result of construction related factors and the presence of methane gas. This means that for these dams actual factors of safety could be significantly less than those calculated by conventional assumptions. This could have significant implications for a number of old dams if the drawdown rates or drawdown level were to be changed. If such changes were to be made to reservoir operations an appropriate site investigtion should be carried out. With limited drawdown and analyses acceptable levels of drawdown and rate could be predicted.

ACKNOWLEDGEMENTS

36. The authors are grateful to Severn Trent Water Limited for their permission to present the findings of the investigation.

The role of instrumentation and monitoring in safety procedures for embankment dams

J. A. CHARLES, P. TEDD and K. S. WATTS, Building Research Establishment

SYNOPSIS. Where the continuing safety of an embankment dam largely depends on visual surveillance, this may be usefully supplemented by the installation and monitoring of appropriate instrumentation. Requirements for successful instrumentation and the effective utilisation of instrumentation results in dam safety assessment are discussed. The limitations of instrumentation need to be appreciated to avoid unrealistic expectations of what can be achieved by the use of instrumentation in safety evaluation.

INTRODUCTION
1. Although most UK embankment dams were built before it was common practice to install instrumentation during embankment construction, many older dams, particularly those showing signs of deterioration, have had instruments installed as part of an investigation. It is also quite common to install instruments in dams where no problems have yet been identified; the routine visual surveillance on which the continuing safety of the dam depends is thus supplemented by the monitoring of appropriate instrumentation. Pore pressure, seepage and settlement are the most helpful indicators of behaviour. This paper deals principally with installing and monitoring instrumentation at older existing dams rather than with installation of instrumentation in new dams during their construction.

ROLE OF INSTRUMENTATION
2. Installing and monitoring appropriate instrumentation should enhance the safety of an embankment dam. It can have a role in several aspects of safety evaluation.
3. General investigation. Little is known about the composition of many old embankment dams and instrumentation may be installed to remedy this lack of knowledge as part of a general investigation which is not related to a particular cause for concern. Suitable instrumentation, usually piezometers, can assist in the assessment of the existing ground conditions.
4. Investigation of particular features of embankment behaviour, problems and damage. Uncertainty about some feature of embankment performance may be resolved by instrumentation.

For example if an embankment has both an upstream clay blanket and a central clay core and it is not known which is the effective watertight element, this may be resolved by installing piezometers to measure the pore pressures in the upstream fill during changes in reservoir level. Thus instrumentation can be used to check the validity of an assumed model of embankment behaviour and an improved understanding of actual behaviour should be obtained. Instrumentation may be installed to investigate a localised situation at which damage has occurred or is suspected. This type of situation could include internal erosion of fill or foundation material or slope instability due to the development of a slip surface.

5. <u>Monitoring of long term safety</u>. Instrumentation may be used to confirm that long term performance is satisfactory and that the structure remains in a safe condition. By monitoring some facet of behaviour it may be possible to predict long term behaviour; measurements of crest settlement over a period may form the basis for predicting the future rate and magnitude of settlement. The predicted behaviour can then be used as a norm against which monitored behaviour can be compared. Changes in measured parameters may be more significant than absolute values. The use of instrumentation for this purpose, particularly routine monitoring of crest displacement, is well illustrated by Robertshaw and Dyke (1990).

MEASUREMENT OF PORE WATER PRESSURE

6. Often pore pressure is the most useful parameter to measure in an embankment dam. Strength and deformation of soil are controlled by the effective stress and this can usually be estimated when the pore pressure is known. The measurement of pore pressure is important in the following situations:
 (a) evaluation of slope stability
 (b) assessment of the effectiveness of the watertight element
 (c) investigation of vulnerability to hydraulic fracture or piping.

7. Piezometers placed in existing dams after construction are generally installed in vertical or near vertical boreholes. The porous element is surrounded by a sand cell which is sealed in the borehole with a plug of grout, such as bentonite. Failure to effectively seal the sand cell can lead to erroneous readings. Generally the porous element should be fully saturated before installation. There are many types of piezometers and recording system (Dunnicliff, 1988; British Standards Institution, 1981). Specially designed piezometers can be pushed or driven into the ground but in stoney fills, driving and jacking can be difficult or impossible.

Standpipe piezometers

8. Standpipe piezometers are the most common type of piezometer installed in existing dams since they are relatively cheap, simple to install, easy to read and their long term reliability is usually good. Despite their simplicity there are a number of problems that occur either because of their

slow response time or as a result of poor installation practice and faulty or badly designed equipment.

9. The design at the top of the standpipe requires attention. Standpipes on dam crests are terminated below crest level for protection and water will pond around the top of the standpipe if drainage has not been provided. Failure to prevent ingress of surface water into the top of a standpipe leads to erroneous piezometric levels being recorded when the piezometer tip is installed in a low permeability material such as puddle clay. Gravel in the top 0.5m of the borehole which is drained to the downstream side of the dam should solve the problem. If drainage between readings cannot be maintained, a waterproof protective cap should be fitted to the top of the standpipe. Ideally the cap should allow the piezometer to be vented as significant errors could occur if the piezometer is in a low permeability material and there are large changes in piezometric level. If the top of the piezometer is covered in water, the smallest hole in a cap for venting air will allow water to be drawn into the piezometer when the piezometric level is falling.

10. The water level in a standpipe is usually measured with an electric dip meter whose operation depends on the electrical conductivity of the water. To avoid premature readings due to condensation nearly all dip meters are now manufactured with one of the electrical contacts located centrally in the probe. Dirty contacts and low batteries can also lead to erroneous results. Other measurement methods are available such as a bubbler system or sealing a transducer into the standpipe (Dunnicliff, 1988). The latter reduces the response time of the piezometer.

11. Standpipe piezometers are valuable for carrying out permeability tests, the results of which can be used to assess the effectiveness of the clay core as a watertight element or the effectiveness of downstream fill as a filter to halt any erosion of the core (Tedd et al, 1988). A summary of some of the problems associated with pore pressure and permeability measurement has been provided by Tedd et al (1989a).

Other piezometers

12. Pneumatic and electrical (vibrating-wire and electrical resistance) piezometers comprise a porous stone and a diaphragm on which the water pressure acts. The systems for reading them are well known (Dunnicliff, 1988; British Standards Institution 1981). They have a rapid response time provided the porous stone and piezometer are properly deaired. More care is required to deair these piezometers which use high air entry (ie low permeability) porous stone. Pneumatic and vibrating-wire piezometers have been used in the upstream shoulders of dams due to the impracticality of access to standpipe piezometers. They cannot be used for permeability tests.

13. The long term performance of pneumatic piezometers is inferior to standpipe piezometers with failure occuring when the diaphragm valve cannot be opened by the pneumatic pressure

or fails to close after operation. Very long tubes increase the time of reading and dirt entering the pneumatic lines can prevent satisfactory operation. Vibrating-wire piezometers have a good long term stability record and are capable of being read remotely at great distance and automatically logged. Their main disadvantage is cost, being two or three times as expensive as a pneumatic piezometer.

14. Hydraulic piezometers are frequently installed in dams during construction. Advantages include suitability for remote reading, use in partially saturated soils and long term reliability. They have a limitation in that the tubing should at no point be higher than 5 metres above the piezometric level, which generally prevents their use in existing dams. Other disadvantages include the greater complexity of operation and problems due to freezing in cold climates.

MEASUREMENT OF TOTAL EARTH PRESSURE

15. Total earth pressure is less frequently measured than pore pressure and, more than for any other measurement, care, time and experience are needed when installing the instrumentation. Where hydraulic fracture is suspected in existing dams, earth pressures have been measured in clay cores using push-in pressure cells (Charles and Watts, 1987). The ratio of the insitu total stress to reservoir pressure can be used as an index of vulnerability to hydraulic fracture (Charles and Watts, 1987; Charles, 1989; Charles and Tedd, 1991). Long term monitoring indicates the stress changes that occur due to changes in reservoir level.

16. Measurement of horizontal earth pressure can be made using oil-filled push-in spade shaped pressure cells with a pneumatic read-out unit (Charles and Watts, 1987). A piezometer may be incorporated into the body of the cell. A cell is installed by pushing it 1 metre beyond the base of a vertical borehole which is grouted after installation. The installation of a cell will affect the in situ stresses and subsequent stress changes in the ground will be affected by the presence of the cell. Details of the installation, reliability and interpretation of the measurements of push in pressure cells are described by Tedd et al (1989b).

17. A system has been devised to measure vertical as well as horizontal pressures from a vertical borehole (Watts and Charles, 1988). Miniature oil filled earth pressure cells can be jacked horizontally 0.6 m into a soft clay soil using a special placing device which has been lowered down a 150 mm diameter borehole. A number of pressure cells and piezometer tips can be installed at pre-determined depths within one borehole. Installation involves skilled personnel in a complex operation, but subsequent monitoring is simple.

MEASUREMENT OF SEEPAGE AND LEAKAGE

18. An important indicator of embankment performance can be provided by the measurement of seepage and leakage flows. Particular attention should be paid to the occurrence of new

flows. Turbid water may be carrying soil particles and can signify internal erosion. Analysis can reveal whether the sample contains suspended particles of soil or whether there are dissolved minerals. Comparison of flows with rainfall and reservoir level can give a good indication of the origin of the flow and in particular whether it is reservoir water.

19. Where a flow occurs and is visible it is relatively easy to measure it, but in old dams there may be no arrangements for collecting such flows. Seepage and leakage through the embankment, foundations or abutments are measured at some old dams and the flow in springs and in the stream below some dams is also checked periodically. In some instances a measuring chamber is installed, but in most dams it is adequate to install a simple system, such as a V-notch incorporated in the existing drains. A datum from which measurements at V-notches are taken can assist with readings. Measuring devices should be maintained clear of debris to ensure their accuracy, and details of all cleaning operations undertaken should be kept with the measurement records.

MEASUREMENT OF SURFACE DEFORMATIONS

20. The deformation of an embankment can be an important indicator of field performance. The most common measurement is of settlement or heave of a dam crest and Charles (1986) described a settlement index based on these measurements. Vertical and horizontal movements can be measured by surveying techniques using theodolite, optical level, electro-optic distance meter (EDM) or photogrammetry (Johnston et al, 1990). Accuracy and reliability of surveying techniques depend on the following:
 (a) type of survey instrument
 (b) repeatability of positioning for instruments, levelling staff, targets etc
 (c) stability of pillars and other reference points
 (d) protection of pillars and other reference points against damage
 (e) competence of survey staff
 (f) effects of meteorological conditions
 (g) length of sight.

21. Optical levelling is the most common method of measuring settlement. Measurement of horizontal displacements requires a theodolite or possibly an EDM. On dams with a straight crest, horizontal displacements upstream and downstream can be observed using line of sight techniques. On several major dams, a geodetic network, which can provide three dimensional values for the position of the movement station on the dam, has been used. Modern integrated total station survey equipment may be linked to a computer and can provide readings to an accuracy of a few millimetres.

MEASUREMENT OF SUB-SURFACE DEFORMATIONS

22. The complex instrumentation required to measure internal displacements has rarely been installed in existing dams.

Valuable information has been obtained in relation to horizontal and vertical displacements at new dams and where such installations are in place and functioning, it is worthwhile incorporating them in the routine monitoring system.

23. Magnet extensometers installed in boreholes to detect vertical movement are generally accurate, reliable and easy to use. The extensometer consists of magnets fixed to the sides of a borehole, which move with the adjacent dam embankment or foundation. A reed switch sensor lowered in a central access tube within the borehole, locates the position of the magnets. The accuracy of the readings depends upon the type of equipment and the care of the reader, but accuracies of better than 1 mm can be obtained. The system permits any number of magnets to be installed at different levels in one borehole (Burland et al, 1972). Such a system can be extended or cut down if a dam is raised or lowered without affecting the continuity of the observations. Tedd et al (1990) have described the monitoring of the deformations of an old puddle clay core dam during reservoir drawdown and refilling using this system.

24. Inclinometers or tilt meters can be used to measure displacement in any direction in a vertical plane. The standard inclinometer is an electronic instrument which is lowered down a special tube set into a vertical borehole, and its inclination is measured at consecutive positions. The overall accuracy of measurement can be 2 mm in a borehole of 20 m depth. The base of the inclinometer tube should be deep enough to ensure that it is in a fixed position. The location of the top of the tube may also be determined by surveying.

25. If it is not possible to gain access to a standard inclinometer tube, a permanently installed tilt measuring system may be required and the electro-level (E-L) system which is based on 20 years of development experience at BRE may be suitable. The E-L is a gravity sensing electrolytic transducer that provides an output voltage proportional to the tilt angle. It consists of a small glass sealed tube, partially filled with electolytic fluid and with metal electrodes in contact with the electrolyte. Great care is required in the installation and consequently they are expensive. As with all inclinometer systems where displacements are summated from a series of readings, errors cumulate and one faulty reading affects the measured total displacement.

26. Recently E-L systems have been permanently installed by BRE in two dams during embankment construction. The E-Ls are about 30mm long and 6mm diameter with a range of ±3° (±52mm displacement over a metre length). At Roadford Dam they were installed to measure the deflection of the upstream asphaltic membrane close to the concrete cut-off structure. E-Ls were fixed to the inside of a series of rigid box sections (50mm by 100mm by 0.5m or 1.0m long) connected together by sliding pin joints that allowed rotation in the vertical plane. Sections were linked to extend 14m up the upstream slope from the toe of the dam and were placed immediately beneath the asphaltic membrane. Cables from the E-Ls were taken into the inspection

gallery. Using a portable readout unit, satisfactory readings have been obtained for a period of more than 2 years which has included the initial impounding of the reservoir (Tedd et al, 1991). E-Ls have been installed to measure horizontal deflections in the fill close to the foundation at the reconstruction of Carsington Dam. They were installed at 1m spacing in 8m long vertical inclinometer tubes at 4 locations in the dam. As the instruments were located on both sides of the core, cable lengths of up to 200m were required to connect to the readout instrument at the downstream toe. They have worked throughout the construction period of the dam.

MEASUREMENT OF DYNAMIC PERFORMANCE
27. Britain is an area of low seismicity where the probability of earthquakes causing significant damage is relatively small. Nevertheless earthquakes do occur and it may be considered desirable to monitor the dynamic performance of large dams under seismic loading. Permanent deformations will be measured by the instrumentation and surveying techniques described above. The monitoring of dynamic behaviour during a seismic event requires continuous monitoring by two automatic strong motion accelerographs, one at foundation level, one at crest level (Charles et al, 1991).

REQUIREMENTS FOR SUCCESSFUL INSTRUMENTATION
28. Instrumentation is only likely to be worthwhile if the following conditions are met during installation and operation.
29. <u>Installation</u>.
(a) The types of instrumentation and their location are carefully designed to give relevant information about, for example, perceived hazards and problems.
(b) Instrumentation is purchased in an appropriate way. The cheapest tender is not always an appropriate basis for selection.
(c) Installation should be carefully supervised by qualified personnel who are conversant with the operation of the instrumentation and the principles of geotechnical engineering.
(d) Zero, datum or reference readings should be obtained at the time of installation. Some forms of instrumentation (eg total earth pressure cells) need to be calibrated before installation.
30. <u>Operation</u>.
(a) Instrumentation should be robust and able to survive adverse site situations including vandalism. Instruments should have an adequate working life. Often the simpler types of instrumentation are the more reliable.
(b) Readings should be taken at appropriate intervals, plotted and assessed at that time by qualified personnel. Lack of long term continuity of personnel can be a problem.
(c) There is a growing tendency for instrument manufacturers to develop blackbox technology including data logging and automatic data transfer of results to computers. It is vital that adequate checks are built into the system so that any

malfunctioning is identified at an early stage.

(d) Measurements should be related to the conditions at the time of the measurement. Records of relevant factors such as reservoir level and rainfall should be kept, otherwise the measurements cannot be reliably interpreted.

(e) There should be a specified course of action if readings indicate some unsatisfactory behaviour of the structure.

31. **Reporting**.

(a) The first instrumentation report should describe the type and location of the instrumentation, calibrations, methods of reading etc.

(b) Subsequently reports should be prepared at specified intervals and include an updating of the measurements and a review of the implications of the measurements for embankment performance and safety.

LIMITATIONS OF INSTRUMENTATION

32. It is well to keep in mind the limitations both of instrumentation in general and of the specific scheme that has been adopted in a particular situation.

(a) The cost of instrumentation has to be related to the expected benefits and this limits the amount of instrumentation that can be installed. However the cost of the instruments themselves usually forms only a small proportion of the total cost of instrumentation which includes installation, monitoring and assessment of results. It is unwise and a waste of resources to economise by using low quality instruments.

(b) Some installations present practical difficulties and this may limit the amount of instrumentation that can be used. Access on embankment slopes can be difficult and costly.

(c) Measurements are usually discontinuous in space and often in time. Data from instrumentation will not be comprehensive and the most adverse circumstances may not have been recorded.

(d) Instrumentation may malfunction and give no measurements, or worse, give erroneous readings.

(e) Installation or presence of an instrument may affect soil behaviour. The installation process will inevitably disturb the ground. The presence of an instrument may affect subsequent ground behaviour.

(f) In certain circumstances installation of instrumentation could seriously damage an existing dam. For example a vertical borehole could pass through a sloping clay core and provide a passage for reservoir water through the core.

(g) Instrumentation is not a substitute for an understanding of the likely mechanisms of behaviour of the structure. Some knowledge of embankment behaviour is a prerequisite of successful instrumentation. Generally one only finds what one is looking for.

INSTRUMENTATION AND SAFETY EVALUATION - CONCLUSIONS

33. There are many old embankment dams in the UK which pose a potential hazard to communities downstream. Little is known about the internal composition of many of these dams. To a

large extent their safety is dependent on diligent maintenance and effective surveillance. Installation and monitoring of appropriate instrumentation can be a valuable addition to visual surveillance in developing effective safety procedures.

34. Pore pressures, seepage and settlement can give helpful indications of the performance of the embankment and the development of any potential hazard (eg slips, internal erosion). However the limitations of instrumentation should not be overlooked. Cost, difficulties of installation, restrictions on the resources available to monitor and analyse the results, all limit the amount of instrumentation installed. The most adverse situation may not have been monitored and instruments may malfunction. In some circumstances changes in measured parameters may be more significant than absolute values.

35. After installation, instrumentation should be read frequently until equilibrium has been reached following the disturbance due to the installation process. Regular monitoring can then commence. Once a pattern has been established, less frequent readings may be sufficient for routine monitoring purposes. Where instruments have been installed because of a particular concern, they should be read frequently until the problem is resolved. Readings should be presented in a tabular form and then plotted in a graphical form, so that they may be compared with other important parameters such as reservoir level and rainfall. Measurements should be analysed and assessed and there must be a specified course of action if unsatisfactory behaviour is monitored.

ACKNOWLEDGEMENTS

The work described in this paper forms part of the research programme of the Building Research Establishment and is published by permission of the Chief Executive. The main client for the work is the Water Directorate of the Department of the Environment.

REFERENCES

1. BRITISH STANDARDS INSTITUTION (1981). BS 5930: Code of practice for Site investigations. BSI, London.
2. BURLAND J B, MOORE J F A AND SMITH P D K (1972). A simple and precise borehole extensometer. Geotechnique, vol 22, 174-177.
3. CHARLES J A (1986). The significance of problems and remedial works at British earth dams. Proceedings of BNCOLD/IWES Conference on Reservoirs 1986, Edinburgh, 123-141. Institution of Civil Engineers, London.
4. CHARLES J A (1989). Deterioration of clay barriers: case histories. Proceedings of Conference on Clay Barriers for Embankment Dams, Institution of Civil Engineers, October 1989, 109-129. Thomas Telford, London.
5. CHARLES J A AND WATTS K S (1987). The measurement and significance of horizontal earth pressures in the puddle clay cores of old earth dams. Proceedings of Institution of Civil

Engineers, Part 1, vol 82, February, 123-152.
6. CHARLES J A AND TEDD P (1991). Long term performance and ageing of old embankment dams in the United Kingdom. Transactions of 17th International Congress on Large Dams, Vienna, vol 2, 463-475.
7. CHARLES J A, ABBISS C P, GOSSCHALK E M AND HINKS J L (1991). An engineering guide to seismic risk to dams in the United Kingdom. Building Research Establishment Report BR 210.
8. DUNNICLIFF J (1988). Geotechnical instrumentation for monitoring field performance. Wiley, New York.
9. JOHNSTON T A, MILLMORE J P, CHARLES J A and TEDD P (1990). An engineering guide to the safety of embankment dams in the United Kingdom. Building Research Establishment Report BR 171.
10. ROBERTSHAW A C and DYKE T N (1990). The routine monitoring of embankment dam behaviour. The Embankment Dam. Proceedings of 6th Conference of British Dam Society, Nottingham, 177-183.
11. TEDD P, CLAYDON J R and CHARLES J A (1988). Detection and investigation of problems at Gorpley and Ramsden Dams. Proceedings of Reservoirs Renovation 88 Conference, Manchester, paper 5.1, 1-15. BNCOLD, London.
12. TEDD P, HOLTON I R and CHARLES J A (1989a). Standpipe piezometers: some problems with pore pressure and permeability measurement. Geotechnical instrumentation in practice, 828-830. Thomas Telford, London.
13. TEDD P, POWELL J J M, CHARLES J A and UGLOW I M (1989b). In situ measurement of earth pressures using push-in spade-shaped pressure cells; ten years experience. Geotechnical instrumentation in practice, 701-715. Thomas Telford, London.
14. TEDD P, CHARLES J A and CLAYDON J R (1990). Deformation of Ramsden Dam during reservoir drawdown and refilling. The Embankment Dam. Proceedings of 6th Conference of British Dam Society, Nottingham, 171-176. Thomas Telford, London.
15. TEDD P, PRICE G, WILSON A C and EVANS J D (1991). Use of the BRE electro-level system to measure deflections of the upstream asphaltic membrane of Roadford Dam. Proceedings of 3rd International Symposium on Field Measurements in Geomechanics, Oslo, 261-271. Balkema, Rotterdam.
16. WATTS K S (1991). Evaluation of the BRE miniature push-in pressure cell system for in situ measurement of vertical and horizontal stress from a vertical borehole. Proceedings of 3rd International Symposium on Field Measurements in Geomechanics, Oslo, 273-282. Balkema, Rotterdam.
17. WATTS K S and CHARLES J A (1988). In situ measurement of vertical and horizontal stress from a vertical borehole. Geotechnique, vol 38, no 4, 619-626.

Crown copyright, 1992.

Investigation, monitoring and remedial works at Tiga Dam, Nigeria

H. S. EADIE, MSc, DIC, PEng, MICE, MIWEM, **D. J. COATS,** CBE, DSc, FICE, FEng, FRSE, **and N. LEYLAND,** BEng, MICE, Babtie Shaw and Morton

SYNOPSIS. The embankment of Tiga Dam in Northern Nigeria has been a subject of concern since its construction in 1978. Unusually heavy rains in 1988, and the ensuing failure of an adjacent similar smaller dam built around the same time, further increased safety fears. This paper summarises the investigations undertaken to assess the safety of the embankment, the findings, and describes the short term remedial works carried out by the client's direct work organisation. The further long term remedial works are outlined along with the economic justification.

INTRODUCTION
1. In July 1989 Babtie Shaw and Morton were successful in obtaining a commission from the Overseas Development Administration to provide investigation, monitoring, training, design and supervision services for remedial works on behalf of the Ministry of Agriculture, Water Resources and Rural Development, Kano. The owners at the time being the Hadejia Jama'are River Basin Development Authority (HJRBDA).
2. Tiga Reservoir is situated near Kano, Northern Nigeria, and feeds a major irrigation project. The dam is a zone filled laterite embankment some 6km long with a maximum height of 48m. Ever since its construction in 1978 there have been fears over its safety.
3. The full project entailed:-
    - the collection and review of all previous work on the dam
    - a limited ground investigation
    - design and supervision of small scale remedial works
    - rehabilitation of existing instrumentation
    - assessment of 1988 flood and Probable Maximum Flood
    - costing total dam failure
    - costing of remedial works and their economic justification
    - training of Authorities staff in dam monitoring and maintenance
    - the recommendation of a staffing plan
    - the full time provision of a senior dam engineer for the project period.

RESERVOIR MONITORING AND MAINTENANCE

4. The particular prompt to this work was the unusually heavy rains which occured in Northern Nigeria in late 1988 which resulted in widespread severe flooding and led to the failure of the nearby (about 10km distant) Bagauda earth embankment dam with loss of life and severe infrastructure damage. As both Tiga and Bagauda had been constructed by the Kano State Water Resources Engineering and Construction Agency (WRECA) the previous concerns over Tiga became more acute.

BACKGROUND

5. Although the embankment was built only some 14 years ago construction records were woefully short and despite efforts it proved to be impossible to trace individuals who would admit to being associated with its construction. Details therefore had to be gleaned from original Dutch Consultant's proposals, their subsequent evaluation report after its completion (without their involvement), and a short report during construction.

6. Further information on the likely construction was obtained however by the examination of the breach in Bagauda and an incomplete and abandoned embankment at Challowa Gorge started by WRECA shortly after Tiga. A bundle of undated and unlabled photographs provided by HJRBDA also proved extremely useful.

7. Particular aspects of construction were:-
   i) The 5-7m wide cut off trench was excavated by ripper up to a maximum of 13m deep through laterite and decomposed rock to fresh rock. Sides were generally vertical except in the alluvial deposits of the old river bed. The grouting programme was reported as being ineffective.
   ii) The cut-off was filled with locally won residual soil in layers of 150-300mm deep and compacted with a 24 ton sheepsfoot roller.
   iii) The embankment was constructed in 200-300m lengths working to the centre with a transverse slope of 1 in 4. The core in places at least was constructed in advance of the shoulders and river sand was dozed over the downstream face for the filter.
   iv) The fill had a very low moisture content and there appeared to be limited watering plant. In consequence very low densities were obtained.

8. Since construction there had been three reports on aspects of its condition including one in 1987 by the US Bureau of Reclamation which recommended the creation of a lower emergency 200m wide spillway which was started immediately after the 1988 floods. This report also noted the existence of wet patches and the appearance of 'boils' between Ch 4000 and 4500m at the toe.

9. There were, prior to our involvement, no formal regular inspections by trained engineers nor regular monitoring of piezometers or weirs.

INVESTIGATIONS

10. It was clear from the earlier reports that the dam had not been constructed to the initially proposed crest levels or side slopes and there was no way of monitoring any settlement. An early exercise was therefore a topographic survey which confirmed at locations side slopes approaching 1 in 2.2 (design 2.5) and an average crest width of only 7m, approximately half that of the original design.

11. Within the time and budget available and considering the size of this dam, only an extremely limited ground investigation could be undertaken. It had as its main aims the estimation of
   1) the internal profile of the dam
   2) the internal piezometric surface
   3) the conformation of the properties and quality of the embankment materials.

12. The work consisted of drilling 15 boreholes within the dam and the downstream toe, followed by the installation of piezometers, the excavation of trial pits in selected areas of the toe, and laboratory testing of samples. The work was carried out by Trevi Foundations (Nigeria) Ltd under our direction and supervision.

DAM ANALYSIS AND INTERPRETATION

13. Although little factual information about the foundations of this 6km long dam is available it is thought that the foundation materials are:-

Ch. 1-1200 metres (embankment height 0-30m)	: Sandy lateritic clay overlying moderately weathered granite or gneiss to which the cut-off extends.
Ch. 1200-2500 metres (embankment height 30-35m)	: About 5 m thickness of silty sandy clay overlying alluvium infill to a buried channel centred between 1800 and 2050 where it is up to 18m deep. The cut-off does not quite reach the underlying moderately weathered gneiss.
Ch. 2500-3100 metres (embankment height 35-48m)	: Moderately weathered granite or gneiss : Area of river gorge with old river bed at about 48m below embankment crest.
Ch. 3100-6000 metres (embankment height 0-30m)	: silty sandy lateritic clay overlying weathered granite to which the cut-off extends.

The sandy lateritic clay in the foundations was similar to the fill material for both core and shoulders but the silty alluvium was much finer.

14. The composition of the embankment is highly variable. The core and downstream shoulder contain both cohesive and granular material which is poorly compacted in places. The granular material found in the core is probably chimney material left on the partially completed core (see 7 (iii)

above) and similar pockets were found in the core of the Challowa Gorge dam.

15. Both the core material and the sand filter materials are finer than intended. The permeability of the filter is less than preferable but its grading is probably such as to prevent soil particles passing through the filter voids provided the adjacent soil does not have a high silt content or is alluvium. The thickness of the drainage blanket is much less then intended.

16. The embankment slopes vary, the crest width is not uniform, the water line is distorted, the level of the embankment crest which was up to 1 metre below intended level when the dam was completed in 1978 had since settled about 400mm (the settlement was greatest from ch. 1000 to 2500m). The downstream slope had been badly eroded by rainfall and remedial works were in hand (but only partially successful), the phreatic surface within the dam body was unclear, there were marshy areas immediately downstream of the dam and a nearby dam of similar age had recently failed by piping! Identifying a typical section for analysis was hardly likely!

17. A number of cross-sections with a variety of assumptions were analysed by conventional limit equilibrium methods and the section shown in Figure 1 gave the lowest factor of safety of 1.5 at the downstream and 1.0 at the upstream (on rapid drawdown). These assumed no seismic forces as the site was known not to be subject to earthquakes. When a horizontal acceleration of 0.05g was introduced the factor of safety

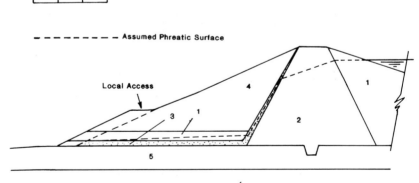

1 Shoulder-Properties: c-0, 0=32, $\gamma' = 18.5 KN/m$
2 Core-Properties: c=0, 0=28, $\gamma' = 19.5 KN/m$
3 Blanket-Properties: c=0, 0=32, $\gamma' = 18 KN/m$
4 Shoulder-Properties: c-0, 0=28, $\gamma' = 19 KN/m$
5 Rock Properties: c-0, 0=36, $\gamma' = 21 KN/m$

Figure 1. Section at Chainage 2750

reduced to 1.3. However, no statement on stability can be made with any confidence until the behaviour of the dam is better understood.

18. It was obvious that the seepage flows contributing to the nearby areas at the toe of the dam had to be investigated. It proved that water in the drainage blanket was under pressure at some places and the remedial works including refurbishment of existing drains as described later relieved this pressure and provided the means of measuring seepage which proved to be modest. It was clear from borehole logs that the cut-off was not always extended to solid rock and sandy silt or gravelly material sandwiched between the core material and weathered rock could allow seepage under the core. The new and refurbished drainage system would also allow examination of seepage flows to see if they were transfering materials since there was a possibility of piping developing and this is probably the condition to which the dam is most vulnerable.

19. The owners of the reservoir were therefore left in no doubt that monitoring of drainage and crest settlement and intelligent interpretation of results including piezometer readings would be of paramount importance.

REMEDIAL WORKS

20. The existing chainage marker system was found to be inaccurate and therefore steel markers (wooden ones being eaten by termites) were established at 100m intervals along the dam crest. A level control was established using the lower spillway as datum.

21. Originally no provision had been made to monitor settlement of the bank and so simple settlement monuments were installed at 50m intervals alongside the roadway on the dam crest. As best as can be judged this work revealed that the embankment has settled at the greatest height by approximately 400mm since 1978.

22. As part of this exercise the opportunity was taken to instruct the HJRBDA staff in good surveying and draughting practice.

23. The embankment drainage has a chimney drain sandwiched between the central core and the downstream shoulder, which is then continued to the downstream toe as a drainage blanket (Figure 1). The original drawings show this as terminating in a rock fill toe but this was not built and the effectiveness of the system was questioned particularly where there was a washdown of the embankment sides which sealed the blanket outlet. Site inspection however revealed toe drains installed in half of the 3km left bank and around a third of the 3km right bank. On the left bank the existing drainage was found to be around 2m deep and with no signs of having functioned. Surrounding water levels were below the drain level. On the right bank seepage between Ch 3000 and ch 4500 had been a source of concern since impounding resulting in a toe drain being installed between ch 3000 and ch 4000.

24. Site inspection revealed 4 outfalls from the system with

RESERVOIR MONITORING AND MAINTENANCE

V notch weirs all discharging clear water, and 22 manholes of which 21 had deep stagnant water to within 1m of ground level. Later 2 further manholes were found with chambers broken just below ground level and full of silt and cobbles, the damage resulting from previous downstream slope erosion repair work. Additionally moist areas were evident downstream of the drains. The overgrown V notch weirs were rehabilitated and downstream channels cleared. The drainage systems were manually cleaned in turn progressing upstream with around $1m^3$ of cobbles removed from each manhole along with other assorted debris. After unblocking, the flow at the outfalls increased by around 50%. The general ground water level downstream of the drain also fell by over 1m resulting in the moist areas drying up.

25. Between ch 4000 and ch 4500 an extremely soft marshy area remained some 150m wide along the toe of the bank where sand boils had been reported during periods of high reservoir levels. It was therefore decided to install a toe drain starting in January 1990 to give both a positive outlet to the sand blanket and to intercept and control underseepage.

26. At intervals of 50m, finger drains were excavated into the dam embankment to connect with the existing sand blanket under the downstream shoulder. (Figure 2). These consisted of 5m wide excavations filled to a minimum depth of 2m with coarse river sand extending the sand blanket to a position 15m downstream of the toe to accommodate any future stability berm. These are drained by a 225mm perforated pvc pipe surrounded by fine gravel and coarse sand filters into manholes located on a carrier drain running parallel to the bank. The carrier drain is of similar construction but with 2 No. pipes and with most of the excavation backfilled with coarse sand. V notch weirs are located in all manholes to enable recording the discharge from each finger and in the main carrier drain.

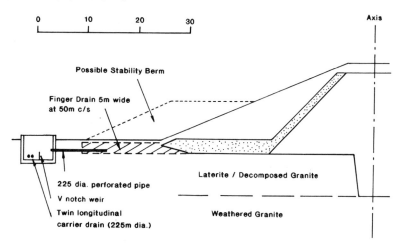

Figure 2. Remedial Drainage Works Chainage 4000 - 4500m

27. The carrier drain trench was excavated from a temporary roadway constructed through the marshy area along the line of the drain. As the excavation progressed upstream surrounding ground water levels fell rapidly from around 0.5m artesian to 2.0m below ground level. On completion of the drain the marshy area had dried completely and baked allowing easy movement of plant.

28. The new drain works were undertaken as the reservoir fell in level, with the resulting head being much reduced. The limited excavations into the drainage blanket yielded only small flows thought to be due to the preceding excavation of the main carrier drain and its effect on ground water in the vicinity.

GENERAL PROBLEMS/SUCCESS

29. Although the HJRBDA have a large workforce the overall staff size is small and structured essentially towards the administration of the irrigation project served by Tiga Dam. The small engineering section is mainly engaged in minor maintenance works which are carried out both by direct labour and outside contractors, with other works designed externally and also constructed using contractors. The construction of the toe drain on the right bank was a far larger project than previously attempted by direct labour and in house resources, and required some $6000m^3$ of coarse river sand, $600m^3$ of fine gravel and some 1.2 km of 225mm perforated pvc pipe. The few staff assigned to the works were inexperienced in site works and required our senior dam engineer to take on the normal role of the contractor's site staff, rather than the usual supervisory duties.

30. A source of coarse river sand was readily located in a tributary of the Kano River 15 km from the dam but fine gravel proved more difficult due to the remote location. Eventually a quarry near Kano some 100km away was found producing single size crushed granite aggregrate thus requiring the purchase of two sizes which were combined by mixing stockpiles on site. Perforated pipe was obtained by hand drilling 6mm holes at 50mm centres for half barrel, into 1.2km of pvc water pipe already owned by HJRBDA!

31. HJRBDA own a reasonable amount of construction plant, but owing to the lack of funds for spare parts and subsequent poor maintenance little was serviceable. This resulted in very slow progress until more reliable plant was hired from BEC, a French contractor operating some 300km away. The delay resulting from slow sourcing of materials and plant stoppages, caused the work to extend through April which in Northern Nigeria is exceptionally hot with daytime shade temperatures commonly in the mid forties, and coincides with the Moslem feast of Ramadan. With an almost wholly Moslem workforce unable to drink or eat during the day, production during April was limited. It was a relief to all involved when the works were completed with only minor disruption from early rains at the onset of the rainy season.

RESERVOIR MONITORING AND MAINTENANCE

MONITORING

32. A dam supervising section was set up within the HJRBDA and the allocated staff given a series of lectures. Assisted by the Senior Dam Engineer, the staff prepared records for the monitoring system, and over the months of the commission took readings regularly. In total 21 existing servicable observation wells were found which were supplemented by the 15 piezometers installed during the site investigation. The 4 existing V notch weirs were supplemented by 15 installed in the new right bank drainage. These supplemented reservoir levels and reservoir discharge records already being kept by HJRBDA, and rainfall records taken by WRECA on behalf of HJRBDA.

SPILLWAY CAPACITY

33. Tiga, as constructed, was seriously deficient in freeboard and spillway capacity as the original design was for less than the Probable Maximum Flood and the embankment crest over 95% of its length was surveyed as up to 1.2m lower than design. The present temporary bypass channel (para 8 above), dug some 4m below the original crest level, was found to be adequate for the Probable Maximum Outflow flood of $2042m^3/sec$, however its cill level is such that storage now falls well below the original design and prevents the development of the full irrigation command area. The 1988 flood which spilled over the original weir was estimated at $650m^3/s$ with a return period of 1 in 16 years.

LONG TERM REMEDIAL WORKS

34. Assuming the dam not to be restored to its original design levels and profile recommended remedial works comprise:-
   1) Improvements to the dam core in the upper zones
   2) Making permanent the temporary spillway with a slightly elevated crest and inflatable weir to further increase storage
   3) Construction of a wave wall
   4) Upgrading of the undersized rip rap on the face
   5) Improvements to the main valve chamber and sealing on existing 0.6m diameter ventilation pipe which runs up the upstream face of the dam
   6) Fit upstream valves to the auxiliary outlet pipes
   7) Install further toe drainage

35. A cost : benefit analysis was undertaken assuming the costs to arise from a total breach of the embankment (as Bagauda). The areas of inundation, damage and crop loss were identified using the DAMBRK program. A cost : benefit ratio of 13.0 was obtained - a very high figure making an overwhelming case for the safety expenditure against the real risk of an embankment failure.

Vegetation and embankment dams

C. G. HOSKINS, BSc(Eng), MICE, RKL Geotechnical, and
P. R. RICE, BSc, ARK Associates

SYNOPSIS. A substantial amount of information is available on the effects of vegetation on structures, but the consequences and benefits on embankment dams have received little attention to-date in papers and the technical press. Many engineers have actively discouraged the presence of all bar a grass cover on slopes, but recent events suggest that this approach may be misguided and larger trees and vegetation on certain parts of the embankment are not detrimental and may improve the safe functioning of a dam. Increasingly, environmental considerations require the planting or maintenance of a more substantial vegetation cover and thus there is likely to be an increasing move towards more vegetation in the future. A case history of dam instability following tree removal is given.

BACKGROUND AND CONTEXT
1. Many of the earliest dams that are still functioning today were planted deliberately with trees. In the 1700s Capability Brown, Humphrey Repton and others remodelled country estates and planted large numbers of trees with planting schemes that extended onto and over the embankment. Other dams on the lowland areas from that time onwards were also planted or allowed to develop a tree covering. Subsequently in the late 1700s and for the next two hundred years, substantial numbers of reservoirs have been built for water supply or augmentation purposes or as canal feeders. Water availability, amongst other facts, have resulted in many of these being sited in upland areas. The embankments were often steeper and of a greater height than those earlier and, with their location in less favourable areas for the subsequent development of a tree cover, their vegetation has been limited typically to grass plus sparse trees and shrubs where these could develop in the more sheltered positions. This allied with the increase in the number of dams in less hospitable locations, appears to have fuelled a general belief that extensive vegetation cover, particularly trees, is detrimental to the well-being of the embankment and, in many instances, any developing trees have been removed. This approach has continued to the present time and many instances

are known where tree cover that has been in place for many years or decades has been cut down at the request of an Inspecting Engineer. In some instances, this has led to consequent problems with the embankment and an example is given later in this paper.

2. The current trend towards an enhanced environmental awareness will necessitate a greater justification for the future removal of mature trees and the loss of the associated woodland habitat. The effects of the trees, whether detrimental or beneficial, must now be considered more fully before any action is taken and their removal shown to be necessary. The increasing use of existing reservoirs for amenity purposes is likely to continue and this will exert further pressure to retain and improve their wooded cover. Reservoir construction in the future is likely to include a significant number of reservoirs for amenity, landscaping and recreational purposes whose size would put them at the lower end or just beyond the scope of the Reservoirs Act 1975, and thus will necessitate a more balanced view.

EFFECTS OF VEGETATION ON EMBANKMENT DAMS

3. The major influences of vegetation on embankment dams are hydrological and mechanical. Whilst all vegetational features have an effect, to some extent, the root system is of particular relevance in dam engineering. They vary from very fine fibrous hair roots through to larger branched roots, frequently with a near vertical tap root. Occasionally near vertical sinker roots often link up the various root levels. Shallow roots form a near surface mat which extracts mineral nutrients from the enriched areas just below the litter layer. Deeper roots form the anchorage and provide the main water extraction facilities. The larger roots are perennial whereas many of the fine roots, whether for mineral or water extraction, are subject to seasonal dieback and regrowth. Much of the root system develops solely to extract water and thus the root pattern is largely dependent on the soil-moisture relationship for any given type. Well drained soils, particularly with a poor nutrient content, will lead to a sparse development of deep roots; wetter soils will necessitate a less extensive growth to obtain the same amount of water. A water table or a densely compacted soil or barrier at shallow depth will result in a more shallow lateral root spread covering a wider area. Much of the root system below a grass cover is within the uppermost 50mm and takes some years to develop. The roots tend to be highly branched and fibrous, but the parts of the root system which remain active in winter are greatly reduced. Most roots from other vegetation are found within a few hundred millimetres of the surface below herbaceous vegetation, but may typically extend to 3m depth below trees. Some roots may be found at greater depth, but these normally

form only a small proportion unless the water table is very low.

4. Two frequent misconceptions are often believed with regard to tree roots. The first is that the root spread is equal to the size of the crown, i.e. the branch spread, and the second is that tree roots are either shallow rooted or are essentially a tap root. Both are incorrect and the roots will develop as required to abstract sufficient water and ensure stability of the tree. Thus the lateral extent of tree roots can be considerable and can extend well beyond their crown area as the hair roots seek to extract water in unfavourable conditions. The extent of influence for various species has been investigated by several authors (eg Ward (Ref. 1), Driscoll (Ref. 2), Greenway (Ref. 3)), and the rank order of water demand tree species given. These have not been entirely consistent but some general comments can be made. The water loving species, willows, alders and poplars have particularly extensive root systems and a high water demand. These tend to be well known in this respect, but other trees can develop moderately deep roots to satisfy their water needs. Included in this list are oak, horse chestnut, hawthorn, rowan/service tree, cherries/plums and, in some instances, sycamores/maples.

5. The roots have a tensile strength which will vary considerably with species, age, size and time of year; a value of about 30 MN/m^2 has been quoted for the Pedunculate Oak. Following felling, the strength of the roots is slowly lost; falling to perhaps 50% after one year but retaining some strength for several years. Roots on a slope, particularly those of the larger shrubs and trees, tend to grow in an upslope direction to counteract gravity forces and thus tend to remain perpendicular to the ground slope. The application of forces to the roots causes them to thicken and thus the thickest roots will be on the upslope side of a tree, particularly those running obliquely to the slope and acting as anchors.

6. Roots have great difficulty in penetrating soils which have strengths greater than 2 MN/m^2 (ie weak rock), but a high bulk density can also limit penetration. Densities above which growth is restricted are approximately 1.4 t/m^3 for clay soils and 1.7 t/m^3 for sandy soils. Clay content also has a significant effect on root penetration and an increasing content tends to retard penetration. The structure, voids content and presence of organic matter and existing roots all tend to influence root penetration. Roots tend to follow the easiest route and thus backfilled trenches and holes offer least resistance to spreading; this has been confused in the past with roots seeking water from the pipe in the drain. Where a fracture does occur, the root will gain entry and spread along depending on the amount of flow.

Opening up of the fracture and increased root development is then likely. Certain trees are more prone to affecting drains and a summary by Cutler and Richardson (Ref. 4) suggests that cherries/plum trees, horse chestnut and sycamores/maples are more likely to affect drains in addition to willows and poplars. Root growth can be sufficient to affect adjacent structures and walls, but this is normally very localised and results from the development of large lateral roots immediately adjacent to the trunk.

7. The effect of vegetation results in an overall increase in the strength of a soil mass as a result of soil reinforcement by the roots and a soil moisture deficit by evapotranspiration. The latter is seasonally variable and thus soil moisture levels will build up in winter and may approach the field capacity. The soil moisture deficit implies suctions in the soil which can be considered as a reduction in pore pressure in a saturated soil. In a partially saturated soil, the effect of a soil suction can be more readily considered as an increase in cohesion. Soil suction increases as the size of voids and capillary channels decrease; its effects are therefore greater in fine grained soils. Methods of assessing the soil strength by allowing for the presence of an additional cohesion, c_r, due to the root reinforcement and a further cohesion, c_s, resulting from the soil suctions have been proposed (Walker and Fell (Ref. 5)). Where the permeability is low and the rate of replacement of water is not large, the development of soil suctions will result in a lowered water table. This can be considered directly as an appropriate way of allowing for the effect of vegetation on soil strength, in preference to adjusting the cohesion.

8. The principal benefits of established vegetation on embankment dams are improved slope stability, surface erosion control against flowing water or abrasion effects and, on the upstream slope in some instances, wave erosion control. Other lesser benefits, in some instances, may include noise attenuation, wind shelter and access barriers. The vegetation may also act as an indicator and show colour changes, more lush growth or changes in inclination which may be indicative of variation from the existing conditions.

9. Detailed methods for quantifying the effects of these benefits are given by Coppin and Richards (Ref. 6) but the most useful process is in controlling slope stability. The beneficial effect of vegetation develops over a long period and may compensate for a gradual loss of soil strength with time due to weathering. Thus there is a tendency for the effects to be underestimated. Often it is the sudden removal of vegetation which can best demonstrate its effectiveness as illustrated by the onset of slope instability after a wooded slope has been cleared. This will normally take the form of

shallow instability or steady creep, but sometimes deeper seated movement can occur. Failure does not normally occur immediately after felling but typically take a few years to occur as the stability gradually decreases as soil moisture deficits are lost and roots rot and lose strength. The most widespread form of failure in the British Isles is in over-consolidated clays where typically a shallow slip up to two metres deep develops. Whilst these typically occur on the relatively steep slopes adapted for motorway embankments, they have been noted on some dam slopes where these materials have been used. Tree and shrub roots extend down typically to two to three metres and thus there is scope for utilising vegetation to reduce the risk of future instability of these slopes.

10. Many shallow slope failures tend to occur during the early spring when the vegetation is only just beginning to start its regrowth and ground water levels are at their highest. Root die-back is also at a maximum at this time. Thus any stability analyses should consider conditions at this time rather than other times of the year when more favourable conditions are present.

11. The response to storm conditions of strong winds, heavy rain or snowfall is clearly dependent on the vegetation species and its shape and height above ground level, assuming the roots are adequate to maintain stability. This may involve severe stresses in the trunk and branches of trees and lead to damage and loss of moderate sized branches. In extreme conditions, toppling or snapping of the main trunk may occur as seen so vividly in south-eastern England after the October 1987 hurricane. The likelihood of toppling or major branch loss is clearly increased where the tree is already dead or generally unhealthy, badly out of balance or mishapened, or has a shallow root system. Experience from the hurricane suggests that sloping ground had little effect due to the increased root development and that toppling or snapping of the trunk was essentially dependent on the species. Most conifers and ash trees tended to snap several metres above the ground whilst most other broad leaved trees toppled. Oaks were not generally felled but, in the worst instances, tended to have all bar the major branches removed. It should not be forgotten, however, that oak trees may shed quite substantial branches in relatively small storms.

12. The potentially adverse effects of trees on embankment slopes which can arise through lack of management are summarised by Johnson et al (Ref. 7). These include uprooting of trees in storms which may disrupt surface protection layers leaving voids vulnerable to erosion, penetration of slope protection and membranes by root systems, concentration of erosion at the water line between trees and blanketting of the surface making surveillance less

effective. Trees on the upstream slope, the crest and uppermost levels of the dam are clearly the greatest potential threat. Roots, however, are unlikely to penetrate into the core unless they are particularly close or have a high water demand. Other authors, e.g. Cronin (Ref. 8), Merivale (Ref. 9) and Page (Ref. 10) warn against trees on dams or give examples where substantial root growth into the core has occurred. In all cases, however, unless the trees had been allowed to develop close to the core, only roots of alder and willow had penetrated into the core. Bishop (Ref. 11) also reported upon root entry into a puddle core and attributed the root growth to a long-term lowering of the water level. The roots in this case were reported to be herbaceous and primarily from docks. Thus potential problems should be minimised if trees and shrubs, and certain of the larger herbaceous plants, are not allowed to develop near the crest of the dam and the more extreme of the water loving species, namely willows, poplars and alders are excluded.

13. Where works on a well wooded embankment or tree removal are planned, it is essential that the effects of the vegetation are fully considered at any early stage. The type, size and general condition of each tree should be recorded together with the positions and details of any stumps. A series of exploratory holes sunk adjacent to and remote from the vegetation should give some visual indication of the effects of dessication. Samples should be recovered to assess the ratio of insitu moisture content to the Atterberg Limits and the degree of dessication assessed. Clearly the time of year must be considered. Roots should be recovered where possible and identified to assist in estimation of the root spread. Where no roots are encountered, this should also be recorded. Once the type and possible extent of the vegetation is more clearly known, the effects of removal can be considered. This must include not only an assessment of the long-term stability following removal, but the effects of removal on other trees and vegetation and the likely increase in runoff. Where excavation is planned adjacent to trees, care should be taken to minimise the effect on the tree, both from the excavation itself, from construction plant or by ill-considered stockpiling around the tree (Ref. 12).

14. Any investigations on dams where removal of vegetation in the past has led to problems should also seek to recover any roots. Several laboratories hold reference samples and most roots can be readily identified. Knowledge of the extent of past vegetation, which may become evident from the root spread, and the type may help assessment of the problem.

15. Vegetation, and particularly trees, will not always continue to perform their necessary function and may develop various weaknesses. The recognition of hazardous trees is an

important aspect of any visual inspection and any risks should be assessed. The Forestry Commission (Ref. 13) have published a useful guide for recognising hazards and recommended courses of actions. In certain instances, mature trees may have reached the end of their useful life and will require felling. Failure to fell mature trees when they cease to perform their intended function or become a hazard may be as serious as ill-considered removal. It should be noted that a felling license from the Forestry Commission is required and certain trees may be protected by a tree preservation order or a blanket protected woodland order. Where weaknesses are noted and felling is required, it is essential that the major roots adjacent to the trunk are removed and the hole backfilled with compacted material of similar type to the rest of the embankment. If the tree appears to be carrying out some beneficial function to the embankment, a suitable replacement should be planted which should be as large as possible to enable the effects of removal to be minimised.

16. Similarly, any trees which have toppled, heaving up a large disc of earth and larger roots, should be removed and the hole inspected. Loosening is possible down to two to three metres, depending on species, but the hole typically may be less than one metre deep. Where this is away from the crest and upper levels of the slope, the hole can be refilled with compacted material. Adjacent to the crest, all disturbed material should be removed to sound fill and the hole refilled; this may require temporary protective measures against inflow from the reservoir. Alternatively, if the damage is limited to certain areas or if the tree is unbalanced, selective pruning or pollarding on the appropriate species may be sufficient to ensure a healthy tree.

17. It should be noted that trees damaged or weakened by natural or man-made agencies do not die off immediately. They will tend to linger on with little new growth and be prone to increased pest and disease attack. Thus they will gradually weaken and die, typically four or five years after their real cause of decline.

18. Disease is often a secondary effect following physical damage, extreme climatic conditions, insect or animal attack, pollution or general poor health. Older trees and clumps of single species are more prone to attack, but most diseases and pests attack a specific species, i.e. Dutch Elm Disease. All trees have a natural span which may range from a few decades for birches to several centuries for the oaks and ultimately all trees must die. Toppling is associated with trees of all ages, but an older tree is more likely to be affected as are trees left adjacent to felled areas which may be subject to substantially changed wind loadings.

RESERVOIR MONITORING AND MAINTENANCE

ALDENHAM RESERVOIR - REMOVAL OF TREES

19. Aldenham Reservoir was formed about 1795 to supply compensation water for mill owners in conjunction with the construction of the Grand Junction Canal. The embankment was built of random fill of London Clay and was raised soon after construction. Problems with stability occurred soon after construction and have continued intermittently to the present day. The present maximum height is 7.6m with a length of 399m. The slopes are variable with a typical slope of 1V:5H on the upper levels of the downstream slope and a variable steeper slope over the lower levels. Despite minor instability on the downstream face, a good covering of mature trees and undergrowth was known to be present by the beginning of this century.

20. Hertford County Council leased the reservoir in the early 1970s and carried out various remedial works following recommendations under the Reservoir (Safety Provisions) Act 1930. This included removal of all the trees and undergrowth over the upper levels of the upstream slope and marginal raising of the crest to increase the freeboard. Following drawdown of the reservoir for other remedial works in 1975, major slips occurred on both the upstream and downstream slopes. A ground investigation was carried out and the slipped area reinstated. Wet areas developed subsequently in the late 1970s and a series of shallow localised drainage trenches were installed. In 1982, signs of incipient instability were noted in the form of severe cracking and deformation of the crest track in the central length of the embankment. Investigation showed a shear surface was present at about one metre depth, and thus remedial works in the form of two metre drainage trenches at five metre spacing were installed and these successfully controlled the instability. Further cracking and deformation subsequently developed to both sides of the initial length in the late 1980's and thus the remedial works were extended to a total length of 75% of the embankment. To-date these have successfully controlled the instability and no further cracking or deformation has been evident.

21. Apart from one short section of embankment which was affected by both the 1975 and 1982 failures, the areas of instability have developed progressively along different lengths of the embankment. This suggests that some time dependent effect is reducing the stability of the downstream shoulder, and an increase in moisture content in a just stable slope appears to be the controlling mechanism in this instance.

22. Unfortunately, during both of the earlier investigations, little consideration was given as to the effect of the previous vegetation on the insitu moisture content and thus no special testing or studies were carried

out. Reconsideration of the index properties and insitu undrained strengths obtained from these investigations, however, suggests that the fill material had a high moisture content and low strength. Thus any previous moisture content deficits that would have been expected beneath the vegetation cover had been lost or substantially reduced by that time. Piezometers at shallow depths of one to three metres have also indicated some evidence of increased water levels and this would further support this idea. Unfortunately, the piezometers have been vandalised and no information is available since installation of adjacent deep drainage trenches.

23. Thus the downstream slope appears to have been only marginally stable, and ill considered removal of the vegetation, possibly assisted by the placement of fill to the crest, was sufficient to allow instability to develop. The vegetation appears to have a two-fold function in assisting stability; pore water pressures would have been reduced, whilst the roots would have helped to hold the superficial layers of the slope together.

CONCLUSIONS
24. The effects of vegetation must be considered as an integral part of the design of a new dam or a continuing facet of the operation and maintenance of existing structures. Some factors that affect the performance of the vegetation are within the control of the designer, whilst others such as climate are clearly not. Thus, early discussion with a landscape architect are essential to develop an acceptable planting proposal. Similarly, the operation and maintenance controls on a dam can affect the health and performance of existing vegetation.

25. Vegetation may amount to a way of adding an extra margin of confidence to stability and reduce the likelihood of future maintenance or remedial work. It may also help guard against long term changes or local variations in conditions on the embankment.

26. Certain species, namely willow, poplar and alder have a high water demand and a very extensive root system. Experience has shown that such species should be excluded on dams at all times. Other species, namely oak, horse chestnut, hawthorn, rowan/service tree, cherries/plums and in some instances, sycamores/maples, also have a moderate water demand and may develop a relatively extensive root system. Their use, together with most other trees, is not considered detrimental on dams, provided they are excluded from certain parts of the embankment. These include the crest and upstream slope, within a few metres of the crest on the downstream shoulder, a similar distance from any drains or ditches and, clearly, not on any auxiliary spillways or

immediately adjacent to structures. Where mature trees have developed on an existing dam, however, they should be retained and maintained in a healthy condition. Where they are too large, out of balance or unhealthy, selective lopping and pruning should be carried out as required, whilst dead trees should be removed.

REFERENCES
1. WARD W.H. The effects of fast growing trees and shrubs on shallow foundations. Journ. Inst. Landsc. Archit., 11, 4, 1947.
2. DRISCOLL R. The influence of vegetation on the swelling and shrinking of clay soils in Britain. Geotechnique, 33, 2, 1983, 6-18.
3. GREENWAY D.R. Vegetation and slope stability, chap. 6. slope stability, 187-222. John Wiley and Sons, New York, 1987.
4. CUTLER D.F. and RICHARDSON I.B.K. Tree roots and buildings. Construction Press, London, 1981.
5. WALKER B.F. and FELL R. Soil slope instability and stabilisation. Proc. extension course, Sydney, A.A. Balkema, Rotterdam, 1987.
6. COPPIN N.J. and RICHARDS I.G. Use of vegetation in Civil Engineering. CIRIA/Butterworths, London, 1990.
7. JOHNSON T.A., MILLMORE J.P., CHARLES J.A. and TEDD P. An engineering guide to the safety of embankment dams. BRE Report. Building Research Establishment, 1990.
8. CRONIN H.F. The planting of trees on embankment banks. Journ. IWE, vol. v, 1951.
9. MERIVALE J. Amenity lakes and the reservoirs act. Journ. Landscape Design, Nr. 205, 19-20, Nov. 1991.
10. PAGE J.C. Engineering aspects of pond construction. The Creation and Management of pond fisheries. Institute of Fisheries Management, 1991, 13-17.
11. BISHOP A.W. The leakage at a clay core wall. Water and water engineering, 349-366, 1946.
12. British Standards Institution. Code of practice for trees in relation to construction, BS 5837, 1980.
13. Forestry Commission. The recognition of hazardous trees. Leaflet published by Forestry Commission/Department of the Environment, 1990.

Resume of maintenance contracts on hydroelectric reservoirs

C. K. JOHNSTON, MICE, and N. M. SANDILANDS, BSc, MICE, Scottish Hydro-Electric plc

SYNOPSIS

1   Several significant papers have been written over the past twenty years which describe the many facets of the operation and maintenance carried out on the seventy-six registered large dams and numerous smaller water retaining weirs and structures owned and operated by Hydro-Electric.

2   This paper reviews a selection of particular projects carried out during the last ten years which demonstrate the variety of reservoir maintenance tasks.

INTRODUCTION

3   The performance of Hydro-Electric's reservoirs and associated structures remains very good. Deterioration of the various dams is not significant. The major structures will continue to give many decades of further effective and reliable service. To secure the future of its reservoirs Hydro-Electric recognise that a well structured and managed reservoir surveillance and maintenance policy is essential.

4   The paper describes works of particular interest in varying areas which have been carried out over the past ten years. The selection of the maintenance work described attempts to demonstrate the diversity of tasks which require to be carried out.

MONITORING AND RESOURCES

5   Over the past decade Hydro-Electric have committed an annual planned maintenance budget of around £2M on civil works, of which approximately 50% is spent on its dams and reservoirs.

6   Occasionally particular projects are required associated with major improvement or refurbishment works which are funded outwith the planned budget.

7   Surveillance is carried out by the Company's appointed Supervising Engineers and local staff contribute by identifying defects which are then incorporated into an Annual Civil Engineering Maintenance Programme.

8   The total engineering effort which is committed to Civil Engineering Maintenance and the operations averages around nine man years/annum.

9   Priorities are carefully considered to optimise the cost effectiveness of the overall programme.

TYPES OF MAINTENANCE WORKS

10 Maintenance works can be categorised as follows:

   A   Concrete Repairs.

   B   Pitching and Rip Rap Reinstatement.

   C   Reservoir Siltation and Debris.

   D   Hydraulic Equipment Refurbishment.

   E   Stability Improvement Works.

A   Concrete and Pitching Repairs

11 The approach to concrete repairs on dams is well recorded in various previous papers and was fully described in the 1986 BNCOLD Lecture "Experience with the Concrete Dams of the North of Scotland Hydro-Electric Board" and will not be repeated here. The authors in associating themselves with the techniques described and the conclusions drawn in this paper confirm the "stitch in time" philosophy although effective must be moderated with consideration of the high unit cost of cosmetic repairs on isolated sites.

12 Proprietary concrete repair systems suitably adapted to take account of particular site conditions have been used successfully at a number of structures to repair and restore the effects of temperature induced

concrete deterioration. Experience points to the skill and competence of the man being more influential in the success of a repair system than the basic materials employed.

13 It is anticipated that as the structures age the incidence of spalling and general deterioration of reinforced thin concrete sections will increase. Access arrangements will be a major proportion of the cost.

B Pitching Repairs

14 There are many references to the performance of the various methods of providing wave protection to structures subject to wave action. Hydro-Electric's experience is that heavy rip rap generally provides a satisfactory method of protection and allows maintenance to be carried out cheaply and effectively following storm damage.

C Reservoir Siltation and Debris

15 The location of Hydro-Electric's reservoirs in upland glacial valleys has mitigated against severe silt deposition in its reservoirs. At some locations however problems are encountered in the management of floating and submerged movement of trees. The methods used to identify and deal with both types of problems at two particular sites are described.

ERICHT DAM

16 At Ericht Dam, constructed in 1932, a problem had existed with the scouring facility for many years. Shortly after construction the scour gate had jammed out of its guides and emergency action had to be taken to prevent the reservoir emptying. The scour arrangement was subsequently remodelled with a culvert and downstream discharge regulator installed. A lack of confidence in the scour gate remained however and scouring was carried out very infrequently. A considerable accumulation of silt and gravel built up upstream of the scour gate and by 1980 this had reached a depth of 6m behind the scour gate with a total quantity in excess of 2000 cu.m.

17 The main source of the silt was from the Allt Glass Intake which discharges into the reservoir on the north bank of the reservoir immediately upstream of the dam.

18 After considering possible options the decision was taken to dredge.

Work was planned for the month of June and the reservoir was pulled down to the minimum draw-down level by operation of Rannoch Power Station. To gain access from the reservoir banks to the areas around the scour culvert a causeway was constructed using local material from the north bank of the reservoir. A tracked excavator was used which travelled to the end of the causeway and loaded the spoil on to dump trucks. The material was tipped in a reception area at the north-west corner of the dam.

19 The work was carried out over a period of three weeks in June 1981 at a cost of £40,000.

20 To minimise the problem of silt being transported by the Allt Glass into the reservoir a gabion weir has been constructed in the river upstream of the diversion to drop the velocity and settle out a large proportion of water transported gravel before it is diverted into the reservoir. The area behind this weir which was located with good machine access is cleared out conveniently and cheaply as required. This technique has been used successfully on a number of upland catchments to reduce the impact of silt and gravel transportation on diverted aqueduct systems.

MULLARDOCH DAM

21 Since the mid-seventies remote operated underwater television cameras have been used to assist in the assessment of the condition and general state of submerged structures.

22 In February 1989 serious problems were encountered with the operation of the 72" DIA Needle Valve. The tube roller emergency gate was dropped but no seal could be obtained. A desk review pointed to debris being the problem and local knowledge of the catchment indicated submerged timber as the likely cause.

23 To eliminate the risk to divers it was decided to call to the site a Mini-Rover underwater vehicle. The survey was completed in one day and confirmed that there was a build up of debris - mainly trees - around the upstream screen and one large branch had migrated through the screen and fouled the gate guide and sill.

24 The needle valve was eventually cleared and a four-man diving team cleared the debris over a period of fourteen days in September 1989. A work boat with pneumatic winches was used to lift the tree branches to the surface and transport them to the shore to be cut up. The reservoir bottom was cleared for a radius

of 15m around the screens.

25 Due to the depth of water - around 20m - and the altitude of the site working time was restricted to just over one hour per man. A decompression chamber was placed on the dam deck with a winch to lift the diver from the water to the deck in the event of any problems.

26 Hydro-Electric use ROV's for selective survey duties on many of our structures and have found them to be an excellent surveillance vehicle which allow the experienced engineer to formulate his own opinion based on a visual presentation. With the video record developing situations can be monitored.

27 Considerations of safety, effectiveness and cost mean that ROV's can be an attractive alternative to divers in cold and dark situations, particularly where deep water necessitates decompression facilities and severely limits working times.

D   GATES AND VALVES

28 Hydro-Electric have over 150 major gates and valves associated with their reservoir works comprising tunnel intake gates, scour gates and valves, and flood' gates.

29 Prior to 1980 little major refurbishment was carried out on this equipment other than routine service maintenance. The effects of age were becoming evident however and dictated the introduction of the current policy of major refurbishment at twenty-five year intervals. In many locations little thought was given to future major maintenance requirements and difficult access problems are encountered. Extensive temporary works may be required and in some cases these costs exceed the cost of the actual overhaul.

30 Three different types of major overhaul of gates and valve are described each with their own particular problem.

LOCH ERICHT SCOUR GATE

31 Following the dredging at Loch Ericht described in Section C and with infrequent use of the gate silt began to build up again by 1985. In February 1986 a survey of the gate was made using an ROV. The gate and guides were found to be in poor condition and the decision was taken to refurbish.

32	The gate is a 3m x 3m sliding gate located on the upstream face of the dam and is permanently submerged. A retrofit discharge regulator is located on the downstream end of the draw-off culvert. The installation of a temporary cofferdam was investigated but was found to be prohibitively expensive. Removal and refitting of the gate and guides was therefore planned to be undertaken by divers to minimise the loss of water and resultant revenue from Rannoch Power Station.

33	The alternatives of refurbishing the existing gate or replacing with a new gate were examined and the replacement option turned out to be cheaper. The new gate was designed by the Contractor and specified to be tight under the head of 14m and capable of emergency operation under unbalanced head. In the event the gate design was very similar to the original.

34	The gate was fabricated in mild steel and coated with a coal tar epoxy. New bearing bars were manufactured in aluminium bronze and fixed by pinning. The bearing paths on the fixed frame were replaced in aluminium bronze and the sealing bars in stainless steel.

35	The existing guide strips were retained and painted under water using underwater epoxy. The guide strip covers were replaced in stainless steel fixed with stainless steel studs.

36	The gate and frame are completely enclosed by screens running the full height of the dam. A removable section was dismantled by divers to allow access to the working area. The gate was lifted to above water level using a chain block in the gatehouse and then pulled sideways out of the screen structure. It was then floated using airbags to a position where it could be lifted to the dam deck using a specially designed lifting frame and loaded on to transport. The new gate was fitted using the same procedure.

37	The existing lifting gear was overhauled and reused.

38	The new bearing paths were fixed by studs and epoxy resin grout. This was a difficult operation underwater but was carried out with complete success.

39	Independent divers were employed to witness the commissioning tests and inspect the underwater works.

40	The cost of the 1985 work was £50,000.

DISCHARGE REGULATORS

41   A number of major valves have now been refurbished. The contract strategy which has proved to be successful provides for removal of the valve from site and refurbishment in the Workshop where full machining facilities are available.

42   The following describes works carried out on the 60" needle valve on the scour culvert at the forty-year-old Cluanie Dam.

CLUANIE DAM NEEDLE VALVE REFURBISHMENT

43   During the statutory inspection of the dam carried out in 1985 the general condition and operation of the valve was unsatisfactory. Refurbishment was therefore arranged for 1986.

44   To gain access the scour culvert was dewatered using a bulkhead gate. This bulkhead is common to the four culverts at Cluanie and the nearby Loyne Dam and is fitted using a mobile crane. The dam roadway is narrow, 3.6m wide, and the placing of the gate proved to be very difficult.

45   The valve is mounted on rails to allow removal from the valve house and it was necessary to demolish a partition wall and remove the main doors to remove the 11 tonne valve from site.

46   The sealing of the valve was by cast gunmetal rings and due to the amount of wear, new sealing rings were fitted. The main rubber seal at the back of the piston was also replaced.

47   The valve body was split at the works and both parts were shot blasted and coated with a coal tar epoxy. The piston and body liners were polished to reduce friction.

48   The operating gear was stripped and inspected and the main operating shaft was found to be damaged and was replaced in stainless steel.

49   A Rotork 'A' range Syncropak 1600 actuator complete with multi-position limit switches and remote indication was fitted to replace the original manual operation.

50   The main causes of the poor condition of the valve were lack of lubrication, lack of exercising and age.

RESERVOIR MONITORING AND MAINTENANCE

51   The existing lubrication points on the valve were inconveniently situated and as a result lubrication was very infrequent. A new system of lubricating pipework was fitted to marshall all the systems to one convenient point.

52   The work carried out on the valves at Cluanie is representative of the extent of refurbishment which has been found to be necessary on other similar valves.

FLOOD GATES

53   The refurbishment of flood gates at Kinloch Rannoch regulating weir is now reviewed.

54   Kinloch Rannoch regulating weir is a major structure constructed over fifty years ago. The gates had operated continuously since construction and although various partial maintenance works were carried out no full refurbishment had taken place.

55   The location of the structure in an area of high natural amenity and the requirement to maintain normal operation as far as possible dictated that the work was carried out behind cofferdams with the reservoirs fully operational.

56   Flows from Loch Rannoch to the headpond at Dunalastair are controlled by three flood gates at the Kinloch Rannoch Weir. Each of these is 12.2m wide by 3.4m high. The gates are of the free roller type with moving roller trains held by chains running on rocking roller paths mounted on the gates. Since the gates are used to regulate flows in the part-open position, the duty is severe.

57   In 1983 an inspection of the gates was arranged from a hanging scaffold. The inspection revealed that one rocking path was damaged and the rollers were heavily grooved. A major refurbishment of the three gates was planned for 1984.

58   The programme for the works allowed six months to complete the refurbishment of the three gates, with work being carried out on each gate in turn. Removable cofferdams were designed using vertical larch needles supported by steel waling beams and braced steel columns. The steel members were fixed to the concrete aprons using stainless steel resin anchor bolts. The timber needles were seated in a bottom channel section and sealing in this channel and at the end posts was by neoprene strips. A crawler crane was used for lifting the steel sections and timber needles and erection was

carried out by a four-man diving team using a small workboat.

59   Once the cofferdam was erected on the first gate the complete roller trains, staunching tubes, sill sealing bars and seals and rocking paths were removed and inspected. The cast iron rollers were heavily grooved, particularly the bottom rollers, which were subject to the heaviest load. The cast iron rocking paths also showed heavy wear. Wear on the rollers was up to 12mm. One rocking path was also cracked, with a section missing. The mild steel staunching tubes were heavily rusted.

60   A survey was carried out on the cast iron fixed roller paths on the weir piers. The wear close to the bottom of these paths was in excess of 20mm.

61   Allowance had been made in the Contract for replacing the complete roller train assemblies, chains, staunching tubes and rocking paths in stainless steel and in view of the condition of these components this was done.

62   To reinstate the fixed paths a 10mm stainless steel plate was to be pinned to the existing guides after machining of the faces. A corresponding reduction of 5mm was made in the radius of the new rollers to maintain clearances. The slight change in the hanging position of the gate was not considered to be significant. In view of the exceptional wear on parts of the path faces these were initially machined using an *in situ* milling machine and a 15mm mild steel plate was fixed to the lower section of the guide face in the milled recess. A skim was then made over this plate to provide a plane surface and the stainless steel plate was then fixed over the complete length of the guide. Finally the paths were grouted with a resin grout.

63   The complete fixed frame was sand blasted with a silica free sand and coated using a coal tar epoxy.

64   The gates were suspended on Reynolds chains and these were given a detailed inspection and found to be in excellent condition. The gearbox assemblies were also stripped and inspected and found to be in good order.

65   On completion of the overhaul the gate was tested dry and the cofferdam was then flooded and wet commissioning tests

carried out. The cofferdam was then dismantled and erected on the next gate. The condition of all three gates was found to be broadly similar.

66 The cost of the refurbishment was £190,000 and the work on the three gates took eight months.

67 A similar refurbishment was carried out on the two 7.6m x 7.6m gates at Dunalastair Dam in the following year.

68 Notwithstanding the normal evidence that gates appear to be operating satisfactorily, all HE's experience indicates that the ageing process, corrosion and wear is taking place and to date we have considered that the often extensive measures and related costs of maintaining key mechanical elements on dams has been fully justified.

STABILITY PROBLEMS

69 The major works at Mullardoch and Torr Achilty Dams are now well documented and these are not considered here. In 1986 prior to these major works a similar but more severe stability problem on a much smaller scale was tackled at Gorton Dam.

70 Gorton Dam is a masonry structure with a crest length of 44.3m and a maximum height of 5.2m. The downstream face

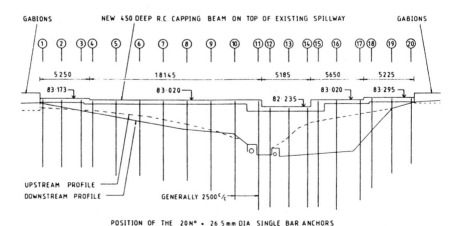

Fig. 1. Elevation of downstream face of Gorton Dam

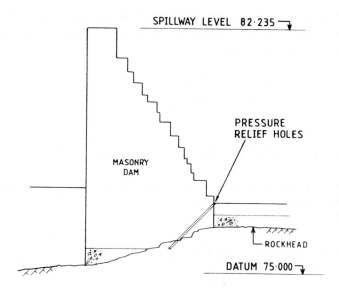

Fig. 2. Cross section through Gorton Dam

of the dam is very steep at 1-in-1.5 and inspection procedures identified evidence of tension cracks on the upstream face. An analysis using Hydro-Electric's dam stability programme indicated a factor of safety against overturning for probable maximum flood of 0.59.

71 The average annual income of the scheme is only £15,000, and the cost of solving the problem was a particular consideration.

72 Access to the site is very difficult. All public roads within 20 km of the site are single track and there is an unsurfaced track 2 km long from the end of the public road to the power station. From here it is a further 1 km to the dam with a climb of 90m. There is no road over this section.

73 Abandonment of the scheme was considered but this offered no cost advantage over remedial works and created other problems.

74 Earlier surveillance of the structure led to a geological investigation being carried out by the BGS in 1968 and cores were taken in 1971 to confirm the density of the masonry. In view of the cost constraints no further investigation was carried out prior to refurbishment.

75 A stressed anchor scheme was selected utilising Dywidag 26.5mm single bar anchors of steel grade 835/1030. The scheme used twenty anchors with design anchor loads varying from a maximum of twenty-two tonnes to five tonnes. Anchor lengths varied from 11.5m to 7m.

76 A 450mm x 750mm wide reinforced concrete capping beam was constructed along the top of the existing spillway. On the lowest spillway section the masonry was trimmed by 450mm to maintain the existing spillway level but on the remainder of the dam the capping beam was constructed on top of the existing masonry.

77 The earlier geological survey indicated that the dam was founded generally on a psammitic granulite with an intrusion of hornblende-microdiorite at the north abutment. A factor of safety of three was adopted for bond stress and a factor of safety of two for tensile stress in the anchor bars.

78 The original plan to use helicopters for transporting all plant and materials to site was replaced with an overland system utilising Muskeg cross-country tractors. The Muskegs hauled the larger plant including 5 ton compressors on sledges to the site. Helicopters were used to transport the concrete which was delivered by trucks from the Ready Mixed Plant at Fort William.

79 The preparatory work and pouring of the cap beam took two weeks and was followed by rotary percussion drilling of 115mm diameter holes for the anchors. Each hole was checked for watertightness and four which were not tight were grouted and redrilled.

80 A double corrosion protecton system with a plastic sheath and factory grouted annulus was adopted. The sheath was ribbed over the anchor length and smooth over the free stressing length. The anchor head consisted of an anchor plate and domed nut located in a square recess in the capping beam. No future testing regime was envisaged and so the anchor head assembly was wrapped in Denso tape and the recess filled with concrete on completion of installation. The holes were grouted with a cement grout with a w/c ratio of 0.4. No additives were used.

81   Stressing was carried out using a hydraulic jack and hand pump and applied load was measured using a pressure gauge and load cell. Loads of up to 150% of working load were applied in two or three load cycles using load increments and periods of observation as recommended in DD81. Four selected anchors were subjected to on-site suitability tests with loading over three cycles and continuous monitoring of load over a ten-day period. The remaining anchors were acceptance tested over two load cycles. All load-extension data was logged and plotted graphically. The behaviour of anchors during test was very close to the predicted figures.

82   In addition to the anchoring works the joints on the downstream face were raked out in an attempt to reduce pore pressure. The cracks on the upstream face were also repaired using a polymer modified mortar.

83   Weather conditions throughout were appalling. In one weekend 150mm of rainfall was recorded at a nearby weather station. The dam spilled almost continually causing considerable difficulties with the work. As a result of this and difficulties in the drilling, work was delayed by three weeks and was completed in December 1986.

CONCLUSION

84   Hydro-Electric currently engineer and place around twenty contracts annually to maintain their reservoirs. In addition local staff carry out minor repairs as preventative maintenance. The engineering works are of a varied nature and require innovation in planning and in finding the optimum engineering solutions for both temporary and permanent works. This interest compensates for the routine nature of many of the works carried out and provides a varied portfolio of tasks.

85   As reservoirs and associated structures age it is predicted that the planned nature of the programme will minimise the detrimental effects of the ageing process.

Performance of blockwork and slabbing slope protection subject to wave action

H. T. LOVENBURY, BSc(Eng), PhD, MICE, and R. A. READER, BSc(Eng), FICE, FIWEM, Rofe, Kennard and Lapworth

SYNOPSIS. As part of a joint Hydraulics Research - CIRIA research project, a survey has been made of the performance of pitching, blockwork and slabbing as wave protection to UK embankment dams. The general findings are reviewed and factors influencing slope protection stability considered.

INTRODUCTION
1. During the last 30 years there have been several incidents of storm damage to the upstream slope protection of British dams. Attention was drawn to these problems by Carlyle in his paper to the 1988 BNCOLD Manchester Symposium (ref. 1). The Department of the Environment has since funded a project to produce a guidance document for the design, maintenance and repair of pitching, blockwork and slabbing protection. The project has been carried out by Hydraulics Research in conjunction with CIRIA, to whom Rofe, Kennard & Lapworth have acted as research contractor. The guide will be published by Hydraulics Research. This paper discusses the performance of these forms of slope protection.

SLOPE PROTECTION
Pitching
2. Over 80% of the earth dams in the UK are protected by stone or rough-dressed masonry pitching or blockwork. This traditional means of protection is well tried, though its use has generally been confined to dams with a reservoir fetch of less than 2 km. Pitching stone typically ranges from 300mm to 700mm in length, 225mm to 300mm in width and 225mm to 450mm in thickness, though in some instances thicknesses up to 600mm have been used. The pitching is generally laid in courses on granular bedding, 225mm to 450mm thick, and retained at its toe by a row of keystones set into the slope. With the exception of the larger heavier masonry blocks, pitching can be placed by hand, which facilitates its repair. The pitching stones are closely placed, the frictional contact between adjacent faces providing a degree of interlocking. Nevertheless there is sufficient open space between the stones to allow ready drainage during wave downrush. In some cases, the pitching appears to have

tightened up as a result of embankment settlement. The thickness used has been determined by experience.

3. Pitched slopes have generally performed satisfactorily. Damage from wave action is normally caused by a proportion of the underlying bedding material being washed out through the joints. Local collapse leads to the removal of the pitching stones. In some instances voids are formed beneath the pitching with the stones left "bridged" above. Once stones are dislodged, and if the damage is left unattended, wave action soon leads to the instability spreading. A commonly reported cause of initial damage is the removal of individual blocks either by vandals or, more frequently, by fishermen. It is generally accepted by owners that routine vigilance and maintenance is essential to forestall the development of more extensive instability problems.

Blockwork
4. Since 1945, pre-cast concrete blocks and cast in-situ concrete slabbing, as well as rip-rap, have frequently been adopted for economic reasons. The dam faces have generally been exposed to longer reservoir fetches. With sizes of up to 500mm wide, 1200mm long and between 300mm and 380mm thick on slopes ranging from 1 on 3 to 1 on 2.5, the placement of square-faced precast concrete blocks has been crane-assisted. Positive gaps, ranging from 12mm to 19mm wide, were introduced at joints partly to allow room for the insertion of lifting tackle should any block need to be replaced at a later date and partly to accommodate settlement of the embankment. The blocks were laid in courses, though in several cases a set random-type pattern was employed with even larger blocks being incorporated at intervals to enhance the appearance. The details appear to have followed masonry practice, but with the features largely dictated by architectural requirements. The block thickness was based on pitching practice. The blockwork is normally retained at intervals by longitudinal cast in-situ concrete beams keyed into the shoulder.

5. Whereas many dams with this form of slope protection have performed satisfactorily, there have been several incidents of instability frequently where the reservoirs are aligned in a westerly direction. At Turret, Backwater and Kielder dams (refs. 2-4) downslope blockwork movement led to opening of some longitudinal (horizontal) joints allowing the underlying bedding material to be washed out. There was also some differential settlement and slight tilting of individual blocks. At Kielder dam, several blocks were ejected during severe storms in January 1984. In all cases the repairs appear to have successfully stabilised the blockwork. The instability problems can be attributed to an underestimation of the maximum wave height, the introduction of positive gaps at joints, and an inadequately graded bedding material. The

positive gaps were a departure from the well-established principles for pitching construction.

6. At Selset dam, the blockwork consisted of varying-sized sawn masonry laid in a random pattern with relatively narrow joints filled with pea gravel or no fines concrete. This protection was dislodged in localised areas during a severe storm in February 1962. Some blocks were lifted out whilst limited collapse took place at other positions. The problems re-occurred over the years and in 1989 the blockwork was replaced by rip-rap. The development of the instability was attributed partly to lack of interlock between the blocks due to their edges being undercut, rather than square, and partly to the washing out of fines from the bedding material through the joints. The block thickness of 203mm was evidently underestimated.

Slabbing

7. Until the early 1960's, cast in-situ concrete slabbing was laid in panels with butt joints. The slab thickness was between 125 and 225mm on slopes ranging from 1 on 5 to 1 on 2.5. Panel sizes were between 1.8 metres and 2.4 metres square, though at one site 4.6 metre square panels were used. The transverse (vertical) joints were typically staggered in alternate rows. The joints were usually open, up to 25mm wide, so as to provide flexibility during the subsequent settlement of the embankment. In some cases, the joints widened as a result of post-construction thermal contraction. The toe of the slabbing was supported by a concrete retaining beam keyed into the slope.

8. Since the mid-1960's, shallow 'V'-shaped interlocking joints have been introduced on several dams as a result of model testing carried out by the Hydraulics Research Station (ref. 5) for the 1 on 5 upstream slope of Diddington (now Grafham Water) dam. The tests, using uniform wave trains, showed that the introduction of 'V' joints in one or both directions gave a marked improvement in stability over panels with butt joints all round. The need to relieve the uplift pressure in the underlayer beneath the slabbing during wave downrush was also recognised. Since the tests showed that the 200mm thick panels should be as large as practical, a size of 3.96 metres square was adopted for the dam. The longitudinal joints were 'V'-shaped whilst narrow drainage slots were incorporated in the transverse joints. Every fifth transverse joint was a butt joint containing a filler to allow for expansion. The slabbing was placed on a gravel bed. Four lines of no-fines concrete blocks were placed at the toe of the slabbing to aid the relief of uplift pressure.

9. On subsequent dams, various details and arrangements were tried, apparently with success, but no preferred method has evolved. The main variations involved the introduction

of dummy joints - slots in the top surface of the slabbing - in place of butt joints and the provision of an impermeable membrane between the bottom of each expansion joint and the drainage underlayer to prevent loss of material. The use of 'V'-shaped interlocking joints was not followed at all dams, however, though expansion joints were included. In two cases, all the transverse butt joints were filled with a cork board filler with the slabbing being retained at intervals by longitudinal beams keyed into the slope. There appear to have been two contrasting approaches to pressure relief. In most cases, drainage was provided across the whole area of slabbing via the transverse joints. In the few cases where transverse dummy joints were used, the slabbing was essentially sealed with drainage being restricted to the toe of the face.

10. Significant displacement of the slabbing has occurred at three sites within the last five years as a result of severe storms: Bewl Water, Hanningfield and King George's, Chingford (refs. 6-7). In every case, the gravel underlayer was washed out and the slabbing collapsed. Interlocking joints were not incorporated in any of these faces. At Bewl Water dam, it is evident that the 127mm thick slabbing was too thin. Similar behaviour was experienced at Blithfield dam (ref. 8) during a severe storm in February 1962 when several isolated panels in the stepped section at the top of the slabbing were dislodged by rotational movement.

DISCUSSION
11. The mechanism of failure by wave action for pitching, blockwork and slabbing is essentially the same. Localised loosening of an individual 'block' by wave impact will lead to loss of frictional interlock enabling the 'block', if of insufficient weight, to be lifted out directly by the uplift pressure acting on its base during wave downrush. Further slope movement will allow 'block' overturning to occur. In either case, erosion of the bedding material and undermining of adjacent 'blocks' will follow. Alternatively, slope movement will lead to an increase in joint width enabling the finer fraction of the bedding material to be removed and cavities to form beneath the blockwork with eventual collapse of the 'blocks'. Unevenness of the surface, especially at the joints, caused by differential settlement of the shoulder fill may render the blocks more susceptible to movement by wave action. Disturbance of the bedding due to shock loading under breaking waves may also initiate block movement. Blockwork with joints filled with fine gravel will move once the gravel is removed by wave action. In the case of pitching, the loss of support caused by removal of individual stones will enable the adjacent stones to be dislodged.

12. The instability problems experienced since 1945 can be attributed partly to underprediction of the maximum wave height and partly to shortcomings in construction detail. The dams were not endangered by the essentially superficial damage caused during the short-duration storms, but repairs are expensive and cause temporary loss of storage.

13. The predicted maximum wave height is derived from the maximum over-water wind speed and the reservoir fetch. The wind speed depends on satisfactory extrapolation of data from the nearest weather station to the reservoir site (which is usually at a higher altitude), the storm duration, and the return period assumed. On a typical UK reservoir the wave conditions in the reservoir are fully developed within a duration of 20 minutes. A storm return period of at least 50 years is regarded as appropriate to the lifetime of the dam. However, the available wind speed data may underestimate the severest conditions. For example, the highest speeds recorded during the October 1987 storm in S.E. England were greater than those predicted from past records. The wind speed at the dam site may be increased by the topography of the reservoir basin, particularly where funnelling occurs.

14. Wave height prediction has been based on methods derived for coastal water conditions with modifications to allow for the limited fetch length, shorter wind duration and deeper water at inland reservoirs. The main modification involves the use of an effective fetch; a reduction in the maximum direct fetch to take account of the reservoir configuration (length/width ratio and irregularity of shoreline) and its alignment relative to the prevailing wind direction. The effective fetch value used must relate to the particular wave prediction method adopted. Most of the incidents of instability experienced have been at reservoirs where the direction of maximum fetch is in relatively close alignment with the prevailing wind direction. In the UK this direction is generally confined to the south-west sector. There is a view amongst dam designers that a value near to the maximum direct fetch should be used in such cases. There is still no single consistent method of predicting wave height applicable to reservoirs of any configuration or topographical setting.

15. Other factors which can increase the wave height of a dam are the slope of the dam face, wave reflection from a landform or structure such as a spillway weir adjacent to the dam and wave diffraction from a structure (the valve tower or overflow) positioned upstream of the face. Wave diffraction may have contributed locally to the wave height at Bewl Water, Kielder and Turret dams. If the reservoir has more than one arm directly exposed to the dam face, the resultant of the individual wave trains will produce an increased wave height. Turbulence may be increased locally at the face by

the presence of protrusions or steps in the face. A curved transition between the top of the slope and the base of the wave wall rather than a vertical step is to be preferred.

16. The predicted wave height is used to determine the minimum thickness of the facing. In the past, the selection of block thickness was mainly a matter of judgement based on rule-of-thumb criteria. However, recent experimental work on unrestrained blocks in Holland and at Hydraulics Research and Strathclyde University (refs. 9-12) has established relationships between wave height and minimum blockwork thickness for different conditions of joint width and underlayer permeability.

17. Consideration must also be given to the various construction details, such as joint type and spacing and the bedding material grading, which also influence stability. Frictional contact and/or interlocking helps to resist uplift pressure, physically prevent block rotation and minimise the loss of bedding material through the joints by erosion. Model testing has also shown that the characteristics of the underlayer have a substantial bearing on the stability of both slabbing and blockwork, stability improving with decreasing underlayer permeability. This is possibly due to the reduction in flow in the underlayer. The grading of the bedding material must also be designed to prevent the loss of the fines through the joints. Unfortunately, these requirements are to some extent conflicting. At one recent dam protected by slabbing, an attempt was made to overcome the problem by placing a narrow strip of coarser material directly beneath the joint openings. The permeability of the underlayer and that of the facing need to be considered together.

18. There is also a strong case for sealing the upper part of the slope above top water level to restrict water entry to the underside of the protection and providing drainage outlets at lower levels to allow escape of water on wave downrush to relieve uplift pressures. When applied to pitching, care should be taken to ensure that the joints are sealed to their full depth, ideally by pressure grouting, to avoid removal of the filling by wave action. Grouting also improves interlocking.

19. In the repairs at Bewl Water, Hanningfield and Kielder, the replacement slope protection thickness was based on the maximum predicted wave height. In that it is probably the few extreme waves that are likely to initiate damage, this approach appears justifiable. Slabbing has the advantage that the short duration shock pressures imparted by breaking waves which could affect the stability of an individual block, can be dissipated better beneath the larger slab contact area. In this respect, it is interesting that

at Selset dam the very largest blocks, although of the same thickness as the other blockwork, were not disturbed by wave action.

CONCLUDING REMARKS
20. Although attention has been drawn to those failures that have occurred, it is considered that blockwork and slabbing protection have generally performed well. Whilst recent research studies have focussed on the stability of open jointed blocks, the trend in design has been towards the use of larger sized panels, thereby reducing the overall surface permeability of the slope protection layer, and interlocking joints. This trend is particularly evident in the approach to repairs. At Bewl Water, for example, the damaged protection was replaced by thicker slabbing with impermeable joints. At Kielder, where the repair involved resetting the blocks on concrete panels (ref. 1), the protection now essentially comprises slabbing with the blockwork as an architectural feature providing additional weight. The same solution of setting stones on a concrete bed has also been adopted at a number of sites where it has been felt necessary to strengthen the pitching protection. The use of grouting to seal the joints in pitched slopes is a variation of the same approach. The need for flexible protection on new dams seems of lesser importance in view of the improved compaction provided by modern earthmoving plant whilst at older dams settlement will have largely ceased. For any new dam, sealed concrete slabbing would on balance appear to be the more suitable method of slope protection.

ACKNOWLEDGEMENTS
The response of the many owners and engineers in contributing to the performance survey is gratefully acknowledged. The paper is published by kind permission of CIRIA, Hydraulics Research and the Department of the Environment. The project forms part of the DoE sponsored Reservoir Research Programme.

REFERENCES
1. CARLYLE W.J. Wave damage to upstream slope protection of reservoirs in the U.K. Proceedings of the BNCOLD Symposium on Reservoir Renovation, University of Manchester, 1988, Paper 6.3.
2. MILNE G.A. Discussion: Technical Session 6. Proceedings of the BNCOLD Symposium on Reservoir Renovation, University of Manchester, 1988, p. D6/4.
3. ROCKE G. Damage to concrete blockwork at Kielder Dam, England. Transactions of the 15th Congress on Large Dams, Lausanne, 1985, vol. 5, 557-562.
4. ROCKE G. Discussion: Final Session. Proceedings of the BNCOLD Symposium on Reservoir Renovation, University of Manchester, 1988, DF/2-4.

5. Diddington Reservoir: Report on model investigation of embankment protection against wind-generated wave action. Hydraulics Research Station, Wallingford, 1964, Report EX 253.
6. SHAVE K.J. Discussion: Technical Session 6. Proceedings of the BNCOLD Symposium on Reservoir Renovation, University of Manchester, 1988, D6/1-2.
7. HORSWILL P. Discussion: Technical Session 6. Proceedings of the BNCOLD Symposium on Reservoir Renovation, University of Manchester, 1988, D6/7-8.
8. LEACH J.A. Discussion: Session 5. Proceedings of the BNCOLD/University of Newcastle-upon-Tyne Symposium on Inspection, Operation and Improvement of Existing Dams, Newcastle-upon-Tyne, 1975, p. D5/8.
9. PILARCZYK K.W. Closure of tidal basins. chap. 2.4.13, Delft University Press, 1984.
10. Wave protection in reservoirs: hydraulic model tests of blockwork stability. Hydraulics Research, Wallingford, 1988, Report EX 1725.
11. TOWNSON J.M. A lower bound for the thickness of revetment blocks. Proceedings of the Institution of Civil Engineers, Part 2, 1987, June, vol. 83, 397-407.
12. NUR YUWONO. The stability of revetment blockwork under wave action. Ph.D Thesis, University of Strathclyde, 1990.

Surveillance and monitoring methods for Italian dams

Eurlng D. MORRIS, MSc, FICE, MASCE, FGS, Consultant

SYNOPSIS As a result of morphology and demography, Italy has a great number of large dams situated near to and lying above centres of population. Legislation has ensured safety and automatic instrumental monitoring systems, which continuously measure the status of the structure and report any anomalies as they occur, file the data electronically and prepare reports have been developed and are described.

PRESENT DAM SAFETY LEGISLATION

1. Some 66 years ago (1925) Italy developed regulations regarding the safety of dams. Since then, the regulations have been regularly updated in line with advances in technology. The most recent amendments, in the form of technical standards passed in 1982. In 1931, a dam was defined as a 'structure retaining water to a depth greater than 10 m above natural ground level, or impounding a volume of water greater than 100 000m^3, irrespective of height. This definition still applies today.

2. The basic philosophy behind dam safety in Italy, is to remove any risk of failure and economic loss associated with dams and to reach a degree of safety significantly above that at which people may have cause to fear that conditions following construction could be worse than foreseen in the design. The rigorous application of the regulations ensures that dams constructed in Italy are built with a degree of safety which does not generate widespread alarm and opposition to their construction. A subjective component, in addition to the principal objective function is, therefore, present or implied in all matters relating to dams.

3. Recent legislation requires all dams to have an 'Emergency Plan' which considers the extreme case hypothesis of collapse of the structure. Preparation of the various 'Plans' is under way and inundation studies are in hand, on a rolling programme, for all existing dams. This will result in a complete series of inundation maps for the whole country.

4. Responsibility for dam safety rests at present with the Ministry of Public Works through its technical services divisions. They are responsible for ensuring that all dams are designed to the current standards, constructed in accordance with the design and

these same standards, and operated in accordance with the relevant regulations. Responsibility is centred mainly with the 'Servizio Dighe' (Dam Service) which has a central office in Rome and local offices in the 10 hydrographic areas into which Italy is divided. They are assisted by other specialist technical sections of the ministry, namely: hydrology and geology.

5. The Dam Service, comprising a nucleus of experienced specialist engineers was originally founded to collect and co-ordinate data on dams, information, test results, etc., and to be a repository of up-to-date technical knowledge of dam design and construction both in Italy and abroad. The Service is one of the most important parts of the system to monitor and control the design, construction and operation of dams in Italy. Through its holding of records of all dams in Italy, it also can and does provide uniform solutions to any problems that arise during construction and operation of dams.

6. The 'Dam Service' is responsible for:
- maintaining appropriate records of each dam;
- examining the feasibility and contract designs for dams to ensure that they conform to the relevant technical standards including the general regulations regarding safety. They provide approvals of the various studies and designs and submit all 'Contract Designs' to the IV Section of the Upper Advisory Committee of the Ministry for their examination and comment. They also prepare the schedule of any conditions imposed by the Upper Advisory Committee after their examination;
- inspecting and approving the foundations during construction;
- inspecting and approving the quality of the materials used and of the construction in general;
- approving minor modifications to the design;
- examining the monthly reports during construction;
- approving the programme of tests on the construction materials and the results of such tests from both the site and the officially appointed laboratories;
- participating in the commissioning tests of the works;
- receiving the monthly reports and the half-yearly reports during operation;
- following any necessary repair works ordered to the dam during the operational phase;

7. The 'Dam Service' has the assistance of two other bodies in this task:
- The 'Genio Civile', the Civil Protection Office (represented regionally), through its specialised hydraulic section, which has responsibility for all matters related to hydraulic resources within their area and for safeguarding the public from floods, both natural and man-made; and
- The IV Section of the Upper Advisory Committee of the Ministry, composed of senior specialists of great experience and drawn from the Universities and Private

Practice. They have the task of co-ordinating hydrologic and meteorologic measurements throughout Italy and the examination and monitoring of the design, construction and operation of dams, in association with the Dam Service.

8. The Ministry, through its various bodies is concerned during all phases of the project's life. The degree of involvement and checking is illustrated below.

9. During the project preparation phase, (feasibility and tender phases) the 'Dam Service has a consultative role and is chiefly concerned with the preparation of a list of conditions imposed by the ministry which must be observed in the design and construction of the project. The conditions include the items listed below in the various sections as well as such items as the number, type and location of instruments to be installed to monitor the dam and the surveillance plan, and the communications links to be installed between the dam and competent offices.

10. During the construction and commissioning phases, checks have the scope of ensuring immediate safety, via analysis of the actual behaviour of the dam and foundations, not always equal to the design assumptions. The results of the analysis determine if the construction proceeds or not. The aspects investigated in this phase are:

- approval of the construction methods and plant and equipment;
- approval of the foundation treatment plan;
- approval of the characteristics and source of the construction materials and of the quality control plan, including the type and number of tests required, (a specialist approved off-site laboratory must be appointed to carry out tests in addition to the site laboratory);
- monitoring of the conditions imposed on the project;
- the supervision of the construction work;
- approval of the experimental partial first filling (by law, dams have to be filled in stages and emptied then filled again with a series of measurements monitoring behaviour of the dam);
- collecting the necessary data for the commissioning;
- approval of the 'Emergency Preparedness Plan.

11. During the operational phase, the dam must be under continuous surveillance by the concessionaire. The basic requirements are:

- monitoring by visual examinations and by a series of measurements of the dam and its foundations using instruments. This monitoring is intended to recognize quickly any early signs of impending dangerous situations so that the reservoir can be emptied in time should this be necessary;
- the preparation at the end of each month of a report containing the results of the readings taken during the period and which is sent to the appropriate authority;
- the testing of all gates and valves at least once every six

months;
- a visit to the dam by the engineer charged with responsibility for the dam at least twice per year and if possible also during periods of the maximum and minimum water level;
- the carrying out by the relevant offices (technical divisions of the Ministry) of periodic inspections of the efficiency of the communications systems at the dam and of the alarm systems installed.

12. For reservoirs already in service, the regulations ensure that at the first sign of trouble, from whatsoever cause, that reduces the degree of safety below the initial value established from the records, the reservoir is emptied to allow the necessary repairs to be carried out. The 'Dam Service' is involved in all such repair work.

13. For existing reservoirs, attention is being concentrated more and more on the monitoring system, the efficiency of the supervising organisation and the overall supervision and regulation. Present techniques are described below.

SURVEILLANCE AND EVALUATION TECHNIQUES

14. In Italy, there are today, more than 500 large dams usually located close to centres of population and between them storing more than 8800 Mm^3 of water of which some 59% is used for power production and the remainder for water supply (domestic, industrial and agriculture). At present more than 35 new dams are under construction or are at the planning stage which will add some 63% to the present capacity. Notwithstanding the legal requirements set out in the various regulations, it became clear that there were insufficient tools available for evaluating adequately the performance of a dam and that the consequent margins of ignorance could lead to excessive and unnecessary prudent criteria for the management of the reservoirs, limiting the exploitation of the resource.

15. In addition, Italy is a country with significant seismic activity. Between 1900 and 1982, more than 8543 earthquakes of over 5 on the MCS scale have been experienced and of these 318 have measured more than 7 on the MCS scale. The south of Italy, which is relatively dry and is subject to seismic activity is also deeply involved in developing water resources.

16. At the beginning of the century, measured data did not really exist and reliance was placed upon simple visual indications of the state of the structure. Later, the assessment of safety of dams in operation was by examination of simple indices of the 'state of health' of the structure by experts analysing 'significant parameters' to verify conditions of normality.

17. Later still, with improvements in instrumentation giving better and more reliable measurements, correlation between 'cause' and 'effect' quantities was added to evaluation. Delays between reading and final plotting and analysis reduced effectiveness, as now. Such delays are a function of logistics and staffing levels.

18. Today, with further improvements in instrument and

computer technology, it is possible to have 'real time' analysis and reporting of the most critical parameters with immediate indication of anomalous readings.

19. In general, evaluation, of safety comes from the visual examination of time series diagrams of the 'cause and effect' quantities and of correlations between them, no matter how the data is gathered and stored. Expressed simply, it is a search for anomalies among measured parameters which gives an indication of a change in state. The quantities measured are those of:

CAUSE	and	EFFECT
producing changes in the state of the structure		describing the evolution of the state of the structure.

20. The use of simple statistical methods (means, moving averages, simple regressions, etc.) to detect anomalies, if done manually, requires great effort and inevitably some delay is involved before competent persons examine the data. In addition, anomalies are usually detected when obvious, and it is possible that the underlying cause could have been detected and corrected earlier with less difficulty. Manual techniques show their greatest limits in times of rapidly changing situations which usually coincide with 'emergency' conditions when, for example, the structure is affected by seismic activity or exceptional floods. Under such conditions, the rapidity and frequency of data acquisition and processing is not compatible with manual techniques.

21. In an effort to overcome these limitations, modern technology using automatic data collection and processing has been introduced in Italy and is presently applied to a number of existing dams as well as new dams under construction. This new approach can be summarised as:

- the collection of measurements via automatic data acquisition systems;
- the filing of the measured data in computerised data bases and their management and processing using specific computer programmes;
- the interpretation of the measured data through mathematical models;
- the real-time checking of the structural behaviour of the dam ('on-line monitoring');
- the periodic detailed checking of significant parameters to identify any deterioration in the state of the structure; and
- periodic fully detailed checks (Certified Check) of the dam by specialists to evaluate the actual situation and define the future surveillance programme.

Automatic Data Acquisition Systems

22. To date, automatic data acquisition systems have been installed in more than 10 Italian Dams. The systems comprise essentially sensors connected to a microprocessor for automation of the readings. The actual arrangement of the instruments depends

upon the type, age, dimensions of the dam, the possible structural problems and the degree of risk associated with the facility. No standard arrangement is, therefore, possible and all are tailored to the installation. On the basis of the experience gained to date the most important system requirements are:
- Flexibility - especially with regard to the need to integrate any automatic system with instruments already installed;
- Expendability - the need to add to the system in time to monitor situations which develop later such as changes following earthquakes, floods, landslides, etc;
- Reliability - in terms of safety of the facility this is of utmost importance. Many Italian dams operate at elevations over 2500 m, close to the limit of perpetual snows. In addition, overvoltages from electrical discharges are a serious and constant danger.
- Simplicity of Maintenance - although modern electronics allow systems to be designed which have no risk of breakdown, the redundancy necessary would be very onerous and modular standardized systems reduce maintenance times and costs to the minimum.

23. A simplified block diagram of a typical system is shown in Figures 1 and 2.

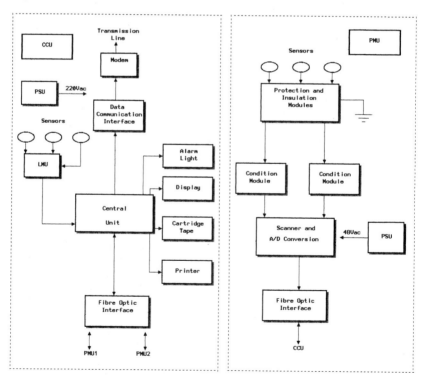

Figure 1 Monitoring System: Block Diagram

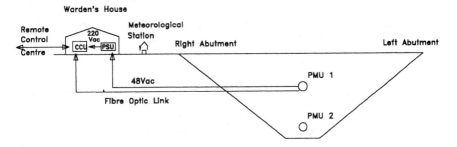

Figure 2 Monitoring System: Typical Layout

The components are:

- measurement sensors such as water level meters, plumblines, settlement meters, piezometers (in dam body and foundations), thermometers and meteorological station;
- peripheral measurement units (PMU) to collect signals coming from the various sensors and transform them into digital form;
- central control unit (CCU) comprising a computer with command and display unit, a printer, magnetic storage and modem for data transmission;
- uninteruptable power supply (PSU).

24. The main functions of the system are:

- reading of measurements at short intervals (10 minutes);
- periodic gathering of measurements at longer intervals (6-8 Hours);
- accelerated reading of measurements activated in special circumstances such as earthquake, exceptional flood, landslide, etc;
- comparison of readings with preset threshold values;
- transmission of data to a local centre for filing in computerised data bases.

Computerized Data Management

25. Although the automatic collection of data is limited to only a small number of dams, the computerized management of collected data (including that collected manually) is now common practice. The major dam owners have long since set up computerized data bases for the storage of measured data. Some systems are used to manage the data for one or two dams and the MIDAS system developed by ISMES is used to manage the data from more than 130 dams, scattered throughout Italy. These data bases allow fast and reliable analysis and overcome the limits of manual analysis such as the interpretation limited to a few fundamental quantities, the discarding or non use of measurements difficult to interpret visually and the long delay between collection and interpretation of measurements.

26. Data, manually or automatically collected are transmitted to a central office either via modem, magnetic disk, or in hard copy. The centre is equipped with PC, printer, tape unit and plotter. This equipment has been found by experience to constitute the best compromise between cost and performance.
Currently, data bases and interactive systems allow:
- storage of original measurements and their transformation into engineering units;
- grouping, printing and plotting measurements to predetermined formats;
- performance of consistency checks and preliminary analyses (moving averages, Fourier analysis, etc.) to verify trends;
- searching for correlations between 'cause' and 'effect' quantities and the development of regression models;
- comparison of measured values with corresponding values computed from 'reference models'.

27. The automatic processing of data has shown the following advantages:
- analysis of all quantities measured, the verification of their usefulness and the rationalization of measurement networks;
- immediate visualization of any quantity of interest and easy verification of the performance of the dam via models.

Interpretation through Mathematical Models

28. Over the last twenty years, a number of Italian Research Groups have sought a methodology for interpreting measurements, better than the traditional simple visual examination of time series diagrams. In this latter method, judgment of safety relies on the experience of the individual operator, whose deductions are not always easy to demonstrate. The use of 'reference models' allows the introduction of more objective criteria, based on comparisons between the real behavior of the dam and theoretical predictions from the model. In this connection, 'reference model' refers to either a deterministic or a statistical mathematical algorithm that, on the basis of known 'cause' quantities is able to predict 'effect' quantities. When the difference between the two is within acceptable limits the structure is considered to be within safety limits. If, on the other hand, after checking for instrument malfunction, errors, or imperfect calibration of the mathematical model, the differences continue to increases with time, then this may mean that some new, possibly risky, situation is developing and that further, deeper investigations are required to identify the actual cause.

29. Two mathematical models are presently widely used in Italy:
- "A posteriori" regressive models; and
- "A priori" deterministic models.

30. **"A posteriori" regressive models** are simple to set up, do not require heavy processing and can be used to check all 'effect' quantities (displacements, seepage losses, strains, rotations, etc.). These models are widely used because of their low cost and ease of setting up, and can produce good results in cases where no particular problems are present. A sufficiently long chronological period of readings is necessary to set them up.

31. **"A priori" deterministic models** are used when the correlation between 'cause' and 'effect' is determined using structural analysis. These require a high technical and economic effort to set up, but once available, remain valid and provide an objective and reliable method of comparison with reality under normal and exceptional operating conditions. Such models are usually based on FEM models.

On-line Dam Safety Monitoring

32. The advances in instrument and computer technology has allowed 'on-line' monitoring systems to be developed. These systems not only collect the data automatically but also perform a comparison between measured and predicted quantities in real time.

33. This comparison does not have the goal of declaring a dam safe or not, but is used to identify situations where anomalies are recorded. Such anomalies may arise from a large number of causes not all of which are dangerous, and only after clarification of the actual cause is it possible to express any accurate judgment about the safety of the structure. This judgment will be based on the evaluation of absolute values and of the evolution in time of differences between measured and predicted values of any individual quantity. Potentially dangerous situations will emerge if several of the measured quantities show anomalous behaviour, not one alone. The automatic monitoring system is, thus, a 'technical filter' which allows attention to be concentrated on the real anomalies. To date the delegation of the final judgement of safety to the computer with no human intervention has been resisted, only man, with his knowledge and experience has the necessary skill to evaluate the actual risk, if any, arising from any anomalous situation.

34. An example of a typical system lay out, indicative of those adopted in Italy, is given in Figure 3. The main elements of the system are:

- the automatic monitoring system installed at the dam and managed by a microprocessor;
- a local control centre with computer terminals and telemetrically connected to the dam site and connected via land-line for data transmission to a remote centre; and
- a remote centre provided with a main frame computer, data bases and mathematical models.

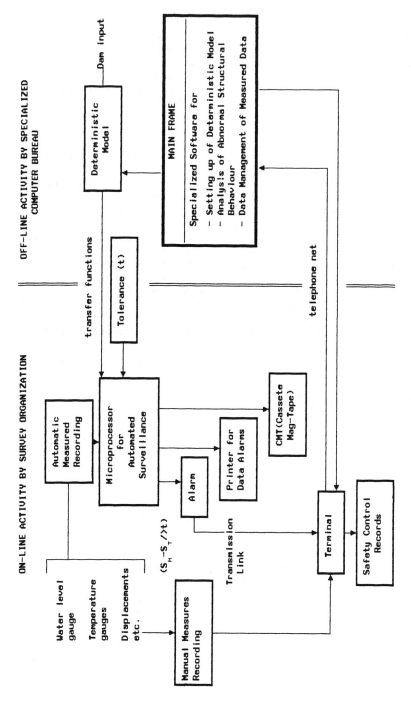

Figure 3. Dam Safety Monitoring: Typical Layout

35. The automatic system installed at the dam performs the on-line monitoring, comparison of measured and predicted values using 'transfer functions' representing the functional relationships between 'cause' and 'effect' quantities previously calculated in the remote centre via mathematical models. In the case of statistical models, the result of linear regressions and in the case of deterministic models, obtained from FEM analyses in which the 'cause' quantities are parametrically varied within their expected limits.

Periodic checks of significant parameters

36. These have the purpose of identifying a small number of physical-mechanical parameters of the dam and foundation and checking their possible deterioration with ageing of the structure. Such deterioration could be due to original defects or to the long term action of external agents. The evaluation of degradation requires various investigations such as: drilling with sampling and in situ methods such as down-hole TV; penetration tests; permeability tests; geophysical tests (cross-hole, seismic refraction, acoustic emission, etc); dynamic excitation etc.

37. The aim of such investigations is:
- to evaluate the progress of any deterioration;
- location and assessment of any cracks or other structural damage;
- check the efficiency of the sealing elements (cut-off, diaphragm, core, deck, etc); and in some cases
- obtaining data for setting up the deterministic reference model.

Periodic Detailed Check (Certified Check)

38. This is a complete check of the structure performed at intervals by independent experts. The aim is to verify the actual safety condition of the dam, to identify any intervention required to restore the structure to its original safety level and to control the routine surveillance programme. The actual details of the check depend upon the type of dam and the problems, if any, present in the structure and foundations. The activities carried out during such a check are:
- examination of the design documentation, of historical records and of all technical documentation produced over the life of the structure;
- interpretative analysis of the dam behaviour on the basis of the measured data via:
  * review of the installed instrumentation network;
  * interpretation of behaviour in time using mathematical models;
  * review of the quality of the data.
- investigations to determine the physical-mechanical properties of the structure and foundation and the degree of degradation, if any, present;
- checks on the stability of the structure, including the review of the design phase calculations and their updating to comply with current regulations;

RESERVOIR MONITORING AND MAINTENANCE

- definition of any improvements necessary to upgrade the monitoring and data processing system;
- definition of any changes required to the routine surveillance programme that must by law be carried out by the owner; and
- checks on the condition of the reservoir including:
  * stability of the reservoir slopes;
  * reservoir siltation in terms of change in storage volume and possible malfunctioning of the outlets;
  * quality and eutrophication of the stored water.
- verification of the impact of the reservoir on the downstream area with the aim of evaluating:
  * effects on fluvial conditions and on ground water recharge;
  * consequences of the propagation of artificial floods produced by sudden opening of the gates. Possible alterations to the morphology downstream or development of the area may have introduced situations of potential danger not foreseen at the design stage. These are investigated by means of numerical simulation of flood propagation.

Site investigation of existing dams

J. M. REID, Babtie Geotechnical

SYNOPSIS. Site investigations for existing dams require to satisfy often conflicting objectives; the need to obtain information about the structure and the need to preserve its integrity. This is illustrated by two recent case studies from Central Scotland. The first, Upper Glendevon dam, is a mass concrete gravity dam with a maximum height of 45 m. The second, Buckieburn dam, is an earthfill embankment with a maximum height of 23m. The investigations illustrate the importance of desk studies, use of existing instrumentation and the need to maximise information from exploratory holes.

INTRODUCTION
1. The investigation of existing dams poses a number of problems for the geotechnical engineer. There are a number of constraints which do not apply to conventional site investigations. For instance many dams are located in upland areas, often some distance from major roads, and this can cause problems of access and loss of working time due to bad weather, especially in winter. Site investigations, like other civil engineering works, should be programmed for the summer months where possible in such areas.
2. At the dam site itself, it may be necessary to erect staging for drilling rigs on the upstream and downstream shoulders of the embankment. If the reservoir level is high, it may be necessary to use pontoons or barges to gain access to the upstream shoulder. All of this adds to the expense and decreases the flexibility of the investigation. It is therefore necessary to give careful consideration to the borehole locations at an early stage and to obtain the maximum possible information from each exploratory hole.
3. A further constraint is the need to avoid causing significant damage to the structure. Injudicious excavations or drilling may seriously exacerbate seepage or leakage and could ultimately threaten the long term stability of the dam. The potential consequences should be borne in mind when methods of investigation are chosen and careful monitoring should be carried out during the investigation. If any problems do develop, remedial action can then be taken

rapidly to preserve the integrity of the dam. The engineer may thus find himself in a "Catch 22" situation: investigations are often required because a dam is showing signs of distress, such as excessive settlement, seepage or leakage; but the methods which would most clearly indicate the cause of the problem cannot be used because of the risk of making the problem worse.

4. The first step should always be to define the objectives of the investigation. Appropriate methods and timescale can then be decided. In emergency situations it may be necessary to proceed immediately with a borehole investigation; for example, in the case of signs of slope instability in an embankment dam it would be necessary to carry out boreholes and laboratory tests and install instrumentation. A desk study would be carried out concurrently with the site work.

5. In many cases the requirement to proceed with site work is less urgent, and a thorough desk study of existing information should first be carried out. A number of sources may be available, including construction records, as-built drawings, instrumentation readings, site investigation reports and records of inspections under the Reservoirs Act (1975). Geological information may be held by the British Geological Survey. Care should be taken when assessing construction records for old dams; information is often scanty and vague, and the terms used may have different connotations now from those intended at the time. For example, many old embankment dams' are described as having a "puddle core"; this may not mean a core of puddle clay as such, but simply the same material as in the shoulders with water added.

6. In many modern dams instrumentation has been installed, either during construction or subsequently. Instrumentation records should be carefully studied to look for correlations between reservoir level, rainfall, drainage flows, seepage and water levels in piezometers. This can yield much information about the behaviour of the structure. It is always advisable to check instrumentation to ensure it is working properly. Standpipe piezometers may become blocked with time and hydraulic piezometers need regular de-airing. Testing can yield other useful information; for example carrying out falling head tests in standpipe piezometers not only demonstrates whether they are blocked but can give an indication of the permeability of the strata.

7. Even if instrumentation records are available, the frequency of readings may be too low to allow any conclusions to be drawn, and it may be necessary to install further instrumentation. It is particularly helpful to monitor instrumentation through a cycle of drawdown and refilling, if this can be arranged.

8. Having carried out the desk study and assessment of instrumentation, a suitable programme of boreholes and

laboratory testing may be carried out. Light cable percussive boring may be used in embankment dams and superficial deposits (ref. 1), and rotary core drilling in bedrock and concrete dams. Hand dug inspection pits can be useful, but large machine excavated trial pits should not be carried out on dams that are in service for safety reasons; slope instability might be triggered by large excavations, or major seepages may be initiated which could not then be stopped. The risks are less with boreholes, but a close watch should be kept on all seepages or signs of settlement during drilling, and if necessary the borehole should be abandoned and grouted up.

9. The techniques used will depend on the circumstances of each dam and the objectives of the investigation, but two general principles apply. First, a high standard of workmanship is required. Because of the difficulties of access and safety constraints, the number of boreholes is likely to be relatively small. Therefore it is doubly important that the quality of drilling, sampling and logging is high, so that a clear indication of the materials within the dam is obtained. Poorly executed boreholes only cause uncertainty, and can be worse than no boreholes at all. These problems can be minimised by careful selection of experienced site investigation contractors for the tender list and close supervision of the work by a suitably qualified engineer or geologist.

10. Second, the maximum possible information should be obtained from each borehole. This may mean continuous sampling in light cable percussive boreholes, for example, and the use of in-situ testing where possible. Instrumentation is frequently installed in boreholes so that further information can be obtained in the future. Investigation of existing dams can be expensive, so every effort should be made to obtain value for money.

11. The reasons for carrying out investigations in dams are varied. There is also a wide range of types of dam and methods of investigation, and each case should be assessed on its merits. This can be illustrated by case studies of two recent investigations of existing dams in central Scotland, Upper Glendevon and Buckieburn (Fig. 1).

UPPER GLENDEVON DAM

12. Upper Glendevon dam is a mass concrete gravity dam with a length of 390 m and a maximum height of 45 m. It is composed of 24 monoliths with a central spillway and was completed in 1955. The dam lies near the headwaters of the River Devon in the Ochil Hills north east of Stirling (Fig. 1). Spillway level is 341.48 m OD. The reservoir is used to regulate the water levels in Lower Glendevon and Castlehill reservoirs further down the valley, from which water is abstracted to the public supply in Fife. The reservoirs are owned and operated by Fife Regional Council.

RESERVOIR MONITORING AND MAINTENANCE

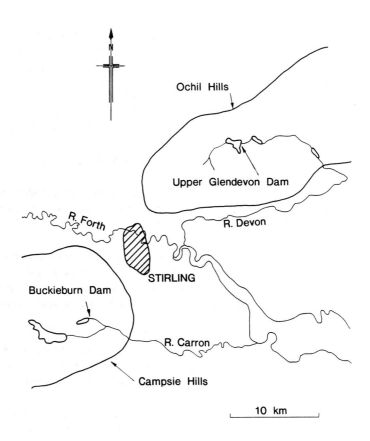

Fig.1. Location Plan

13. There were problems with leakage through and below the dam from early in its history (ref. 2,3) and a major programme of grouting of the dam and bedrock was carried out in 1959/60. Subsequent concerns about the stability of the dam under earthquake loading (ref. 4) have resulted in the water level being kept 4.3 m below spillway level for many years. Fife Regional Council wish to bring the reservoir back into full use and in 1990 commissioned Babtie Shaw and Morton to carry out a feasibility study of remedial options.

14. Two possible options were identified; a rockfill embankment on the downstream side of the existing dam or a series of rock anchors through the dam into bedrock. Ground investigations were carried out as part of the feasibility study to provide design information to enable the two options to be assessed and costed.

15. Initially a desk study and site visit were carried out. Being a relatively recent dam, there was a considerable

amount of existing information, not only on the construction but on the subsequent remedial works. Of particular value were record drawings showing geological conditions in the cut-off trench and a collection of progress photographs taken during construction.

16. Based on the desk study, proposals for a ground investigation were drawn up. These included surface geological mapping, boreholes and trial pits in the proposed borrow areas, along the line of the proposed new spillway, and in the foundation area of the proposed rockfill embankment. It was also proposed to drill two vertical rotary cored boreholes from the crest road to obtain information about the condition of the dam and bedrock.

17. The site works were originally programmed for November and December 1990, but due to slippage in the programme the contract was not awarded until late December. The successful contractor was Ritchies Ltd of Kilsyth. A commencement date in early January 1991 was agreed; however on the commencement date the site was inaccessible because of heavy snowfalls and access was not possible for a further 10 days. The site is over 4 km from the nearest main road, along a very exposed single track road, at an altitude of 350 m OD in the middle of the highest range of hills in central Scotland; not an ideal place to carry out site investigation work in the middle of winter. The contractor applied for an extension of 10 days because of extreme weather conditions, and this was granted. Once access was established, however, there were few further interruptions due to weather, in spite of hard frosts and further snowfalls. The contractor adopted seven day working and the site works were completed within the revised programme, by the end of February 1991.

18. The rotary boreholes through the existing dam were carried out using diamond tipped drilling bits and water flush to obtain 100 mm diameter cores. Sandbags were used to prevent any spillage of diesel or other contaminants into the reservoir. The concrete was found to be of poor quality in places and there was some loss of core due to disintegration of the cement. This had been anticipated from earlier descriptions of problems with the concrete (refs. 2, 3).

19. The boreholes were drilled in the middle of monoliths 16 and 20 respectively. During drilling of the borehole in monolith 16, water was observed seeping out of the junction of monoliths 15 and 16 about 12 m below crest level on the downstream side of the dam. The water appeared to be drilling flush water from the borehole; it was brown in colour from the drill cuttings, and the driller reported a loss of water returns in the borehole. The seepage was not under pressure and there was no sign of leakage on the upstream side. The water was presumably running along a construction joint between two pours of concrete, which was consistent with the condition of the concrete in the cores and the history of the dam. Nevertheless it was a somewhat

unnerving experience to stand on the 4 m wide crest of the dam, with a full reservoir on one side and 45 m drop on the other and see seepage coming out on the downstream face. The potential hazards involved in investigation of existing dams become very real at such moments! The Inspecting Engineer and Supervising Engineer were consulted and on their advice drilling continued with careful monitoring of the seepage. This got no worse, however, and ceased when the borehole was advanced into bedrock. Both boreholes were completed as planned.

20. The Ochil Hills consist of a thick pile of lavas of Lower Old Red Sandstone age (ref. 5). In the area of Upper Glendevon dam the strata consist of interbedded andesites and agglomerates. The boreholes encountered fresh bedrock immediately below the base of the cut-off trench. The core diameter was reduced to 63 mm in bedrock and a water-foam flush was used to support the sides of the borehole. This proved successful and 100% core recovery was obtained.

21. The depth of the boreholes was determined by the potential depth of the rock anchors below rockhead. This was roughly equal to the height of the dam, and both boreholes were taken to a depth of 80 m. Packer tests were carried out at regular intervals to assess the permeability of the strata and determine grout takes for the rock anchors. The tests were carried out according to BS5930 (ref. 6) and gave permeabilities in the range 10^{-7} to 10^{-6} m/s with little variation.

22. Uniaxial compressive strength tests and tensile strength tests were carried out on samples of core. These, however, can only give values for the samples tested, not for the rock mass as a whole. It had been decided therefore to carry out in-situ high pressure dilatometer (HPD) tests in the boreholes; these provide an estimate of the shear modulus of the rock mass. The tests were carried out in two depth ranges; immediately below the base of the dam to assess the strength of the foundation strata, and near the base of the boreholes for rock anchors. The HPD tests were carried out by specialist sub-contractor Cambridge In-situ Ltd immediately on completion of drilling the test section to avoid relaxation of the strata. A series of unload-reload loops were carried out up to the limit of the test equipment (20MPa). Values of shear modulus ranged from 1300 MN/m^2 to 9300 MN/m^2. The variation in the results may reflect the different positions of the test sections within individual lava flows.

23. On completion of drilling, CCTV surveys were carried out in both boreholes. These confirmed the poor condition of the concrete, with horizontal voids at various depths. Traces of grout were detected near the base of the concrete and in the upper 5 m of the bedrock, reflecting the previous remedial works. Below this depth, the bedrock was tight with no open joints or cavities. This tied in well with the good

condition of the core and the fairly uniform permeability values obtained in the packer tests. The CCTV survey was carried out by the Transport and Road Research Laboratory. Finally, piezometers were installed in the boreholes at the junction between the concrete and bedrock, thus providing ongoing information on water pressures at the base of the dam.

24. The above study illustrates how the amount of information from a limited number of boreholes can be maximised. The results of the investigation were used to assess the feasibility of the two options, and a report was submitted to Fife Regional Council in May 1991 recommending the rockfill embankment option. The client decided to proceed with this option immediately and additional ground investigation work was carried out in June and July 1991. The weather posed no problems for access on this occasion, though the afternoon when 37.5 mm of rain fell in a thunderstorm does live in the memory!

25. The work was carried out by Ritchies as an extension to their existing contract, and involved further boreholes and trial pits in the borrow areas, spillway and embankment foundation. On the dam itself, further investigations were carried out into the condition of the concrete, following the results of the earlier investigation. Cores were obtained from the upstream face of the dam for specialist laboratory testing. Four locations were chosen, at various levels. It was hoped to reach some of the earlier levels of construction, but owing to the vagaries of the Scottish weather the reservoir level was only 4 m lower in early July than it had been in late February.

26. A number of safety precautions were required for the concrete coring. The drilling was carried out from a power-operated cradle suspended from a weighted platform on the crest. A watchman on the crest and a boatman in the reservoir were present at all times. The coring was carried out by specialist sub-contractor Site Services Ltd. An expanding grout was used to backfill the boreholes to ensure that the reservoir water did not penetrate into the concrete. The power cradle was fitted with hand winches in addition to electrical ones; the need for these was demonstrated on one occasion when the electrical cable got caught round a stanchion and pulled out its socket. The cradle had to be hand winched to the dam crest; as it happened, the Resident Engineer was in the cradle, having descended to check that one of the boreholes had been satisfactorily grouted.

27. An extensive suite of tests were carried out on the core by a NAMAS accredited laboratory, Laing Technology Ltd. One of the main purposes of the testing was to confirm that the concrete had not suffered from alkali-silica reaction. The testing confirmed that this had not occurred.

28. Using the results of the ground investigations the rockfill embankment was designed and the tender documents for

the work were issued in January 1992. The total cost of the ground investigations, of which the works on the existing dam were only part, was £190,000. This compares with the tender estimate of £5,300,000 for the civil works, i.e. 3.6%.

BUCKIEBURN DAM

29. Buckieburn dam is an earthfill embankment dam with a length of 200 m and a maximum height of 23 m. It was completed in 1905. The reservoir has a very small natural catchment and is fed mainly by aqueducts collecting the headwaters of nearby streams. The reservoir is owned by Central Regional Council and is used for the public water supply, including a fish farm 1 km downstream of the dam. The reservoir lies in the Campsie hills south west of Stirling (Fig. 1). Spillway level is 248.41 m OD.

30. In November 1970 a slide occurred on the downstream shoulder, and an emergency site investigation of the kind described in paragraph 5 was undertaken (ref. 7). Remedial works were carried out, including a rock fill toe and berm of locally obtained sand and gravel fill.

31. A report by the Inspecting Engineer was issued in January 1990. This included recommendations for a number of measures, including lowering the spillway crest and replacing the rip-rap on the upstream face. Concern was also expressed about possible seepage through the embankment and leakage into the discharge tunnel below the dam. Central Regional Council appointed Babtie Shaw and Morton to design the remedial works and investigate the areas of concern.

32. In contrast to Upper Glendevon, where there was a simple aim of obtaining information, at Buckieburn the nature and extent of the potential problems had to be defined and quantified. A step by step investigation was therefore carried out, with the results of each step being assessed and used to determine the next stage. The investigation was carried over the period October 1990 to April 1991. Initially a desk study was carried out. Little information on the construction of the dam was available, but one longitudinal section along the crest showed a "puddle clay core". The boreholes in 1970 had not detected this; the whole embankment appeared to be formed of locally derived moraine fill (ref. 7). This was confirmed by subsequent boreholes and trial pits.

33. Piezometers had been installed in the 1970 boreholes and water levels had been recorded at monthly intervals since. The water levels were plotted against reservoir level and rainfall; there appeared to be a connection in some cases, but the readings were too widely spaced for any firm conclusions to be drawn.

34. The existing piezometers were all in the embankment fill. In order to assess the seepage it was decided to install eight piezometers in the natural strata. Bedrock consists of basalt lavas of Lower Carboniferous age (ref. 5),

and is overlain by morainic glacial drift deposits. These deposits are highly variable, ranging from sands and gravels to gravelly clays and clayey gravels, and contain numerous large boulders. These conditions caused difficulties for drilling; however it was important to install the piezometers as early as possible in order to monitor water levels through the winter of 1990-91. The work was carried out by Wimpey Geotec Ltd in December 1990 and January 1991 using rotary percussive techniques. There was some disruption of work in January due to snow, but this was not as serious as at Upper Glendevon.

35. Falling head tests were carried out on the existing piezometers in December 1990 and new piezometers in January 1991. A few were blocked, and leakage was occurring in others, but most were still functioning. The permeability of the embankment fill lay in the range 2×10^{-8} m/s to 7×10^{-7} m/s, compared with greater than 1×10^{-5} m/s in the basalt.

36. In order to check the stability of the embankment for the proposed remedial works, boreholes were carried out in the core and upstream shoulder; the 1970 emergency investigation had been confined to the downstream shoulder. The reservoir was drawn down to allow access to the upstream shoulder. Water levels in all the piezometers were monitored daily throughout the period of drawdown and recovery. This yielded very valuable information on the movement of water through the embankment and the natural strata.

37. The boreholes were carried out in March and April 1991 by Norwest Holst Soil Engineering Ltd using light cable percussion methods. Staging was used to gain access to the borehole positions on the upstream shoulder. Continuous undisturbed samples were taken and appropriate laboratory tests carried out to obtain shear strength parameters. Great care had to be taken to avoid contamination of the reservoir, as water was being supplied to the nearby fish farm throughout the site operations. This was achieved using a combination of sandbags, sawdust and careful drilling techniques by the contractor. A geophysical investigation was carried out in the discharge tunnel by G.B. Geotechnics Ltd.

38. The information from this range of investigations enabled the objectives of the study to be met: the proposed remedial works were shown to significantly improve the stability of the embankment; seepage through the downstream shoulder was shown to be negligible, and unconnected to reservoir level; seepage was shown to occur through the fractured basalt bedrock and not to affect the stability of the embankment; the tunnel leakage was shown to be associated with a change in the strata from bedrock to superficial deposits and may reflect difficulties of construction at this point or differential settlement of the tunnel.

39. The cost of the site investigations at Buckieburn was £21,500 compared to an estimate of £550,000 for the proposed

remedial works, i.e. 3.9%. The cost does not include the monitoring and testing of piezometers, which was carried out by personnel from Babtie Shaw and Morton and Central Regional Council. As a result of the investigations the scope of the remedial works was considerably reduced from what had originally been envisaged. The step by step approach thus proved highly effective. The remedial works are programmed to be carried out during 1992.

CONCLUSIONS
40. The investigation of existing dams differs from conventional site investigation in a number of ways. Greater emphasis has to be placed on indirect methods of investigation and less on boreholes and trial pits. The safety of the structure has to be ensured at all times. Provided the objectives are carefully thought out at the start of the project, and limitations accepted, very useful results can be obtained, leading to savings in the cost of subsequent civil engineering works.

ACKNOWLEDGEMENTS
41. The author wishes to thank the Director of Engineering, Fife Regional Council and the Director of Water and Drainage, Central Regional Council for permission to publish this paper and Messrs G.Rocke and A.Macdonald of Babtie Shaw and Morton for advice, encouragement and critical comments.

REFERENCES
1. JOHNSTON T.A. et al. An engineering guide to the safety of embankment dams in the United Kingdom. Building Research Establishment, Watford, 1990.
2. ALLEN A.C. Leakage and stability of the Upper Glendevon dam of the Fife Regional Authority, Scotland. BNCOLD Symposium, Newcastle, 1975, Paper 3.6
3. ALLEN A.C. and BOARDMAN, J. Upper Glendevon dam, Scotland. ICOLD 14th Congress on large dams, Rio de Janeiro, 1982, Paper Q52.
4. CHARLES J.A. et al. An engineering guide to seismic risk to dams in the United Kingdom. Building Research Establishment, Watford, 1991.
5. BRITISH GEOLOGICAL SURVEY. British Regional Geology: The Midland Valley of Scotland 3rd edition, HMSO, London, 1985.
6. BRITISH STANDARDS INSTITUTION. BS5930:1981 Code of practice for site investigations.
7. OSBORN H.D. Buckieburn reservoir, Stirlingshire: failure of downstream slope of embankment dam and subsequent remedial works. BNCOLD Symposium, Newcastle, 1975, Paper 5.9.

Dam ageing

G. P. SIMS, BScEng, PhD, FICE, Engineering & Power Development Consultants

SYNOPSIS. The ICOLD committee on the Ageing of Dams has been established to report on the way in which dams and their related structures change with age and what investigatory procedures are available for the detection and assessment of ageing. The Committee intends to report in 1992 and this is considered an appropriate time to report to the British Dam Society on the Committee's work and the forthcoming bulletin.

INTRODUCTION

1. The study of the deterioration of dams has traditionally been a concern of the International Commission on Large Dams (ICOLD). A Bulletin (ref 1) was published in 1983 to report on 1105 case histories of deterioration. These were collected from a sample of 14700 projects from 33 countries. Known as the CDDR (Committee on Deterioration of Dams and Reservoirs), it is an excellent database for the study of ageing dams and related works. In order to gather information from as wide a source as possible the Committee has also taken into account the information on deterioration contained in 'Lessons from Dam Incidents USA' (refs 3 and 4) published by USCOLD.

2. ICOLD set up the Committee on Ageing of Dams in 1986, at the 54th Executive meeting in Jakarta. Its terms of reference were to report on:

- The changes with time of the structural properties of dams, using illustrative case histories.

- The investigatory procedures for the detection and assessment of ageing.

To achieve its goals, the Committee has established three sub-committees to deal with: concrete and masonry dams, embankment dams, and appurtenant works. Table 1 summarises the classification of structures used which, to comply with the terms of reference, excluded reservoirs and temporary diversion works.

TABLE 1. CLASSIFICATION OF STRUCTURES

Classification	Major Typical Components	Structures
Concrete and Masonry Dams and Foundations	Gravity Arch	
Earth and Rockfill Dams and Foundations	Earthfill	Homogeneous Zoned
	Rockfill	Clay Core Concrete Central Diaphragm
Appurtenant Works	Operational Installations	Power Station Intake Fish pass
	Safety Installations	Spillway Stilling Basin

3. The Committee has not tried to prepare a text book to explain the underlying scientific basis of each example of deterioration. The published literature will continue to serve this purpose. The Bulletin under preparation is rather intended to identify the way in which dams and their associated structures tend to age. In studying the database of examples of deterioration, the Committee has been able to identify a limited number of scenarios to cover the great majority of reported incidents. The Bulletin will introduce each scenario, its causes and development, in sufficient detail to allow a broad understanding of the factors involved. Based on the experience gained from the reported incidents in the database, the Committee has then offered recommendations which, if followed, will reduce the effects of ageing in both new and existing structures. In some circumstances, it may be possible to prevent the occurrence of the scenario. ICOLD intended that the Committee should report after six years, in 1992. With one year to go the format of the report is becoming clear and this seems to be an appropriate time to report to the British Dam Society on the Committee's work.

THE CONCEPT OF AGEING

4. Figure 1 summarises some of the concepts and nomenclature used by the Committee. Deterioration of dams is influenced by their design, construction and operation. It is a requirement placed on those involved to maintain the dam in a safe condition until its decommissioning, and to achieve the performance during its operational lifetime. This, as we in Britain can confirm, can be well in excess of 100 years. Ageing is a class of deterioration. Specifically, it is

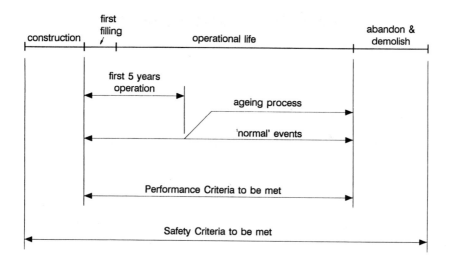

Fig. 1. Dam ageing: concept and nomenclature

defined as deterioration occurring later than the first five years of operation. It is not concerned with the effects of poor design or construction that may become apparent shortly after the first filling, neither does it encompass the influence of extreme events such as floods or earthquakes. Thus ageing, as defined by the Committee is associated with changes in structural materials that are significant to the extent that their properties are worsened. Ageing initially affects the performance of the project; later, if no corrective action is taken, the safety of the project may be put in jeopardy.

5. It is helpful to identify the more important ageing scenarios. These are the processes involved and time-related cause-effect chains leading to structural weakening. The CDDR report has defined 209 deterioration scenarios. A careful study of these, together with the available literature on incidents and accidents (refs 1-5) lead to the identification of the ageing scenarios that are the basis of the Committee's report. They are summarised on Table 2 which confirms that there are 42 scenarios, each of which will be described in detail in the Bulletin. To avoid repetition wherever possible, the ageing of the structural elements of appurtenant works is dealt with in the sections dealing with dams. Thus all scenarios of ageing for appurtenant works are associated with the flow of water and the problems that arise from this.

6. The Committee has structured its account of each scenario in a consistent way to simplify the presentation.

TABLE 2. MAJOR AGEING SCENARIOS

CONCRETE & MASONRY DAMS (432)

Foundations

1. Loss of strength under repeated load (23)
2. Erosion or solution (9)
3. Ageing of grout curtains and drainage systems (18)

Dam Body

4. Chemical reaction resulting in swelling (35)
5. Shrinkage or creep (21)
6. Loss of strength due to reaction with the environment (139)
7. Loss of strength under repeated load (71)
8. Freezing and thawing (93)
9. Ageing of structural joints
10. Ageing of upstream facings (23)
11. Ageing of pre-stressed structures (0)

EARTH & ROCKFILL DAMS (262)

Foundations

1. Deformations (6)
2. Loss of strength (6)
3. Uplift pressure increase (18)
4. Internal erosion (65)
5. Degradation due to chemical processes (9)
6. Change in state of stress (0)

Dam Body

7. Deformations (29)
8. Loss of strength (15)
9. Pore pressure increase (10)
10. Seepage through facings (5)
11. Internal erosion (22)
12. Embankment degradation (12)
13. Permeability change (0)
14. Surface erosion (55)
15. Capacity change of drains (0)
16. Loss of bond between concrete and embankment (10)
17. Ageing of synthetic polymers (0)

APPURTENANT WORKS (167)

1. Local scour (29)
2. Erosion by abrasion (29)
3. Erosion by cavitation (35)
4. Obstruction by solids carried by flow (6)
5. Problems with gates and other discharge equipment (30)
6. Excessive flow (38)

Note: The numbers in brackets are the number of examples identified for the scenario in question.

- Description of the scenario
- Illustrative example
- Analysis of the causes of the ageing, the way it tends to develop, and the consequences if left untreated
- Comments on the possibility of prevention and control of the process, and the way in which it can be monitored

ILLUSTRATIVE SCENARIOS

7. In this paper, an account is given of two scenarios to illustrate the approach. They will be taken from two of the major divisions of the Bulletin: concrete and masonry dams, and appurtenant structures.

Concrete and Masonry Dams

8. The first scenario to be described is that in which chemical and physical reactions reduce the strength of the constituents of a concrete or masonry dam. This scenario has been chosen somewhat arbitrarily; any of the eleven scenarios listed on Table 2 would have served the illustrative purpose equally well. It is however the most frequently reported scenario and British engineers will no doubt have encountered it and will perhaps find it of interest.

Loss of Strength due to Reaction with the Environment

9. The reaction is usually with reservoir or ground water, or air and is influenced by the ambient temperature. Reactions are usually more intense in dams built of poor materials with inadequate thickness, where the increased permeability allows good contact between the reacting components. Of the 139 case histories, 43 relate to the high permeability of the concrete or masonry, 71 to the chemical action of water on these materials, and 25 to the effects of thermal cycling. Sixty cases concern dams commissioned before 1930, 43 cases between 1930 and 1950, and 31 between 1950 and 1970.

10. The example selected by the Committee is the 44m high Cogliandrino Dam in Italy. It is a cellular gravity structure completed in 1926. The outer shell is of cement mortar masonry and the hearting is of cement-lime concrete containing up to 30% of "plums". The upstream face is coated with a 30mm thick layer of sprayed concrete. The dam has no joints, nor any drainage or inspection gallery. It was built in a location where the temperature varies between -5°C and 23°C.

11. Hydraulic lime had been obtained by baking a locally available marly limestone. Thus the cement-lime binder proved to be chemically weak and to be soluble, at least to some extent, in the reservoir water. Hence the permeability of the dam, already high when it was constructed, became higher with time. Between 1966 and 1978 the reservoir was held at its

RESERVOIR MONITORING AND MAINTENANCE

minimum operating level owing to problems elsewhere. When normal operation was resumed, a large leak appeared on the downstream face.

12. Testing confirmed the following properties of the concrete:-

density	$22KN/m^3$
original density	$24KN/m^3$
average compressive strength	13MPa
elastic modulus	8.4 GPa to 11.4 GPa
average porosity	7%
average sonic velocity	3600 m/s

13. It was decided that it was too expensive to grout the entire dam body although this appears to have been the technically favoured option. Instead a reinforced concrete upstream face was anchored to the dam, and substantial drainage incorporated to reduce uplift.

Causes and Processes, Development and Consequences

14. The Bulletin will highlight the importance of the porosity of the concrete in allowing chemical reactions to occur quickly. Transfer is possible when the size of interconnecting pores exceeds 1mm, diffusion being the main mechanism involved. However, in concrete subject to alternate wetting and drying, the mechanisms of transfer include capillary forces and hydrostatic pressure: these effects do not usually extend beyond about one metre. The attack by aggressive water on concrete is an important subject and ICOLD's Bulletin 71 (ref 6) covers this subject. The large majority of attacks have been caused by soft water and this is at the heart of this scenario. Soft water, with a pH of about 7 may dissolve the most soluble of the hydrated constituents of concrete, namely calcium hydroxide. This combines with atmospheric CO_2 to form the calcium carbonate efflorescence so frequently observed.

15. A swedish study (ref 7) confirms that the reaction occurs more rapidly in winter, despite the lower temperature, when joints and cracks tend to open. Thus, one can generalise that damage from this source is more prevalent in porous concrete. Experience also shows that cementitious material suffers more when it is exposed to alternate wetting and drying, or is under high hydrostatic pressure. Exposed concrete is influenced by rapid, say daily temperature variations. The steep temperature gradients established can result in tensile stress large enough to fracture the cementitious material. The consequence is an increase in porosity, leading to the accelerated degradation of the dam.

16. The Committee draws several lessons from this scenario. The importance should be recognised by designers of new structures of the value of high quality concrete and good quality lift joints as providing durability. Engineers concerned, as in the UK, with the evaluation of existing structures will be reminded of the value of an impermeable dam body with good downstream drainage. Often they will find that coring is appropriate to determine the permeability of the dam and that grouting trials may be needed. The Committee suggests that the stability of the structure should be re-evaluated with current estimates of material density and strength, and uplift pressure. The results of this analysis will help to decide what further steps may be needed.

Appurtenant Works

17. A second example of the scenarios studied by the Committee concerns appurtenant works. The scenario 'Problems with gates and other discharge equipment' has been selected not only because of its relative frequency but also because of its potential importance. If gates fail to operate whole installations are placed at sudden and spectacular risk with the obvious additional threat to human lives.

Problems with Gates and Other Discharge Equipment

18. Thirty case histories have been studied of which Picote Dam in Portugal is typical, reference 8. Picote Dam is an arch structure 100m high, 93.5m long, on the Douro River in Portugal. Built in 1958, the dam has a spillway with four tainter gates 20m wide by 8.6m high, located on the dam crest. The spillway has a capacity of $10,400 m^3/s$. The hydrostatic pressure is transmitted through radial arms to steel trunnion girders, secured to the concrete piers by steel bars. The gate position is controlled by chains, secured to the bottom girder of the gate and passing over the front of the skin plate. A chain is provided on each side of the gate, and each is operated by a separate winch powered by an electric motor. The motors are linked by a synchronising shaft.

19. During the flood on 16th February 1966, the gate was being opened under remote control. The lifting chain had lost its ability to articulate, due to the corrosive and other effects of its immersion in the reservoir and to the accumulation of debris. The left hand winch motor failed, leading to damage to the synchronising shaft bearings on that side.

20. Each winch motor was designed to lift the gate unaided, hence the gate continued to lift through the action of the right hand winch motor. However, the gate was not sufficiently stiff to be lifted from one side only, and the consequent

RESERVOIR MONITORING AND MAINTENANCE

warping led to high friction forces on the side seals and guides. The additional load on the right hand winch motor led to its failure. The gate descended out of control, and the trunnion girders were torn free, dragging the gate with them.

Causes and Processes

21. As a result of their study of problems reported within this scenario, the Committee has been able to identify shortcomings in design, construction and operation that contribute to premature ageing. The most frequently encountered errors in design include: i) wide gates with insufficient lateral stiffness to allow them to be lifted safely from one side only; ii) valve bodies with insufficient strength to resist vibrational loading, particularly under prolonged operation; iii) inadequate ventilation downstream of gates or valves discharging into a conduit; iv) failure to site stilling wells in positions that will provide steady operation of the gates without hunting; v) excessive reliance on automatic control equipment; vi) failure to take account of the consequences of a human error by the operators, particularly when they are under stress; and vii) inadequate support to the design through such activities as model or other testing.

22. The construction errors encountered owe much to lack of quality assurance, welding details which are inadequate for fatigue loading, and materials containing defects which are incorporated in the works.

23. The role of good management of the operation and maintenance is to ensure that the equipment and its operators work reliably when needed. The nature of the equipment that is the subject of this scenario is such that its failure often has immediate and obvious consequences. Among the particular points noted by the Committee is the failure to exercise and service the equipment regularly, to ensure that the automatic devices work reliably, seals remain effective, and that corrosion is controlled. In addition the failure to clear silt regularly from the operating equipment, such as control chambers and the failure to provide an adequate back-up to operators under stress are also common features of the incidents reported.

Development and Consequences

24. The development and consequences of problems with gates and other discharge equipment depend largely on the purpose of the equipment: whether it is to control the discharge from a spillway, an intake, or through outlet works. Inadequately designed spillways pose a serious threat to dams, appurtenant works, and to the people living downstream. The remedial

measures have included the replacement of the entire device, where necessary. An example of this is at Campliccioli Dam in Italy, where a siphon was replaced by gates on the dam crest.

25. When designing spillway gates to operate reliably in hostile environments and in appalling conditions of weather, the designer frequently has to foresee that the dam or appurtenant works have to survive a complete failure of the gates to open. He is forced to consider this possibility because of experience of the not infrequent failure of control devices, standby power sources, and human operators at times of great storm.

Monitoring and Control of Ageing

26. The maloperation of gates or other discharge equipment is usually detected by a visual observation.

RECOMMENDATIONS

27. The Committee intends to present its recommendations in the Bulletin for each of the three classifications of earth and rockfill dams, concrete and masonry dams, and appurtenant works. The recommendations will be under two headings: general and specific. The general recommendations for appurtenant works, listed below, will serve as an example of the guidance that will be found in the Bulletin.

General

28. The general recommendations are those that are common to all or most of the ageing scenarios identified by the Committee within the classification of the ageing of appurtenant works. Generally, ageing will be prevented, or its effects minimised through good quality design, construction and operation. Experience shows that many problems will be minimised or even avoided altogether if account is taken of the following:

(a) Hydraulic model studies, whether physical or mathematical are important in understanding the nature of flow and in designing appropriate structures.

(b) Outlet works should be robustly designed and simple to operate. They should incorporate devices to simplify cleaning and repair. They should be designed to be operable even after failure has begun. Spillway gates should, if possible, be designed to fail in the open position.

(c) The maintenance programme should be organised to ensure reliable operation of the works. Regular exercise of electrical and mechanical equipment should be routine.

RESERVOIR MONITORING AND MAINTENANCE

(d) Regular inspection of all components are necessary together with special inspections following heavy floods or earthquakes.

(e) It is frequently beneficial to determine at the design stage the gate operation sequence. This should be complied with during the operation of the appurtenant works.

(f) Facilities should be provided wherever possible for the installation of temporary stoplogs upstream and downstream of appurtenant works.

Specific Recommendations

29. Specific recommendations are made to address the problems underlying each individual scenario. As an example, the specific recommendations to reduce problems with gates and other discharge equipment are listed below:

(a) Gates should be stiff enough to allow them to be lifted from one side only.

(b) Flotation chambers used for the control of gates should be located remote from areas where the flow velocity is high.

(c) Adequate ventilation must be provided in closed conduits downstream of gates.

(d) Structural analysis is required to check the strength of hollow jet discharge valves. The analysis must take account of the vibration commonly encountered with these valves.

CONCLUDING REMARKS

30. This paper has sought to explain to British dam engineers the way in which the ICOLD Committee on Dam Ageing has approached its task. As a member of that Committee, the author has become deeply aware of the value to the engineering community in sharing experience. This however, is not enough. The work of ICOLD underlines the value, if this were needed, of the patient analysis of the collective experience of the profession so that all may benefit. The Bulletin under preparation will, it is hoped assist designers, constructors and operators to produce and care for structures that will defy the effects of ageing better than their predecessors.

Acknowledgement

31. The author acknowledges with thanks the contribution of members of the ICOLD committee on Dam Ageing, particularly its chairman, Mr. J.O. Pedro.

REFERENCES

1. ICOLD. Deterioration of Dams & Reservoirs, Examples and their Analysis - Paris, 1983.

2. ICOLD. Lessons from Dam Incidents, Bulletin - Paris, 1974.

3. USCOLD. Lessons from Dam Incidents, USA, ASCE 1975.

4. USCOLD. Lessons from Dam Incidents, USA II, ASCE 1988.

5. ICOLD. Alkali Aggregate Reaction in Concrete Dams, Bulletin 79, 1990.

6. ICOLD. Exposure of Dam Concrete to Special Aggressive Waters, Bulletin 71.

7. Fristrom G, Sallstrom S. Control and Maintenance of Concrete Structures in Existing Dams in Sweden. ICOLD Congress, Q47, 1967.

8. Lemos F.O, Martins, H.F, Peixeirolc & Leite D.O, An Accident with a Big Tainter Gate of a Spillway. ICOLD Congress QC15, Madrid, 1973.

Development of a three-dimensional computer system for dam surveillance data management

D. M. STIRLING, BSc, City University, London, and
G. L. BENWELL, BSurv, MPhil, LS, MISAust, University of Otago

SYNOPSIS. Dam surveillance is vital for the early detection of possible failures or problems so that remedial action can be taken. In the past the multitude of survey results and instrumentation readings involved have been recorded on paper and laboriously plotted manually. Computer systems using database management systems and automatic plotting routines speed up these processes enabling the surveillance engineer to detect more efficiently possible problems. This paper discusses how the use of three-dimensional computer graphics to represent such data provides an enhanced overall impression of the behaviour of structures than is possible by studying tables of data and two-dimensional plots.

INTRODUCTION
1. The Urban Water Research Association of Australia provided a grant to the Melbourne and Metropolitan Board of Works (MMBW), in cooperation with the Department of Surveying and Land Information of The University of Melbourne, for a programme of research into the development of a computerised data management system with particular emphasis being placed on the use of three-dimensional computer graphics for the improved retrieval and presentation of dam surveillance data. The first stage of this project was a tour of various authorities in Australia and New Zealand to ascertain the current status of the computerisation of dam surveillance data in these countries. Nine authorities were visited and their methods and systems studied (refs 1-2). It was noticeable that organisations which had adopted systems used them to automate the production of the form of graphical display which had previously been produced manually. In this study no attempt was made to duplicate the existing manual output presently employed by the surveillance engineers at the MMBW. Instead differing combinations of data have been combined into single displays so that a greater overall impression of the behaviour of a structure may be obtained.
2. In consultation with the MMBW it was decided to develop a prototype system based on Intergraph's Interactive

Graphics Design Software (IGDS) and Data Management and Retrieval System (DMRS) software. These are now rather old and hardware intensive packages but were used as they were specifically designed to have linkages between IGDS graphical elements and DMRS textual entities. Using these linkages data can be extracted from the database by placing the screen cursor over the graphical element representing the desired instrument instead of having to remember and type in the instrument's actual name or number.

3. It should be noted that the purpose of this research program was to illustrate the possibilities that can be realised using a powerful three-dimensional computer graphics system linked to a textual database. Therefore only a demonstration system was developed as the time and resources available did not permit full system development. Additionally the full potential of the system can only be appreciated using a colour graphics terminal and the black and white line drawings which have been used in this paper are limited in their ability to display the results.

THOMSON DAM

4. The structure selected for this pilot study was Thomson Dam which is the MMBW's newest and largest water storage structure. It is an earth and rockfill embankment dam situated 125 km east of Melbourne and actually consists of two structures, the Main Dam and the Saddle Dam, separated by a rock outcrop. Construction commenced in 1975 and was completed in 1983. The Main Dam has a height of 166 metres and a crest length of 570 metres and the Saddle Dam has a height of 70 metres and a crest length of 575 metres. Filling commenced in July 1983 and the water level had, by October 1989, reached to within ten metres of Full Supply Level.

5. One of principal reasons for the selection of Thomson Dam for this pilot study was the large number, over 500, of survey monitoring points and surveillance instrumentation on, within or around the dam including 153 surface movement/settlement points, 112 pressure cells, 35 hydrostatic settlement cells and 112 piezometers of various types.

THE DATABASE

6. After reviewing the results of the visits to the other water authorities it was decided that, for instrumentation, raw observed data should be stored in the database along with computed results. In this way any future improvements in the mathematical modelling of data reduction or recalibration of instruments could be used to improve previously computed results as well as any new observations. It was also decided that a quality tag should be included for checking for gross errors at the data input stage with each individual instrument having its own quality tag to take into account differing reliabilities. Any newly entered observation

differing from the previous observation by more than the value in the quality tag for that instrument would be queried by the system and the reading confirmed or corrected.

7. After the database structure had been decided upon and data loaded into the database attention was turned to the various options available for data extraction and presentation. Some routines were developed for the production of written reports on the database. For an operational system routines such as these would have to be written to cover all day-to-day query tasks. It is possible for these routines to be interactive and prompt the operator for information such as instrument name and desired date. However this requires the operator to know and remember the names, numbers and positions of all the individual instruments. This is not necessary if a three-dimensional computer graphics model can be attached to the database.

8. The next stage was to create a computer model of the topographic and structural detail of Thomson Dam and its surroundings. 1:500 scale contour plots of the area were manually digitised with index contours at twenty five metre intervals and five metre infill contours. Similar digitising was carried out on 1:500 plots of the foundation excavations. The spillway, inspection adits and the internal structure of the Main Dam were added. Finally colour coded graphical elements depicting all the instrumentation and monitoring points were included in the model.

9. One feature of the IGDS-DMRS interface is the ability to highlight graphical elements whose database attributes meet specified database search criteria. When the command is processed any element which satisfies the search criteria is highlighted. Highlighting normally involves temporarily converting the element colours to white. This facility could, for example, be used to highlight instruments whose latest readings have changed from their previous readings by more than a specified amount, say the instrument's quality tag. Once they have been highlighted more detailed reports could be requested for analysis.

DATA PRESENTATION

10. Various forms of three-dimensional display were developed using a combination of different types of observational data from the database.

11. The first displays developed were for the presentation of results of the surface monitoring surveys. Based on work carried out in the UK (ref 3) it was decided to depict changes between epochs using three-dimensional vectors (Fig. 1). The vectors of movement are exaggerated by a factor of one hundred times compared with the topographic background. To aid interpretation in plan views of the dam circles of settlement were also used. This technique involves placing a circle at the later end of the movement vector with the radius of the circle corresponding

RESERVOIR MONITORING AND MAINTENANCE

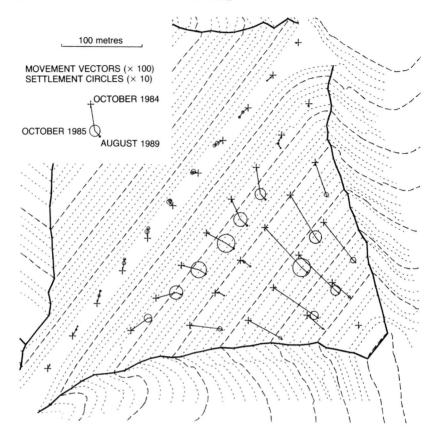

Figure 1

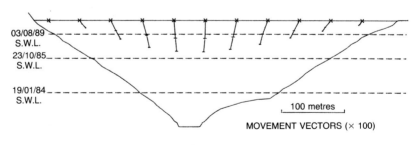

Figure 2

to the amount of vertical displacement with the circle being drawn in one colour if the point has settled and in another colour if the point has risen. The exaggeration factor for the radius of the circles is only ten compared with one hundred for the vectors. Fig. 1 shows quite clearly that the points along the crest of the dam have moved upstream and towards the centre of the crest whereas all the points on the downstream face have moved downstream.

12. As the vectors have been created in three dimensions it is possible to view them from different angles. A facility in IGDS allows the user to set the "display depth" of a view which results in an apparent slice through the dam. The display depth can be reset at any time so that different slices, thick or thin, can be viewed. The display depth for fig. 2 has been set to produce a cross-section looking downstream through the crest of the Main Dam. The corresponding stored water levels of the three epochs have also been superimposed on this view. This view shows the settlement of the crest and the movement of material towards the centre of the crest.

13. By using split screens on a workstation it is possible to display these and other views simultaneously. Each view can be altered, rotated, have details switched on or off and zoomed independently of the others. Therefore, using a series of displays, it should be possible to obtain a better understanding of the mechanics of the movement of points on the surface of the dam.

14. The next area of study was to look at methods for the depiction of piezometer readings for a single epoch. The first method of display was to use coloured squares at the position of each piezometer where the colour of the square indicated the total head reading for that instrument. When this display was first shown to the MMBW's Surveillance Engineers it was suggested that a useful addition would be to have a line drawn from the position of the piezometer to the elevation corresponding to its total head reading. Fig. 3 quite graphically shows a falling-off of pressure through the dam. It is noticeable that the two piezometers in Zone 2A both have negative head readings whereas all the piezometers in Zone 1, the central portion of the dam, have positive readings. The three lowest piezometers have significantly lower readings than the line of piezometers immediately above the grout blanket. Therefore it should be possible to identify rapidly instruments which are giving anomalous readings and then begin an investigation into possible causes for these readings.

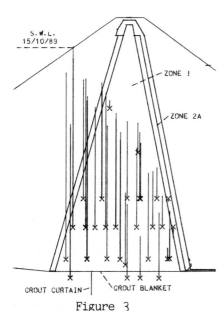

Figure 3

RESERVOIR MONITORING AND MAINTENANCE

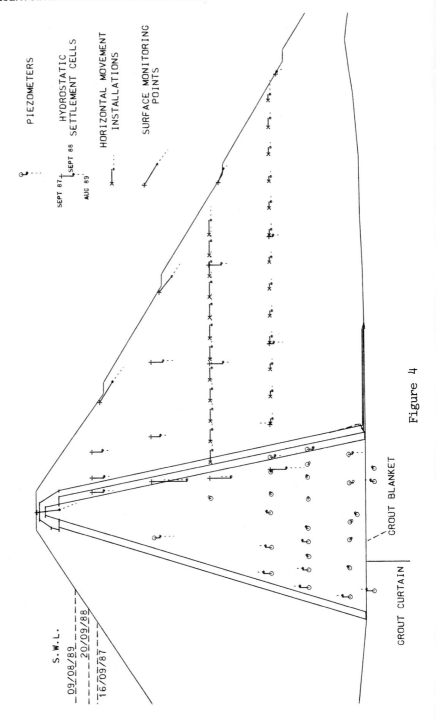

Figure 4

15. Having already produced displays for the depiction of surface movement it was decide to attempt to combine surface movement with information on movement within the internal structure of the dam. The two types of instruments providing these data are hydrostatic settlement cells for vertical movement and horizontal movement installations. Three epochs were selected to be as near to yearly intervals as the information in the database would allow.

16. Fig. 4 depicts changes indicated by hydrostatic settlement cells and horizontal movement installations along a section through the centre of the Main Dam. Because of the relatively small movements indicated by these instruments the exaggeration factor used was two hundred and fifty compared with the rest of the view. The three stored water levels are shown at their correct elevations.

17. The display for horizontal movement appears to indicate that the anchor points in the upper of the two lines of installations are moving more than the anchors in the lower line of installations.

18. Also included in Fig. 4 are three-dimensional vectors for surface monitoring points. Once again the exaggeration factor used was two hundred and fifty. By combining all three types of readings together analysis of the resulting display confirms that both horizontal and vertical movements increase in magnitude with elevation.

19. It was then decided to add changes in piezometer readings to the same view to see if any correlation between changes in water pressure and movements could be discerned. The depiction of changes in water pressure was to be in a similar fashion to the changes in hydrostatic settlement cells i.e. vertical lines from epoch to epoch. This time a factor of one was entered so that the lines indicated true changes in total head readings. These are also shown in Fig. 4. This appears to show that there is increasing water pressure in the upstream portion of the dam gradually changing to decreasing pressure in the downstream portion of the dam.

CONCLUSIONS

20. There can be no doubt that the use of a computerised database system is far more efficient for the day-to-day tasks involved in dam surveillance when compared with the older techniques of paper files and manual plotting. A number of organisations have developed computer systems where the output has basically duplicated what was previously produced manually. Many of these systems are completely automatic in that the range of data selected are automatically scaled to fit the screen or paper and text and other annotations placed automatically. This results in these systems being very easy to use but also in them being restricted in what they can produce. On the other hand a three-dimensional computer graphics system may, at first

glance, appear to be difficult to use but knowledge of a relatively few basic commands and concepts can result in a powerful and flexible system. The user can manipulate the displayed graphics using standard commands, including facilities for scaling each of the three axis by differing amounts, rotating the views around the axes, turning detail on and off, adding, deleting and moving elements and positioning text. This requirement to have at least a working knowledge of the graphics system greatly increases the power and flexibility which can be achieved.

21. The ability to link graphic elements with a textual database also increases the efficiency for data extraction. In this way the required instruments can be selected using the screen cursor, the relevant data automatically extracted from the database and the graphic elements appearing on the screen almost instantaneously.

22. This experimental work has shown that a powerful three-dimensional computer system can be used to produce composite displays of a variety of dam surveillance data which can allow new methods of interpretation to be developed and hopefully lead to a better understanding of the internal mechanics of such structures. This in turn may lead to better or more cost effective remedial works to be designed and undertaken at an earlier date than is possible when using more conventional surveillance techniques. Full details of this work can be found in Ref. 4.

REFERENCES

1. STIRLING, D M and BENWELL, G L. The status of the computerisation of dam monitoring data in Australia. Proceedings of the Surveillance and Monitoring Surveys '89 Symposium. The University of Melbourne, November 1989, 173-182,

2. STIRLING, D M and BENWELL, G L. Surveillance data management - review of practice in Australasia. ANCOLD Bulletin, No. 84, December 1989, 97-113

3. HOPKINS, J K, WICKHAM, D B and STIRLING, D M The use of close-range photogrammetry for reservoir embankment monitoring. Proceedings of The Embankment Dam Conference of the British Dam Society, University of Nottingham, September 1990, 137-142.

4. STIRLING, D M, BENWELL G L and MURNANE, A B. Management and display of dam surveillance data. Urban Water Research Association of Australia Research Report No. 21, April 1991, 112 pages.

The BRE dams database

P. TEDD, I. R. HOLTON and J. A. CHARLES, Building Research Establishment

SYNOPSIS. A computerised database containing information on dams that come within the ambit of the Reservoirs Act 1975 is being developed at the Building Research Establishment. Preliminary analyses of the data are presented including age, height, type, problems, and remedial works.

INTRODUCTION
1. A computerised database is being developed as part of the Department of the Environment's (DOE) research programme on the safety of dams. It currently contains information on some 2300 dams that come within the ambit of the Reservoirs Act 1975. Both concrete and embankment dams have been included, but the present structure of the database is biased towards embankment dams. Dams that have failed or have been discontinued are also included. Basic details on dams such as name, location, height, capacity, and type of construction are recorded together with information on problems, investigations and remedial works.
2. The database will provide a register of all the 2450 dams that come within the Reservoirs Act 1975 and will enable all information sent to DOE in its returns from the Enforcement Authorities to be stored in a readily accessible form. It will also be of use in DOE's reservoir safety research programme in providing background information for particular research projects. It is used by BRE to analyse the occurrence of problems and remedial works in relation to particular features of embankment dams enabling research needs to be identified.
3. This paper describes the structure of the database, the method of storing data and the facilities available for retrieving and analysing the data. The problems involved in setting up such a database are discussed. Some preliminary analyses of the data are presented. It is intended that as the database becomes more comprehensive more detailed analyses of data will be published.

SOURCES OF DATA
4. Information has been obtained from many sources including the World Register of Dams, the Welsh Register, Enforcement Authorities and returns from a questionnaire circulated to dam owners and panel engineers during the preparation of "An

Engineering Guide to the Safety of Embankment Dams in the United Kingdom" (Johnston et al, 1990)[3]. Much of the information on problems and remedial works has come from published literature. A list of published references is also contained in the database. Information is continually being added to the database and small modifications to the structure are being made. Information from any source for inclusion in the database will be appreciated.

STRUCTURE OF THE DATABASE
5. The database uses a computerised information management system called Oracle. The system allows information to be stored, retrieved and interrogated on the mainframe computer at BRE.

Data Storage
6. Information on a particular dam is entered into the database under various headings known as record types. These record types have been divided into three groups as shown in Table 1. Group 1 contains basic information about the dam and its construction. Group 2 relates to problems (including incidents and deterioration), investigations and remedial works. The two records in group 3 store the references.

Table 1. Record types in database

Group 1	Group 2	Group 3
Details	Problems	References
Purpose	Investigations	All references
Damtype	Remedial works	
Sealant		
Foundation		
Cut-off		
Spillway		
Outlet		
Upstream protection		

Group 1: Basic Details and Dam Construction
7. The DETAILS record in group 1 contains basic information about the dam such as name, height, date completed and capacity of the reservoir. The DETAILS record for the failed Bilberry dam is given as an example in Appendix A.
8. Other information about the construction and structure of the dams is stored in the remaining records in group 1. In these records information is categorised as shown in Appendix B. For example, in the DAM TYPE record there are eight categories of dam types, the first six corresponding to those used in the World Register. The number of categories in the records can be increased to accommodate any modification to the database. Categorising the information facilitates storing and searching for particular features. Having specified a particular type more details can then be added in the form of

text. Details of any references cited in the text are stored in the ALL REFERENCE record.

Group 2: Problems, Investigations and Remedial Works
9. For each of these records the categories may be conveniently divided into groups as shown in Appendix C.
10. <u>Problems record</u>. The current list of categories in the PROBLEMS record covers every aspect of deterioration of dams and does not distinguish between the seriousness of the problems. There are seven groups of categories, three of which relate to the major causes of failure of embankment dams i.e. internal erosion, external erosion and slope instability.
11. Difficulties have arisen in selecting the categories in the PROBLEMS record with trying to cover all aspects of deterioration. Some categories are possible indicators of problems e.g. wet patches on the downstream slope, whereas others are serious incidents such as a breach.
12. Performance indicators such as leakage and settlement pose certain difficulties in their interpretation. Whilst leakage is a possible indicator of a malfunction of the dam, it is also likely that some reports of leakage could be the result of poor drainage on a dam and be unconnected with the water from the reservoir. Similarly, settlement of an embankment dam generally only becomes a problem if it is caused by some serious malfunction of the dam such as internal erosion or slope instability or if freeboard has been seriously reduced. Within the settlement category there is the facility for specifying the settlement index proposed by Charles[1] (1986).
13. <u>Investigations record</u>. The categories within the INVESTIGATION record also include instrumentation and monitoring. The groups into which the categories can be placed include the three main areas of monitoring 1) pore pressure 2) leakage and 3) movement (see Appendix D).
14. <u>Remedial works record</u>. As far as possible, the categories have been grouped in a similar way to those in the problems record (see Appendix E).
15. <u>References</u>. Published references relevant to a particular dam are listed by a number in the REFERENCE record. Details of all the references in the database are given the ALL REFERENCE record. An important facility of the database is the ability to search the references.

RETRIEVAL AND ANALYSIS OF DATA
16. There are two ways of retrieving information from the database. Information on any particular dam can be retrieved by keying in the appropriate commands and viewing the available entries as in the example DETAILS record entry in Appendix A. Alternatively a query language can be used to analyse the data. Lists of dams with particular features and which have had certain problems can be obtained. Some examples of analyses are given below.

RESERVOIR MONITORING AND MAINTENANCE

Analysis of the Details Record

17. Analyses of the database reveal a number of interesting features about the U.K. stock of dams. Similar surveys and analyses carried out by others (Moffat 1982[5], Hughes 1988[2] and Millmore and Charles 1988[4]) confirms the validity of the general form of these analyses.

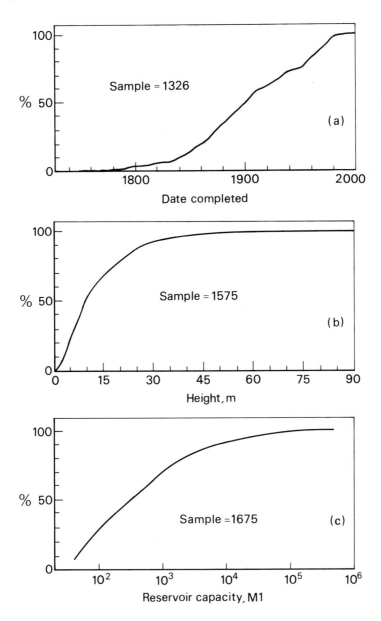

Fig. 1. Analyses of database records

18. Figure 1 shows the results of three analyses that have been carried out on data from the DETAILS record. Figure 1a indicates the rapid increase in dam construction in the U.K. during the latter half of the 19th century. Figure 1b shows that 68% of British dams are under 15m, but the percentage is certainly much greater than this as information on height was available for only 1575 dams and it is likely that the remainder are lower than 15m. Figure 2 shows the percentage of dams built in a height range for particular periods of time. Examination of the data on height and date completed shows that although the maximum height of dams has increased with time, more than 50% of the dams built during the last three decades were under 15m in height.

19. Figure 1c shows the vast range in the reservoir capacities of British reservoirs. It should be noted that the capacity has a log scale. Many of the largest reservoirs are associated with concrete dams in Scotland. Information was available on only 1675 dams and it is likely that those not included will have small capacities.

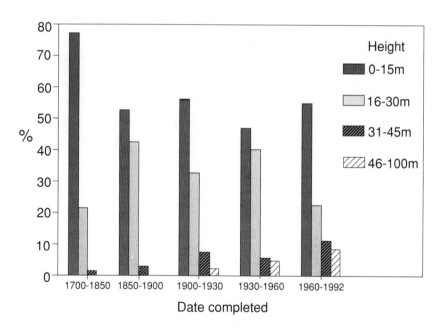

Fig. 2. Heights of dams built during different periods

Analysis of Problems and Remedial Works Records

20. At present the database includes 340 dams with records of problems and 366 dams with records of remedial works. Leakage is the most commonly reported problem with 101 cases being reported in embankment dams. The majority of these have

had related remedial works carried out. The database includes 594 records of remedial works spread over 366 dams. They range from major works such as the core being grouted to relatively minor works like repairing valves.

Table 2. Occurrence of problems and remedial works

	Problems	Remedial works
Internal erosion - leakage	195	111
Slope stability	61	69
External erosion - floods	93	114
Draw-off works	41	119
Wave damage	17	25
Settlement	95	84
Other	85	85

21. Table 2 shows a broad breakdown of recorded problems and remedial works. The categories included in these various groups are shown in Appendices C and D.

AVAILABILITY

22. Requests for searches of the database have come from a variety of organisations including consulting engineers, research workers and insurance companies. Reasonable requests for information from the database have been undertaken free of charge. Many requests have been accompanied by exchanges of information. Any information acquired in confidence would only be given with the consent of the originator. It is not considered practical or desirable to give or sell the complete database to any other organisation.

CONCLUSIONS

23. The database provides a means of storage and retrieval of basic information on the dams in the U.K. The amount of detail on any particular dam varies considerably from only dam name and owner to full details including information on construction, problems, investigations and remedial works and with a comprehensive list of references. Analysis of problems and remedial works provides a valuable tool for the identification of research needs.

ACKNOWLEDGEMENTS

24. The work described in this paper forms part of the research programme of the Building Research Establishment and is published by permission of the Chief Executive. Continued support and encouragement from Mr C E Wright of the Water Directorate of the Department of the Environment is gratefully acknowledged. The co-operation of some Enforcement Authorities in providing additional information to that sent to DOE is appreciated. The support of Mr G J Smith and Mr C J Johnson of BRE in setting up the database is acknowledged.

REFERENCES
1. CHARLES J.A (1986). The significance of problems and remedial works at British earth dams. Proceedings of BNCOLD/IWES Conference on Reservoirs 1986, Edinburgh, 123-141. Institution of Civil Engineers, London.
2. HUGHES A.K.(1988). The Supervising Engineer and the Reservoirs Act 1975. Proc. of Reservoir Renovation 88 Conference. Manchester, Paper 2.5. BNCOLD, London.
3. JOHNSTON T.A.,MILLMORE J.P., CHARLES J.A. and TEDD P (1990). An engineering guide to the safety of embankment dams in the United Kingdom. Building Research Establishment Report BR 171.
4. MILLMORE J.P. and CHARLES J.A.(1988). A survey of U.K. embankment dams. Proc. of Reservoir Renovation 88 Conference Manchester, Technical Note 1, pp.1-5. BNCOLD.
5. MOFFAT A.I.B (1982). Dam Deterioration - A British Perspective. Proc. of 1982 BNCOLD Conference. Paper No 8

Crown copyright,1992.

APPENDIX A. EXAMPLE OF A DETAILS RECORD

REFNO - 1585 DAMNAME - Bilberry (1845) EXISTING - Failed 1852
ENFORCEMENT AUTHORITY -
UNDERTAKER - Holme Valley Commissioners
GRIDREF SE 103070 COUNTY - West Yorkshire NEAREST TOWN - Holmfirth
RIVER - Digley Brook ENGINEER - George Leather CONTRACTOR - Sharp
DATE COMPLETED - 1845 CONSTRUCTION TIME - 6 years HEIGHT - 20m
CAPACITY - 310,000m^3 CREST LENGTH - 90m SURFACE AREA - ...m^2
EMBANKMENT VOL. -...m^3 UPSTREAM SLOPE - 3.0 DOWNSTREAM SLOPE - 2.0
ALTITUDE (TWL) - 250m SPILLWAY CAP.- ...m^3/sec DRAWOFF CAPACITY - ...m^3/sec
IMPOUNDING - Yes RISK CATEGORY - A

APPENDIX B: CATEGORIES IN THE PURPOSE AND CONSTRUCTION RECORDS

PURPOSE RECORD	DAMTYPE RECORD	UPSTREAM PROTECTION RECORD
Irrigation	Earth fill	Masonry blocks
Hydroelectric	Rockfill	Brickwork
Flood control	Gravity	Asphaltic membrane
Navigation	Buttress	Concrete
Water supply	Arch	Riprap
Recreation	Multiarch	Concrete blocks
	Service	Pitching

SEALANT		
Position	Nature	FOUNDATION RECORD
Upstream face	Concrete	Competent rock
Internal	Asphaltic	Fissured rock
Homogenous	Puddle clay	Soft clay
	Rolled clay	Firm clay
	Plastic	Stiff clay
	Earthfill	Ballast

OUTLET RECORD	CUT-OFF RECORD
Pipes through core	Concrete
Pipes in tunnel through core	Puddle clay
Tunnel through abutment	Grout
Pipes through fill	
Culvert in foundation	

RESERVOIR MONITORING AND MAINTENANCE

APPENDIX C: CATEGORIES IN THE PROBLEMS RECORD

Internal Erosion
Leakage
Turbid leakage
Wet patches d/sslope
Breached
Internal erosion
Fractures in discharge pipe
Leak by pipe or culvert
Leak into culvert

External erosion
Inadequate spillway capacity
Over topped
Breached
Spillway damaged
Inadequate freeboard
Inadequate flood capacity

Settlement
General settlement
Localised settlement
Culvert damaged by settlement
Mining

Slope instability
u/s slip in-service
d/s slip in-service
Natural ground slip
u/s slip during construction
d/s slip during construction

Draw-off works
Culvert damage
Scour blocked
No draw-off arrangements
No upstream controls
Tower damaged
Valves inoperable
Leak into valve tower

Wave damage
Wave damage u/s protection
Wave damage to wall

Other
Animal activity
Trees or vegetation

APPENDIX D: CATEGORIES IN THE INVESTIGATION RECORD

General
Construction instrumentation
Trial pits
Borehole sampling
Laboratory testing
Geophysics
Diving surveys
CCTV

Pore and earth pressures
Pore pressures in core
Pore pressures in fill
Pore pressures natural ground
Earth pressure core

Leakage
Infrared
Tracers
Flow measurements

Flood studies
Flood studies
Hydraulic model

Movement
Crest settlement
Crest lateral movements
Internal movements
Slope movements

APPENDIX E: CATEGORIES IN REMEDIAL WORKS RECORD

Internal erosion - leakage
Core grouted
Diaphragm wall placed in core
Sheet piling
Membrane repaired
Extended cut-off
Toe drain
Core raised
Core repaired with clay
Upstream blanket
Foundations grouted

External erosion - floods
Spillway repaired
Spillway enlarged
TWL lowered
Additional spillway
Spillway reconstructed

Draw-off works
Valve tower
Outlet tunnel repaired
Outlet pipes repaired-replaced
Valves repaired/replaced

Slope instability
Downstream slope flattened
Downstream slope drainage
Toe drain
Upstream slope flattened
Berms added
Downstream face repaired

Settlement
Crest restored to original
Wave wall repaired
New wave wall built

Wave damage
Upstream face repaired

Other
Demolished
Total reconstruction
Partial reconstruction
Raised
Lowered
TWL raised
Discontinued

The reservoir safety research programme of the Department of the Environment

C. E. WRIGHT, Department of the Environment, D. J. COATS, Consultant, and J. A. CHARLES, Building Research Establishment, Department of the Environment

SYNOPSIS. The Department of the Environment is responsible for reservoir safety legislation in Great Britain. An outline is presented of the various projects carried out as part of the Department's research programme on reservoir safety from 1983-1991. The overall objective of the programme is to promote adequate, consistent and cost-effective safety standards and in particular to provide Panel Engineers with an appropriate background for carrying out their duties under the Reservoirs Act 1975.

INTRODUCTION
1. Legislation imposes a system of safety checks on reservoir construction and operation in Great Britain. Two Acts of Parliament have been passed in this century for this purpose; the Reservoirs (Safety Provisions) Act of 1930 has now been superseded by the Reservoirs Act 1975 which applies to all reservoirs that are designed to hold or capable of holding more than 25 000 m^3 of water above the natural level of any part of the land adjoining the reservoir.
2. In 1983 the Department of the Environment (DOE) instituted a programme of research to promote reservoir safety in response to a recommendation by a House of Lords Select Committee. This action was linked to a decision to implement the Reservoirs Act 1975 and the realisation that many dams would require remedial works to satisfy modern safety standards. The research has been implemented through external contracts and through work at the Building Research Establishment (BRE) which is an Executive Agency of DOE. The Department has been advised by a Reservoir Safety Research Committee. Some projects have been jointly funded with bodies such as the Construction Industry Research and Information Association (CIRIA), the Science and Engineering Research Council (SERC) and the Water Research Centre (WRC).
3. While the responsibility for the safety of a reservoir generally rests with its owners or the persons having management or control of it, the DOE has an interest in such safety through its sponsorship of the Reservoirs Act 1975. The Department also acts in effect as a proxy customer for the many private owners who are each responsible for one or two

reservoirs since these owners are not organised collectively to fund the necessary research. This proxy customer role was recommended in a report (ECE 512) of the Economic Commission for Europe in June 1989.

4. The main focus of the research is concentrated on the 2500 reservoirs which come within the ambit of the Reservoirs Act 1975, although some of the research also covers smaller reservoirs. The research programme has been principally directed towards the safety of existing dams rather than new construction, since few new reservoirs are being built in the UK. The overall objective is to promote adequate and consistent safety standards for large reservoirs at minimum cost thereby maintaining a satisfactory balance between reducing risk and the expenditure involved. The research is designed to provide Panel Engineers with an appropriate and consistent background for carrying out their duties under the Act. Also it is in the UK interest to promote research and produce reputable guides which could later influence European practice and codification.

5. The research projects are described under the following headings; hydrology, hydraulics, structures, risk. The background to a current assessment of the DOE research programme is described and some thoughts on the future direction of the programme are presented.

HYDROLOGY

6. A significant part of the research programme addresses the hydrological aspects of reservoir safety. This is appropriate because one of the most common causes of reservoir dam failure is overtopping due to inadequate spillway design and incorrect estimation of extreme flood events. In 1975 the Natural Environment Research Council (NERC) published the "Flood studies report", which was followed by the Institution of Civil Engineers' (ICE) "Floods and reservoir safety : an engineering guide" in 1978. These two reports have been widely used as a basis for uprating and improving reservoir spillway capacity. However, further hydrological research has been considered advisable. A number of "Flood studies supplementary reports" have been published by the Institute of Hydrology (IH) at various dates and these have been listed by Reed and Field (1992). The ICE guide is currently under review, as reported in this conference, and DOE funding has supported three further projects.

7. <u>Spatial and temporal rainfall variability</u>. Three data sets were examined to improve the understanding of spatial and temporal rainfall patterns. The main region of study was a 10 000 km^2 area centred on the Hameldon Hill radar site north of Manchester. Daily data was obtained for a large number of sites and these were analysed together with hourly radar-derived data for a number of storm events. Additionally, recording rain-gauge data were used from the upper Dee catchment in North Wales (Stewart, 1992; Stewart & Reynard, 1992).

8. <u>Flood estimation for reservoired catchments</u>. The characteristics of reservoired and gauged catchments in the UK have been compared to assess whether there was any significant bias in the recommendations contained in the 1975 NERC "Flood studies report". IH have examined differences between summer and winter probable maximum flood estimates, the sensitivity of the design peak hydrograph to increasing storm duration and reservoir lag and the effect of snowmelt. The various methods of reservoir flood estimation in Scandinavia, Australia, South Africa and the United States have been reviewed and their relevance to the UK situation noted (Reed and Field, 1992).

9. <u>Joint occurrence of wind and rain</u>. The ICE guide "Floods and reservoir safety" recommended an examination of the combination of various design inputs when testing the adequacy of flood provision at reservoirs, for example the coincidence of wind induced wave run-up with the probable maximum flood. This IH study is on-going and a paper describing some of the work is being given at this conference (Reed and Anderson, 1992).

HYDRAULICS

10. A number of projects covering hydraulic aspects of reservoir safety are aimed to improve the design or reduce maintenance costs. They include field studies, laboratory studies and mathematical modelling.

11. <u>Wave prediction in reservoirs and blockwork protection</u>. Work on wave prediction by Hydraulics Research (HR) has included a literature review (HR, 1987) and a comparison of available methods (HR, 1988a). The blockwork stability study (HR, 1988b) has been followed by a further study which has included an evaluation of cases of failure at selected reservoir sites with three types of upstream protection: pitching, concrete blockwork and concrete slabbing. In 1992 HR are due to publish their report to give guidance on best practice in the design, maintenance and/or rehabilitation of blockwork protection for dams.

12. <u>Performance of wedge-shaped blocks in high velocity flow</u> A review of work on wedge-shaped blocks was completed in 1989 by CIRIA as part of phase I of a study on the use of this technique developed by Russian engineers. The visit of Professor Pravdivet to the UK in Autumn 1991 assisted in the completion of the phase II report which is the design guide for the use of these blocks. If a site suitable for the use of this method is found in the UK, it is hoped that phase III, a full-scale test performance for the benefit of other potential users, will be initiated.

13. <u>Mathematical model for spillway flow</u>. CIRIA technical note TN134 (Ellis, 1989) made available a numerical model for high velocity flows in open channel chute spillways. Older spillways may have features which complicate flow patterns that are often supercritcial and require advanced analytical techniques to reproduce flood profiles. This report provided guidance on the use of these techniques for a range of spillway

features, it contained the results of laboratory studies of cascades in the spillway floor, and guided the reader in the use of numerical modelling and physical modelling solutions. In some cases it is expected that the use of this CHUTE numerical model would provide a more cost-effective analysis than a physical model.

14. Reinforced grassed spillways. CIRIA report R116 (Hewlett et al, 1987) on the design of reinforced grassed spillways incorporated the results from full scale field trials at Jackhouse reservoir in north west England. This report provided guidance on the ability of unreinforced and reinforced grass to withstand unidirectional flow for up to two days duration in steep waterways such as auxiliary spillways on dams, and also on the protection of embankments against erosion by overtopping during extreme flood events. An impetus for producing this guide to good practice for design engineers was given by the full implementation of the Reservoirs Act 1975 in the period 1986-87, which brought about an increase in the number of dams being modified to provide additional capacity to pass extreme flood events.

STRUCTURES

15. Reservoir safety is closely linked to the safety of impounding structures and their ancillary works. The water retaining structure which impounds a reservoir may be formed by a concrete or masonry dam or more commonly in the UK by an embankment dam. In view of the numerical preponderance of embankment dams, the main emphasis in the research programme has been placed on this type of dam while a relatively small part of the research programme has been related to concrete dams.

16. Concrete and masonry dams. Only 20% of the total population of 2500 dams which come within the ambit of reservoir safety legislation are concrete and masonry dams. A report has been commissioned on internal pressures and uplift in massive concrete dams and this should be completed in the near future. In co-operation with CIRIA it is intended to produce an Engineering Guide to the Safety of Concrete and Masonry Dams in the United Kingdom to complement the BRE guide to embankment dams (Johnston et al, 1990).

17. Embankment dams. Of the 2000 embankment dams that come within the scope of the reservoir safety legislation, approximately 70% were built prior to the year 1900 and only 7% after 1960. Consequently less than 10% of the existing stock of embankment dams have been built during the recent period in which the theories of modern soil mechanics have significantly influenced design and modern heavy earth moving plant has been available for placing and compacting fill materials. The majority of these older dams have puddle clay cores and many also have a puddle clay filled cut-off trench (Millmore and Charles, 1988). The long term behaviour of this large stock of old embankment dams is of considerable importance.

18. Research sponsored by DOE on embankment dams has

included chemical deterioration of fill materials (Babtie, Shaw and Morton, 1986), modes of dam failure (Binnie and Partners, 1986) and stress analysis (Dounias et al, 1989). Work on a guide for small embankment type reservoirs for water supply and amenity use is due to be completed in 1992. In addition to these projects, intra-mural research into the safety of embankment dams has been carried out at BRE as described below.

19. <u>Ancillary works</u>. A project has commenced on the assessment of pipes and valves in reservoirs and dams. Access and inspection techniques have been reviewed. Field trials are planned involving the use of CCTV and measurement of wall thickness with an ultra-sonic probe.

BRE investigations into the safety of UK embankment dams

20. Research into the safety of embankment dams has continued at BRE for over 50 years, commencing in 1937 when the Soil Mechanics Section (now Geotechnics Division) of the Building Research Station (now BRE) carried out an investigation into the failure during construction of the embankment of a new reservoir at Chingford (Cooling and Golder, 1942). Since 1983 the BRE work on embankment dams has formed part of the reservoir safety research programme of DOE and the principal emphasis has been on investigating the safety and performance of old puddle clay core dams (Charles and Tedd, 1991). Field investigations at selected dams have examined mechanisms of deterioration and ageing particularly internal erosion. A number of performance indices have been developed to normalise measurements of stress, settlement and seepage. A simple method of classification of erodibility has also been proposed.

21. <u>Stress measurements and hydraulic fracture</u>. The most common cause of deterioration of clay cores is associated with cracking and internal erosion. Hydraulic fracture by the reservoir water may occur due to internal stress transfer caused by differential settlement and there is a risk that hydraulic fracture will occur if the pressure due to the reservoir head [$\gamma_w h_w$] is greater than the total earth pressure [σ] in the core acting on a plane which crosses the core, (ie $\gamma_w h_w$ is greater than either the total vertical stress [σ_v], or the total horizontal stress acting along the axis of the dam [σ_{ha}]). The ratio $\sigma/(\gamma_w h_w)$ has been proposed as an index of susceptibility to hydraulic fracture:

$$[HF]_I = \frac{\sigma}{\gamma_w h_w}$$

If $[HF]_I < 1$ there is a risk of hydraulic fracture. Field measurements of earth pressure in narrow puddle clay cores and cut-off trenches have confirmed the existence of relatively low stresses at certain locations in some dams (Charles and Watts, 1987; Tedd et al, 1987, 1988, 1989; Charles, 1989). Pressure cells pushed into the soil from the bottom of vertical boreholes can only be aligned to measure horizontal stress and the need to measure vertical stress has led to the development of the BRE miniature push-in pressure cell which has been designed to be jacked horizontally into soft clay from a vertical borehole with the cell in an attitude to measure either vertical or horizontal stress (Watts and Charles, 1988).

22. <u>Erodibility of puddle clay</u>. If a water filled crack does occur in the clay core, soil adjacent to the crack will swell in the presence of free water and effective stresses will approach zero. When the puddle clay in an embankment or its foundation is unprotected, the internal stability of the soil when subjected to drag forces from seepage and leakage is crucial for the long term performance of the embankment. All clays will erode under severe conditions, but it is important to know whether erosion resistance plays a significant role in the behaviour of puddle clay core dams. The erosion resistance is governed by true cohesion, that is the strength at zero effective stress. The erodibility of British puddle clays has been investigated in laboratory tests; the cylinder dispersion test (Atkinson et al, 1990) has been developed for this purpose and clays can be classified as follows:
      type N; non-dispersive, cohesionless
      type C; non-dispersive, cohesive
      type D; dispersive

23. <u>Filter properties</u>. The filter properties of the selected fill, which was typically placed on both sides of the puddle clay core, have been examined to see if erosion of the core would be prevented by such fills (Tedd et al, 1988). Particle size distribution and permeability criteria for filters have been used in the assessment. It seems that many typical selected fills would act as filters and halt internal erosion of the core. The situation in clay filled cut-off trenches may be more critical. The trench may have been excavated through jointed and fissured rock which may permit internal erosion of the puddle clay subsequently placed in the trench.

24. <u>Deformation measurements</u>. Crest settlement is the most common measurement made at old embankment dams and can be an important indicator of embankment performance. It can be caused by a variety of processes some of which relate to the normal behaviour of an embankment dam (eg secondary consolidation of puddle clay, stress changes during reservoir drawdown and refilling) others to mechanisms of deterioration and failure (eg development of slip surfaces, internal erosion). It is necessary to have some method of interpreting the results in order to assess the performance of the dam. Charles (1986) proposed a settlement index S_I that was analogous to the coefficient of secondary consolidation for a clay soil:

$$S_I = \frac{s}{1000 \cdot H \cdot \log(t_2/t_1)}$$

where s is the crest settlement in mm measured between times t_2 and t_1 after the completion of embankment construction, and H is the height of the dam in m. If the settlement of the embankment was entirely due to creep of a granular fill or secondary consolidation of a clay fill, it is unlikely that S_I would be larger than 0.01 and probably somewhat smaller. A value of 0.02 is suggested as a baseline above which other explanations need to be found for the movements. Recent investigations have indicated that large reservoir drawdowns will cause significant movements (Tedd et al, 1990) and the settlement index will need to be adapted to take cognisance of this.

25. <u>Measurement of seepage and leakage</u>. Work has continued on methods of identifying the source of seepage flows in the downstream slope of embankment dams (Tedd and Hart, 1988). This has included an evaluation of chemical analyses, temperature measurements and correlations of measurements of seepage flow with reservoir level and rainfall. An index which may be of use in this context is the seepage and leakage index:

$$Q_I = \frac{q}{1000\ A\ i\ k}$$

where q is the flow through the watertight element in litres per second, A is the area of watertight element in contact with the reservoir water in m^2, k is the maximum acceptable permeability of intact watertight element in m/s, i is the mean hydraulic gradient across the watertight element. If all the flow through the watertight element was due to steady seepage, then it would be expected that $Q_I<1$. Leakage through imperfections, discontinuities and cracks may give much greater values. In practice it will be difficult with many clay core dams to assess q. Some old dams may have no seepage collection system. Where a collection system does exist, some measured flows may not have come through the watertight element but may be due to rainfall or may have come from the surrounding hillsides. Alternatively some leakage may pass into the foundations and not be detected by the monitoring system.

26. <u>An engineering guide to the safety of embankment dams in the United Kingdom</u>. Prepared by BRE in collaboration with consulting engineers Babtie Shaw and Morton, the guide (Johnston et al, 1990) provides guidance to those with responsibility for reservoir safety, particularly in the application of the principles of geotechnical engineering (soil mechanics) to dam safety. It has been based on the experience of panel engineers and also applies the results of BRE research findings.

RISK

27. The evaluation of hazard and risk is an important facet of reservoir safety which involves a knowledge of hydraulics, hydrology, geotechnical engineering and a number of other subjects as well as methods of risk analysis. Projects primarily associated with risk assessment have been grouped together under this heading.

28. **Probabilistic risk assessment.** Following a recommendation by the House of Lords Select Committee on Science and Technology (1982), DOE and WRC (representing water industry funders) jointly commissioned a study of the feasibility of applying probabilistic risk assessment (PRA) to reservoir safety in the UK. The work was let to the Safety and Reliability Directorate (SRD) of the United Kingdom Atomic Energy Authority. SRD were commissioned to work with Binnie and Partners. The feasibility study concluded that there were no fundamental reasons why PRA should not be applied to reservoir safety. It recommended that a detailed risk assessment of a selected reservoir should be carried out (SRD UK Atomic Energy Authority, 1985). Following this detailed study, it was concluded that risk assessment based on the existing data base could not be relied on to quantify the risk of dam failure and the following actions were recommended (Cullen, 1990):
   (a) retrieval and collation of information on dam incidents should be continued in order to expand the database
   (b) geotechnical research into UK embankment dams should continue and possibly expand with widespread dissemination of significant findings
   (c) continued cautious development of fault tree methods applied to dams in conjunction with expert system approaches should await expansion of the database to provide sufficient information.

Some of the issues arising from the study were discussed by Parr and Cullen (1988).

29. **Modification of DAMBRK program.** A widely used computer program, DAMBRK, available from the US National Weather Service has been modified to suit typical UK conditions. It can be used to evaluate the formation and propagation of a dam failure wave along a valley. The modified program has addressed the analytical problems caused by changes from sub-critical to super-critical flow in steep upland catchments, has been made more user friendly with the option of using metric units, has been verified using data from known cases of dam failure, and tested for changes in parameters such as speed of embankment failure and valley steepness using sensitivity analyses. The computer program, manual and report are available from Binnie & Partners (1991a).

30. **Assessment of flood damage following dam failure.** This project reviewed the methods available for evaluating the damage caused by the depth of inundation and the velocity of flow following a dam failure. Observed damage in cases of dam failure was correlated to hydraulic characteristics and a methodology developed to allow both the extent of flooding and an assessment to be made of the areas where partial and total structural damage would occur, and where loss of life would be likely. A methodology was provided to enable flood damage costs to be evaluated for UK conditions (Binnie and Partners, 1991b), with an input from the Flood Hazard Research Centre at Middlesex Polytechnic.

31. _Reservoir hazard assessment_. This project was a literature review of reservoir hazard indices, with an assessment of their value for the United Kingdom (Binnie and Partners, 1992). There are few hazard indices which are based solely on the hazard which may strictly be understood as the capacity of a reservoir to cause damage. Therefore the report included those indices which also involve an assessment of risk or probability of a dam failure and a consequent release of the reservoir contents. The methods available have been classified under three levels of complexity, ranging from the relatively subjective method given by the Institution of Civil Engineers (1978) for the purpose of spillway sizing to more complex and expensive methods.

32. _Design of flood storage reservoirs_. A technical update of the CIRIA design guide TN100 (Hall and Hockin 1980) has been completed and provides the practising engineer with guidelines for the design of flood storage ponds for flood control in partly urbanised catchment areas (Hall et al, 1992). Sections deal with the causes and prevention of flooding in urbanised drainage areas, design flood estimation, flood routing, water quality considerations, detailed engineering design and the operation and maintenance of ponds. It has been estimated that more "large raised reservoirs" as defined by the Reservoirs Act 1975 are being constructed in the UK for flood storage than for any other purpose. Although hydrological methods presented can be used with care down to areas as small as 50 ha, the design of storage facilities for smaller developments served by stormwater sewerage systems may best be tackled using the Wallingford procedure (National Water Council, 1981) or similar package.

33. _Regional flood risk_. IH published a report (Dales and Reed, 1989) which described a statistical analysis of rainfall records used to develop a methodology for assessing the risk of an extreme rainfall event at one of a group of reservoirs. It concluded that if, for example, 22 reservoirs occupied an area of 249 km^2, the risk of a design exceedence of a 1:10 000 year event was only 1/6th of what it would be if it was assumed that each of the reservoirs experienced independent rainfall events.

34. _Seismic risk_. An engineering guide to seismic risk to dams in the United Kingdom has been produced by BRE in collaboration with consulting engineers Sir William Halcrow and Partners, and includes embankment, masonry and concrete dams (Charles et al, 1991). Although Britain is an area of low seismicity, the seismic risk to UK dams is not negligible and should be addressed. It is proposed that the seismic safety evaluation of a dam and its ancillary works should be based on a safety evaluation earthquake and the safety of the dam against catastrophic failure should be ensured under the level of ground motion produced by this earthquake. Recommended peak ground accelerations for the safety evaluation earthquake are related to four dam categories based principally on the downstream hazard in the event of failure and three zones of

seismicity level into which it is suggested that the UK can be divided.

35. <u>Computerised data base</u>. Detailed studies of published accounts of serious incidents and failures of UK embankment dams have been undertaken by BRE (Charles and Boden, 1985; Charles, 1986; Charles, 1989). A computerised data base of reservoirs with their dimensions and other characteristics together with incidents, deterioration, inadequacies and remedial works at British earth dams has been compiled and now includes details of some 2000 dams and reservoirs. The list of dams entered on the data base is being extended so that it is comprehensive. It is described by Tedd et al (1992) in a paper to this conference.

ASSESSMENT OF RESEARCH PROGRAMME

36. The Department has been advised by a Reservoir Safety Research Committee which has included leading UK dam experts. The research programme has been reviewed annually at the meetings of this committee. In addition steering groups have been appointed for individual projects.

37. All government departments which fund research are encouraged to practise research assessment to ensure that research is suitably directed and demonstrates value for money. In September 1991 DOE appointed an independent consultant, Dr D J Coats, to act as research assessor for the Reservoir Safety Research Programme, 1983-1991. It is anticipated that Dr Coats will report to the Department prior to this conference and will present his findings at the conference.

FUTURE DIRECTION OF RESEARCH WORK

38. Following the publication of the assessment, the future direction and development of the research programme will be re-assessed particularly with regard to areas of research which have been identified as requiring considerable further attention. In addition the following aspects of the programme will be considered:
 (a) assessment of scale of effort required in terms of both money and duration
 (b) allocation of priorities
 (c) if necessary, modification of mechanisms for monitoring, appraisal and control.

39. Until the completion of the current assessment of the research programme it is difficult to give details of the future programme. However certain features can be identified:
 (a) a substantial effort will continue to be directed towards the basic hydraulic, hydrological and geotechnical aspects of reservoir safety
 (b) the BRE computerised data base will be used to identify and quantify areas of concern where further research is required; updating and analysis of the data base of UK dam records will be continued
 (c) publication of appropriate guidance to the dam industry on aspects of reservoir safety is seen to be of major

importance
(d) a systematic study of other national European legislation and regulations, with particular emphasis on assessing safety trends within the European community, will be undertaken by BRE.

ABBREVIATIONS
BRE	Building Research Establishment
CCTV	Closed circuit television
CHUTE	Numerical model for flows in open channel spillways
CIRIA	Construction Industry Research and Information Association
DAMBRK	Computer program for simulation of dam failure
DOE	Department of the Environment
HF	Hydraulic fracture
HR	Hydraulics Research
ICE	Institution of Civil Engineers
IH	Institute of Hydrology
NERC	Natural Environment Research Council
PRA	Probabilistic risk assessment
SERC	Science and Engineering Research Council
SRD	Safety and Reliability Directorate of UK Atomic Energy Authority
UK	United Kingdom
US	United States of America
WRC	Water Research Centre

REFERENCES
1. ATKINSON J H, CHARLES J A and MHACH H K (1990). Examination of erosion resistance of clays in embankment dams. Quarterly Journal of Engineering Geology, vol 23, 103-108.
2. BABTIE SHAW and MORTON (1986). Chemical deterioration of fill material in earth dams in the United Kingdom. Department of the Environment Contract PECD 7/7/193.
3. BINNIE and PARTNERS (1986). Modes of dam failure and flooding and flood damage following dam failure. Department of the Environment Contract PECD 7/7/184.
4. BINNIE and PARTNERS (1991a). Dam break flood simulation program, DAMBRK UK. Department of the Environment Contract PECD 7/7/271.
5. BINNIE and PARTNERS (1991b). Estimation of flood damage following potential dam failure: guidelines. Department of the Environment Contract PECD 7/7/259. Foundation for Water Research.
6. BINNIE and PARTNERS (1992). Review of methods and applications of reservoir hazard assessment. Department of the Environment Contract PECD 7/7/309.
7. CHARLES J A (1986). The significance of problems and remedial works at British earth dams. Proceedings of BNCOLD/IWES Conference on Reservoirs 1986, Edinburgh, 123-141. Institution of Civil Engineers, London.
8. CHARLES J A (1989). Deterioration of clay barriers: case

histories. Proceedings of Conference on Clay Barriers for Embankment Dams, Institution of Civil Engineers, October 1989, 109-129. Thomas Telford, London, 1990.
9. CHARLES J A and BODEN J B (1985). The failure of embankment dams in the United Kingdom. Proceedings of Symposium on Failures in Earthworks, 181-202. Institution of Civil Engineers, London.
10. CHARLES J A and WATTS K S (1987). The measurement and significance of horizontal earth pressures in the puddle clay cores of old earth dams. Proceedings of Institution of Civil Engineers, Part 1, vol 82, February, 123-152.
11. CHARLES J A and TEDD P (1991). Long term performance and ageing of old embankment dams in the United Kingdom. Transactions of 17th International Congress on Large Dams, Vienna, vol 2, 463-475.
12. CHARLES J A, ABBISS C P, GOSSCHALK E M and HINKS J L (1991). An engineering guide to seismic risk to dams in the United Kingdom. Building Research Establishment Report BR 210.
13. COOLING L F and GOLDER H Q (1942). The analysis of the failure of an earth dam during construction. Journal of Institution of Civil Engineers, vol 19, 38-55.
14. CULLEN N (1990). Risk assessment of earth dam reservoirs. Department of the Environment Contract No PECD 7/7/191. Water Research Centre Report no DOE 0002-SW/3.
15. DALES M Y and REED D W (1989). Regional flood and storm hazard assessment. Department of the Environment Contract PECD 7/7/135. Institute of Hydrology Report no 102, 159pp.
16. DOUNIAS G T, POTTS D M and VAUGHAN P R (1989). Numerical stress analysis of progressive failure and cracking in embankment dams. Department of the Environment Contract No PECD 7/7/222.
17. ELLIS J R (1989). Guide to analysis of open-channel spillway flows. Technical note 134. CIRIA, London.
18. HALL M J and HOCKIN D L (1980). Guide to design of storage ponds for flood control in partly urbanised catchment areas. Technical Note 100. CIRIA, London.
19. HALL M J, HOCKIN D L and ELLIS J B (1992). Design of flood storage reservoirs. Department of the Environment Contract PECD 7/7/240. CIRIA-Butterworth-Heineman, London.
20. HEWLETT H W M, BOORMAN L A and BRAMLEY M E (1987). Design of reinforced grass waterways. Report 116. CIRIA, London.
21. HMSO (1930). Reservoirs (Safety Provisions) Act, 1930. HMSO, London.
22. HMSO (1975). Reservoirs Act 1975. HMSO, London.
23. HYDRAULICS RESEARCH (1987). Wave prediction in reservoirs: a literature review. Department of the Environment Contact PECD 7/7/187.
24. HYDRAULICS RESEARCH (1988a). Wave prediction in reservoirs: comparison of available methods. Department of the Environment Contract No PECD 7/7/187.
25. HYDRAULICS RESEARCH (1988b). Wave protection in reservoirs: hydraulic model tests of blockwork stability. Report No EX 1725. Department of the Environment Contract PECD

7/7/187.
26. INSTITUTION of CIVIL ENGINEERS (1978). Floods and reservoir safety: an engineering guide (reissued 1989).
27. JOHNSTON T A, MILLMORE J P, CHARLES J A and TEDD P (1990). An engineering guide to the safety of embankment dams in the United Kingdom. Building Research Establishment Report BR 171.
28. MILLMORE J P and CHARLES J A (1988). A survey of UK embankment dams. Proceedings of Reservoir Renovation 88 Conference, Manchester, Technical Note 1, 1-5. BNCOLD, London.
29. NATURAL ENVIRONMENT RESEARCH COUNCIL (1975). Flood studies report (5 vols). NERC, London.
30. NATIONAL WATER COUNCIL (1981). Design and analysis of urban storm drainage: the Wallingford procedure. DOE/NWC Standing Technical Committee report no 28.
31. PARR N M and CULLEN N (1988). Risk management and reservoir maintenance. Journal of Institution of Water and Environmental Management, vol 2, no 6, 587-593.
32. REED D W and FIELD E K (1992). Reservoir flood estimation - another look. Department of the Environment Contract PECD 7/7/181. Institute of Hydrology report no 114, 116pp.
33. REED D W and ANDERSON C W (1992). A statistical perspective on reservoir flood standards. Paper in this conference.
34. SAFETY and RELIABILITY DIRECTORATE UK ATOMIC ENERGY AUTHORITY (1985). A feasibility study into probabilistic risk assessment for reservoirs. Water Research Centre External Report No ER 188E.
35. STEWART E J (1989). Areal reduction factors for design storm construction: joint use of raingauge and radar data. Proceedings of Symposium on New Directions for Surface Water Modelling. IAHS Publication No 181, 31-40.
36. STEWART E J (1992). Spatial variations of extreme rainfall events in upland areas. Department of the Environment Contract PECD 7/7/190. Institute of Hydrology Report in preparation.
37. STEWART E J and REYNARD N S (1991). Rainfall profiles for design events of long duration. Proceedings of British Hydrological Society Symposium held in Southampton, September 1991, 4.27-4.36.
38. STEWART E J and REYNARD N S (1992). Temporal variations of extreme rainfall events in upland areas. Department of the Environment Contract PECD 7/7/190. Institute of Hydrology Report in preparation.
39. TEDD P, CHARLES J A and BODEN J B (1987). Internal seepage erosion in old embankment dams. Proceedings of 9th European Conference on Soil Mechanics and Foundation Engineering, Dublin, vol 1, 507-510.
40. TEDD P and HART J M (1988). The use of infrared thermography and temperature measurement to detect leakage from old embankment dams. Proceedings of International Symposium on Detection of Subsurface Flow Phenomena by Selfpotential/ Geolectrical and Thermometrical Methods, Karlsruhe.
41. TEDD P, CLAYDON J R and CHARLES J A (1988). Detection and

investigation of problems at Gorpley and Ramsden Dams. Proceedings of Reservoirs Renovation 88 Conference, Manchester, paper 5.1, 1-15. BNCOLD, London.

42. TEDD P, POWELL J J M, CHARLES J A and UGLOW I M (1989). In situ measurement of earth pressures using push-in spade-shaped pressure cells; ten years experience. Proceedings of Conference on Instrumentation in Geotechnical Engineering, Nottingham. Geotechnical Instrumentation in Practice, Purpose, Performance and Interpretation, 701-715. Thomas Telford, London.

43. TEDD P, CHARLES J A and CLAYDON J R (1990). Deformation of Ramsden Dam during reservoir drawdown and refilling. The Embankment Dam. Proceedings of 6th Conference of British Dam Society held in Nottingham, September 1990, 171-176. Thomas Telford, London, 1991.

44. TEDD P, HOLTON I R and CHARLES J A (1992). The BRE dams database. This conference.

45. WATTS K S and CHARLES J A (1988). In situ measurement of vertical and horizontal stress from a vertical borehole. Geotechnique, vol 38, no 4, 619-626.

Crown copyright 1992. The views expressed in this paper are those of the authors and not necessarily those of the Department of the Environment.